Review of
Optical Manufacturing
2000 to 2020

Review of Optical Manufacturing 2000 to 2020

Aizhong Zhang
Richard N. Youngworth
EDITORS

SPIE PRESS
Bellingham, Washington USA

Library of Congress Control Number: 2021938747

Published by
SPIE
P.O. Box 10
Bellingham, Washington 98227-0010 USA
Phone: +1 360.676.3290
Fax: +1 360.647.1445
Email: books@spie.org
Web: www.spie.org

Copyright © 2021 Society of Photo-Optical Instrumentation Engineers (SPIE)

Cover art courtesy of Optimax Systems, Inc.

All rights reserved. No part of this publication may be reproduced or distributed in any form or by any means without written permission of the publisher.

The content of this book reflects the work and thought of the author. Every effort has been made to publish reliable and accurate information herein, but the publisher is not responsible for the validity of the information or for any outcomes resulting from reliance thereon.

Printed in the United States of America.
First Printing.
For updates to this book, visit http://spie.org and type "PM334" in the search field.

SPIE.

Table of Contents

1	**Introduction**		**1**
	Aizhong Zhang and Richard N. Youngworth		
	References		6
2	**Optical Materials**		**11**
	Ralf Jedamzik, Uwe Petzold, Frank Nürnberg, Bodo Kühn, and Gordon von der Gönna		
	2.1	Introduction	11
	2.2	Optical Glass Production	13
		2.2.1 What is optical glass?	14
		2.2.2 Raw materials	17
		2.2.3 Melting and coarse annealing	18
		2.2.4 Fine annealing	23
		2.2.5 Refractive index measurement	26
		2.2.6 Refractive index homogeneity measurement	28
		2.2.7 Refractive index homogeneity: striae	28
		2.2.8 Transmittance measurement	29
		2.2.9 Stress birefringence measurement	33
		2.2.10 Bubble and inclusion measurement	34
		2.2.11 Other properties	35
	2.3	Fused Silica Production	35
		2.3.1 Properties of fused silica	35
		2.3.2 Making of fused quartz and fused silica	42
		2.3.3 Homogenization of quartz glass	45
		2.3.4 Shaping and forming of quartz glass	46
		2.3.5 Doping quartz glass	48
		2.3.6 Quartz glass for applications in the near-infrared	49
		2.3.7 Laser-induced damage threshold of quartz glass	50
		2.3.8 Applications	53
	2.4	Crystalline Materials Production	54
		2.4.1 History	54
		2.4.2 The crystalline state	54
		2.4.3 Crystal growth	55

		2.4.4	Optical ceramics	56
		2.4.5	Post-growth heat treatment	57
		2.4.6	Properties and qualities	59
	2.5	Optical Material Trends		63
	References			65
3	**Optical Fabrication**			**71**
	Jessica DeGroote Nelson			
	3.1	Introduction		71
	3.2	Traditional Fabrication Methods		72
		3.2.1	Cast iron lapping	72
		3.2.2	Conventional or pitch polishing	76
	3.3	CNC Optics Manufacturing		77
		3.3.1	Spherical CNC generation	77
		3.3.2	Deterministic small tool polishers	79
		3.3.3	Mid-spatial frequency smoothing methods	87
	3.4	Special Considerations for Aspheres and Freeforms		89
		3.4.1	Aspheres	89
		3.4.2	Freeforms	90
	3.5	Concluding Remarks		92
	References			92
4	**Metrology**			**105**
	Daewook Kim, Isaac Trumper, and Logan R. Graves			
	4.1	Introduction		105
	4.2	Standard Opto-Mechanical Metrology		107
		4.2.1	Coordinate measuring machines	108
		4.2.2	Machine vision	112
		4.2.3	Structured light projection	113
		4.2.4	Laser tracker	115
		4.2.5	Infrared scanning system	116
	4.3	Precision Process Guiding Metrology		119
		4.3.1	Contact stylus profilometry	120
		4.3.2	Non-contact slope sensor scanning	124
		4.3.3	On-machine metrology	125
	4.4	High-Precision Quality Check or Verification Metrology		126
		4.4.1	Birefringence measurements	127
		4.4.2	Swing arm profilometry	129
		4.4.3	Deflectometry	130
		4.4.4	Null interferometry	133
		4.4.5	Instantaneous dynamic interferometry	138
		4.4.6	Microscopic white-light interferometry	139
	4.5	Concluding Remarks		141
	References			142

5 Optical Coatings — 151
Ronald R. Willey
- 5.1 Introduction — 151
- 5.2 The New Century — 156
  - 5.2.1 Better understanding — 156
  - 5.2.2 Design — 158
  - 5.2.3 Materials — 160
  - 5.2.4 Equipment — 165
  - 5.2.5 Masking for uniformity — 169
  - 5.2.6 Monitoring and control — 170
  - 5.2.7 Processes — 176
  - 5.2.8 Index versus thickness — 184
- 5.3 Conclusions — 188
- References — 189

6 Infrared Optical Systems — 195
Adam Phenis and Jason Mudge
- 6.1 Introduction — 195
- 6.2 New Applications — 197
  - 6.2.1 Industrial imaging — 197
  - 6.2.2 Defense — 199
  - 6.2.3 Science — 200
- 6.3 Component Development — 201
  - 6.3.1 Infrared sensing — 201
  - 6.3.2 Infrared sources — 214
- 6.4 Acknowledgments — 216
- References — 216

7 Polymer Optics — 223
Robert Parada, Jr., Douglas Axtell, and Dan Morgan
- 7.1 Introduction — 223
- 7.2 Polymer Materials — 223
- 7.3 Optical Design Considerations — 228
- 7.4 Small-Volume Manufacturing — 230
- 7.5 Large-Volume Manufacturing — 232
- 7.6 Metrology Considerations — 236
- 7.7 Ancillary Services — 240
  - 7.7.1 Gate vestige removal — 240
  - 7.7.2 Optical thin film coating — 241
  - 7.7.3 Alignment and joining — 242
  - 7.7.4 Stress reduction — 244
- 7.8 Emerging Techniques and Applications — 245
  - 7.8.1 Micro-optics — 245
  - 7.8.2 Photonics — 246

	7.8.3	Asymmetric form factors	246
	7.8.4	Novel materials	247
	7.8.5	Healthcare	248
	References		248

8 Optical Fibers and Optical Fiber Assemblies — 263

Devinder Saini, Kevin Farley, and Brian Westlund

8.1	Optical Fibers		263
	8.1.1	Preform manufacturing	264
	8.1.2	Fiber draw	273
	8.1.3	Fiber testing capabilities	280
	8.1.4	Advanced fibers	283
8.2	Optical Fiber Applications and Assemblies		285
	8.2.1	Applications	285
	8.2.2	Mode mixing and de-speckling	286
	8.2.3	Anti-reflection technologies	287
	8.2.4	Assembly design considerations	292
	8.2.5	1D and 2D arrays	293
	8.2.6	High-power design considerations	296
	References		298

9 Diffractive- and Micro-structured Optics — 303

Tasso R. M. Sales and G. Michael Morris

9.1	Introduction		303
9.2	Fabrication of Surface-Relief Masters		307
	9.2.1	Single-point diamond turning (SPDT)	308
	9.2.2	Optical and E-beam lithography	309
	9.2.3	Laser pattern generation	311
	9.2.4	Reactive ion etching	314
	9.2.5	Nickel electroform tooling	316
9.3	High-Volume Manufacturing Processes		317
	9.3.1	Polymer-on glass wafers	318
	9.3.2	Roll-to-roll manufacturing	319
9.4	Diffractive Lenses in Broadband Imaging Systems		321
	9.4.1	Diffractive lens fundamentals	321
	9.4.2	Diffractive/refractive (hybrid) achromatic lenses	325
	9.4.3	Multi-layer diffractive optical elements	328
	9.4.4	Multi-order diffractive lenses	329
	9.4.5	Multifocal diffractive ophthalmic lenses	332
9.5	Markets for Micro-Structured Optics		339
	9.5.1	Gesture recognition and 3D imaging/sensing systems	339
	9.5.2	Display and image projection systems	342
	9.5.3	Solid-state lighting	344

9.6 Summary		346
References		347

10 Illumination Optics — 355

Henning Rehn and Julius Muschaweck

10.1	Introduction	355
10.2	Fields of Application	355
	10.2.1 Indoor lighting	356
	10.2.2 Outdoor lighting	357
	10.2.3 Automotive lighting	357
	10.2.4 Medical lighting	358
	10.2.5 Airfield lighting	358
	10.2.6 Entertainment lighting	359
	10.2.7 Data and video projection	360
	10.2.8 Flat panel displays	360
	10.2.9 Smartphones and smart watches	361
	10.2.10 Solar	361
	10.2.11 Freeform optics	362
	10.2.12 Trends after 2000	363
10.3	Light Sources and Their Fabrication	363
	10.3.1 Legacy light sources	363
	10.3.2 Rise of LED technology	364
10.4	Optical Components for Illumination and Their Fabrication	366
	10.4.1 Standard lenses	366
	10.4.2 TIR lenses and related collimators	367
	10.4.3 Fresnel lenses	367
	10.4.4 Compound concentrators	368
	10.4.5 Reflectors	369
	10.4.6 Light guides	369
	10.4.7 Homogenizers	370
	10.4.8 Sheets	371
10.5	Prototype Technologies	372
10.6	Outlook	372
References		373

Chapter 1
Introduction

Aizhong Zhang
Optimax Systems Inc., 6367 Dean Pkwy, Ontario, NY, 14519, USA

Richard N. Youngworth
Riyo LLC, PO Box 2325, Boise, ID, 83701, USA

In the history of technology, scientific ideas for systems and components were often conceived in theory well before practical implementation was possible. However, manufacturing capability often sets the practical limit of scientific exploration. Manufacturing technological development has always been the key measure for research and commercial implementation, especially in our field of optics and photonics. Visionary leaders in the field, technical and business, might want to keep up with the advancement of manufacturing capability to effectively invent, develop, and make decisions.

Optics and photonics are often a critical part of the solution to many challenges facing humanity, whether it the constant need for energy sources, medical solutions (as we write this introduction, the world is coping with the Covid-19 crisis), or numerous other industrial, consumer, and defense applications. With this background, it might be a proper time to look back at the industry of optical manufacturing at this critical time in human history, and prepare for the challenges ahead.

During the period of 2000–2020, the industry of optical manufacturing witnessed dramatic changes in key areas for hardware implementation such as optical materials, fabrication methods, and metrology, to name a few. Some of these technologies have incubation periods back to the 1990s or even earlier. This review aims to summarize some of the critical changes that have impacted the way optical systems are manufactured. It evaluates a number of new manufacturing techniques with growing popularity, and points out some trends in optical manufacturing in the future. Several fundamental questions that we try to elucidate in this review book include:

(1) What are the newly employed techniques in the optics industry in the past two decades?

(2) What in optical manufacturing has had an order of magnitude or more change in the past two decades? The changes could be in precision, scale, cost, and package size, etc.

(3) Looking to the future, what manufacturing techniques are not widely used yet but may become popular in the next 20 years?

To answer these questions, we divide the review book into ten chapters, each with a focusing theme. These chapters are not exhaustive to fully describe the current optics industry, but rather we aim to provide a snapshot of the current time so that the readers may quickly grasp the basics of current manufacturing capabilities and promising new techniques.

On optical materials, in the 1990s, a major change was to develop lead- and arsenic-free glass types due to environmental concerns. In the past two decades, there is still a demand for new glass types, and the requirements on tolerances of refractive index, dispersion, homogeneity, stress birefringence, striae, bubbles and inclusions, etc. are getting ever tighter [1]. Chalcogenide and fluoride glasses have become more widely used. With fine annealing, the refractive index of a glass material can be adjusted with precision up to the fifth digit after the decimal point and homogeneities up to the sixth digit [1]. Advanced measurement methods such as spectral goniometer measurements of refractive index and temperature dependence of refractive index, homogeneity interferometer measurements, stress-birefringence measurements based on differential interference contrast microscopy, etc. have provided deeper insight in glass material properties and significantly improved the production quality. The production process of glass ceramics with extremely low thermal expansion coefficient, such as Zerodur, has been further optimized, and is widely used in ground-based and space-based telescopes [1,2].

On optical fabrication, computer numerical controlled (CNC) optics manufacturing has become the mainstream optical fabrication method widely used in modern optical machine shops. Aspheric optical components fabrication has become more common, and the cost is significantly reduced. Precise compression molding of glass preforms has become more widely used, especially in mass production of aspheric lenses in consumer cameras [1]. Freeform optics manufacturing has become more mature [3], and the corresponding freeform design and freeform metrology are also rapidly evolving. Magnetorheological finishing (MRF) [4] as a deterministic sub-aperture polishing process is gaining popularity in high-end precision optical fabrication. Additive manufacturing or 3D printed optics is an active research field that may become more influential in industry in the near future. In astronomy, with proper segmentation, extremely large astronomical telescopes with primary mirrors up to tens of meters are being constructed [5,6].

Modern optical metrology employs a series of tools with different accuracies and dynamic ranges, including instruments such as coordinate measuring machines, profilometers, and interferometers, to name just a few. These tools are used for different applications to monitor surface figure, form, and waviness as they evolve from grinding to polishing. Some tools are optimized for molded optics or optical coating inspection as well. Subaperture stitching interferometry allows precise measurements of large optical surfaces [7]. Modern metrology could achieve nanometer accuracy for large astronomical telescopes several meters in diameter [8]. Metrology for extreme ultraviolet (EUV) lithography used for integrated circuits routinely investigates defects at individual atomic level.

Although the thin film optical coating methods of physical vapor deposition (PVD), chemical vapor deposition (CVD), and atomic layer deposition (ALD) have all been developed prior to 2000, better understandings on heat, stress and materials have led to more precise control of the coating process. In the past two decades, several exciting new coating technologies have been introduced. For example, plasma-etched organic layers for antireflection coating on polymers and glasses has been developed by the Fraunhofer IOF [9,10]; Lawrence Livermore National Laboratory (LLNL) has developed multilayer dielectric coatings for petawatt-class lasers and advanced hard x-ray multilayer coatings [11,12,13] With the growing use of freeform optics, advanced optical coatings for applications with broadband and wide-angle antireflection are also more common.

Since its introduction by Theodore Maiman in 1960, laser has fundamentally revolutionized the landscape of optics industry. A large variety of lasers and numerous applications have been developed. In the period of 2000–2020, industrial laser sales have increased dramatically. In 2020, the total laser market revenue is about 16 billion, with ~40% from diode lasers [14]. In comparison, the total laser market revenue was about $5.6 billion in 2001 [15], almost 300% growth in the past two decades. Particularly notable during this period is that fiber laser output power has been steadily growing, and its market share keeps increasing. High-power fiber lasers have replaced conventional CO_2 lasers in many applications, thanks to the compact design with simpler optics, longer lifetime, and flexibility to use. With compact, reliable and affordable lasers, laser remote sensing has expanded from defense applications into civilian applications. For intelligence, surveillance and reconnaissance, the frontier of defense LiDAR systems is evolving from airborne to satellite-borne [16]. Meanwhile LiDAR for autonomous vehicles have been commercially manufactured by many suppliers. Large laser systems aiming to initiate nuclear fusion, such as the Omega Laser Facility at the University of Rochester, the National Ignition Facility at LLNL [17], the

Măgurele Laser near Bucharest, Romania, (part of the European Extreme-Light Infrastructure (ELI) Project), and the Laser Megajoule (LMJ) currently under construction near Bordeaux, France [18] could all deliver petawatt power picosecond beams to a target with chirped pulse amplification.

Optical instruments beyond the visible spectrum have been further explored. On the ultraviolet (UV) side, driven by the ever-increasing demand from the semiconductor industry, lithography is pushing the frontier of human precision engineering. The latest ASML NXE:3400C platform employs EUV light wavelength of 13.5 nm, and it has a resolution of 13 nm, an NA of 0.33, and a throughput capability reaching 170 wafers per hour. Further, ASML and Carl Zeiss SMT are working on the next generation EUV exposure tool with an NA of 0.55 [19]. On the infrared side, thanks to drastic miniaturization of infrared systems, many previously expensive defense IR systems are now modified to become more compact and commercially available. Infrared systems including oximeters, heart rate monitors, proximity sensors, and biometric systems based on fingerprint, face, iris, or vein recognition have become widely used in consumer electronics and wearable medical devices. On the far infrared end, terahertz imaging and spectroscopy have wide applications in security screening, industrial inspection, etc. [20,21]. Broadband optical systems from illumination to detection, including supercontinuum lasers, hyperspectral imaging systems, low-coherence interferometry systems and optical coherence tomography (OCT) systems, etc. have become more commonly used.

Due to the ubiquitous use of consumer electronics, such as computers, tablets and smart phones, polymer optics have undergone explosive growth in the past two decades, and have become the most common type of cameras in our daily life [22,23]. Image taking and video shooting whenever and wherever have drastically transformed the way people communicate and interact. Large volume polymer optical cameras are being produced hundreds of millions of units per month. Driven by the volume production and diverse applications, the manufacturing process of fabrication, coating, and metrology of polymer optics has rapidly matured and is still under continuous improvement. Polymer optics-based cameras have become one of the key data generators of the Internet. Coupled with the power of social media, the content recorded by these polymer cameras has far-reaching effects, as manifested by many videos that had gone viral and influenced societies and governments.

The low-loss optical fiber was invented by scientists from Corning Inc. in the early 1970s [24]. Since then, billions of kilometers of optical fibers have been installed to wrap around the globe. Together with long-lifetime semiconductor lasers and erbium-doped fiber amplifier (EDFA), low-loss optical fibers have become the key infrastructure of the Internet to enable people around the world to communicate at the speed of light. In 2020, more than 60% of the global population have Internet access, and there are more than 20 billion networked devices [25]. The global IP traffic is now more than 250 exabytes per month [26], or more than 100 terabyte per second (TB/s). In comparison, the global IP traffic was about 84 petabytes per month in 2000, hence a 3000-fold increase in two decades. In the past two decades, optical fibers have experienced significant improvements of optical preforms, fiber draw, and fiber coatings. Advanced fibers such as multicore fibers, photonic crystal fibers, etc. have become more mature. The application of optical fibers has extended from telecommunications to medical diagnosis, fiber sensing, and high laser power delivery.

Diffractive and micro-optics could enable optical systems with large aperture, light weight, and affordable cost. In the past two decades, micro-optics have been developed with ever-smaller feature sizes, reaching nano-optics realm. Other than conventional micro-optical materials, such as plastics, fused silica and silicon, new materials such as CaF_2, InP, GaAs, etc. are increasingly used, partly driven by monolithic integration with other optical components [27]. Surface relief masters of micro-optics can be fabricated by single-point diamond turning, optical or electron-beam lithography, laser writing, and reactive-ion etching, etc [28]. Large-volume replications of diffractive and micro-optics could be produced from the surface relief masters by techniques of casting, hot embossing, and injection molding, to name a few [29]. Computer generated holograms (CGHs) have been widely used in measuring surface figures of aspheres and freeform optics [30]. Full-color, high-definition CGHs for static 3D displays have been developed recently [31]. However, real-time, dynamic 3D holographic displays without vergence-accommodation conflict are still under development [32]. Diffractive and micro-optics are used in consumer electronics for applications of 3D imaging and sensing, illumination systems, display and projection systems, among others. Further, they also find applications in space telescopes [33] and medical systems, especially in intraocular lenses (IOLs) [34].

In the field of illumination, light-emitting diodes (LEDs) have become the dominant light sources to illuminate our world, both indoor and outdoor, replacing incandescent and fluorescent light sources. Other core illumination components, such as total internal reflection (TIR) lenses, Fresnel lenses, reflectors, light guides, homogenizers, etc. are all redesigned for illumination systems with LED sources. Illumination systems must find a balance between efficacy, luminance and cost. The projection and display markets have risen rapidly, thanks to the development of technoloiges, such as liquid crystal display (LCD), digital light processing (DLP), organic light-emitting diode (OLED), etc. Displays come in various sizes, from large outdoor displays of several meters to microdisplays of millimeters in length. New design principles with freeform surfaces such as simultaneous multiple surfaces (SMS) method and tailoring, etc. have become widely used in illumination design [35,36,37,38,39]. 3D printed optics have started to gain popularity in illumination optics, since the required accuracy is relatively lower than that in imaging.

Since its inception in 1880s in Jena, Germany by pioneers of Carl Zeiss, Ernst Abbe, and Otto Schott, modern optics industry has evolved on the shoulders of generations of giants. Today, from flat panel displays to home theater projectors, from cellphone cameras to augmented reality goggles, from LED headlights to LiDAR guided self-driving cars, from contact lenses to intraocular lenses, optical products have become so ubiquitous in our daily life that are often overlooked. On the industrial level, from telecommunication optical fibers to astronomical telescopes, from medical microscopes to refractive eye surgery, from EUV lithography to terahertz security screening, from integrated photonics to quantum computing, from laser fusion labs to directed-energy weapons, optics has been proven an enabling technology integral to the global economy and national security. The rapid advancement of the optical manufacturing techniques in the first two decades of the 21st century is truly remarkable, but in a more densely webbed world connected by optics-enabled Internet of things, a faster evolution of optical technologies may be yet to reveal in the coming decades.

References

[1] P. Hartmann, R. Jedamzik, S. Reichel, and B. Schreder, "Optical glass and glass ceramic historical aspects and recent developments: a Schott view," *Applied optics* **49**(16), D157–D176 (2010).

[2] T. Westerhoff, T. B. Hull, and R. Jedamzik, "ZERODUR manufacturing capacity: ELT and more," in *Astronomical Optics: Design, Manufacture, and Test of Space and Ground Systems II*, **11116**, 1111612, International Society for Optics and Photonics (2020).

[3] J. Nelson, "The evolution of freeform fabrication and testing: lessons learned and the roadmap to higher precision," *EPIC Meeting on Precision and Freeform Optics at WZW-OPTIC* (2019).

[4] S. D. Jacobs, D. Golini, Y. Hsu, B. E. Puchebner, D. Strafford, I. V. Prokhorov, E. M. Fess, D. Pietrowski, and W. I. Kordonski, "Magnetorheological finishing: a deterministic process for optics manufacturing," in *International Conference on Optical Fabrication and Testing*, **2576**, 372–382, International Society for Optics and Photonics (1995).

[5] G. H. Sanders, "The thirty meter telescope (TMT): An international observatory," *Journal of Astrophysics and Astronomy* **34**(2), 81–86 (2013).

[6] R. Tamai, M. Cirasuolo, J. C. González, B. Koehler, and M. Tuti, "The E-ELT program status," in *Ground-based and Airborne Telescopes VI*, **9906**, 99060W, International Society for Optics and Photonics (2016).

[7] P. Murphy, G. Forbes, J. Fleig, P. Dumas, and M. Tricard, "Stitching interferometry: a flexible solution for surface metrology," *Optics and Photonics News* **14**(5), 38–43 (2003).

[8] C. J. Oh, A. E. Lowman, G. A. Smith, P. Su, R. Huang, T. Su, D. Kim, C. Zhao, P. Zhou, and J. H. Burge, "Fabrication and testing of 4.2 m off-axis aspheric primary mirror of Daniel K. Inouye Solar Telescope," in *Advances in Optical and Mechanical Technologies for Telescopes and Instrumentation II*, **9912**, 99120O, International Society for Optics and Photonics (2016).

[9] U. Schulz, C. Präfke, C. Gödeker, N. Kaiser, and A. Tünnermann, "Plasma-etched organic layers for antireflection purposes," *Applied optics* **50**(9), C31–C35 (2011).

[10] U. Schulz, P. Munzert, F. Rickelt, and N. Kaiser, "Breakthroughs in photonics 2013: Organic nanostructures for antireflection," *IEEE Photonics Journal* **6**(2), 1–5 (2014).

[11] R. Soufli, R. Qiu, T. Pardini, C. Burcklen, J. Alameda, J. Robinson, C. Walton, E. Spiller, P. Mirkarimi, R. Negres, et al., "Advanced multilayer coatings for national security," tech. rep., Lawrence Livermore National Lab.(LLNL), Livermore, CA (United States) (2019).

[12] C. J. Stolz, R. A. Negres, and E. Feigenbaum, "Trends observed in 10 years of thin film coating laser damage competitions," in *Optical Interference Coatings*, ThA–1, Optical Society of America (2019).

[13] S. Romaine, "Further support of the development of advanced multilayer optics coatings for x-ray imaging," tech. rep., Lawrence Livermore National Lab.(LLNL), Livermore, CA (United States) (2020).

[14] C. Holton, A. Nogee, J. Hecht, and B. Gefvert, "Laser markets navigate turbulent times," *LASER FOCUS WORLD* **56**(1), 46–61 (2020).

[15] S. G. Anderson, "Review and forecast of the laser markets. Part I: Nondiode lasers," *Laser focus world* **37**(1), 88–110 (2001).

[16] D. R. Shedd, "It's time for a tactical LiDAR satellite," *Defense Systems* (2021).

[17] V. Goncharov, S. Regan, E. Campbell, T. Sangster, P. Radha, J. Myatt, D. Froula, R. Betti, T. Boehly, J. Delettrez, et al., "National direct-drive program on OMEGA and the National Ignition Facility," *Plasma Physics and Controlled Fusion* **59**(1), 014008 (2016).

[18] J. Miquel, C. Lion, and P. Vivini, "The laser mega-joule: LMJ & PETAL status and program overview," in *J. Phys. Conf. Ser.*, **688**(1), 012067 (2016).

[19] E. van Setten, G. Bottiglieri, J. McNamara, J. van Schoot, K. Troost, J. Zekry, T. Fliervoet, S. Hsu, J. Zimmermann, M. Roesch, et al., "High NA EUV lithography: Next step in EUV imaging," in *Extreme*

Ultraviolet (EUV) Lithography X, **10957**, 1095709, International Society for Optics and Photonics (2019).

[20] J. El Haddad, B. Bousquet, L. Canioni, and P. Mounaix, "Review in terahertz spectral analysis," *TrAC Trends in Analytical Chemistry* **44**, 98–105 (2013).

[21] J. P. Guillet, B. Recur, L. Frederique, B. Bousquet, L. Canioni, I. Manek-Hönninger, P. Desbarats, and P. Mounaix, "Review of terahertz tomography techniques," *Journal of Infrared, Millimeter, and Terahertz Waves* **35**(4), 382–411 (2014).

[22] M. Schaub, J. Schwiegerling, E. Fest, R. H. Shepard, and A. Symmons, *Molded optics: design and manufacture*, CRC press (2011).

[23] W. S. Beich, "Injection molded polymer optics in the 21st-century," in *Tribute to Warren Smith: A Legacy in Lens Design and Optical Engineering*, **5865**, 58650J, International Society for Optics and Photonics (2005).

[24] F. Kapron, D. B. Keck, and R. D. Maurer, "Radiation losses in glass optical waveguides," *Applied Physics Letters* **17**(10), 423–425 (1970).

[25] Cisco, "Cisco annual internet report (2018–2023) White Paper," tech. rep. (2020).

[26] Cisco, "Cisco visual networking index (VNI)," tech. rep. (2018).

[27] T. J. Suleski and R. D. TeKolste, "Roadmap for micro-optics fabrication," in *Lithographic and Micromachining Techniques for Optical Component Fabrication*, **4440**, 1–15, International Society for Optics and Photonics (2001).

[28] D. C. O'Shea, T. J. Suleski, A. D. Kathman, and D. W. Prather, *Diffractive optics: design, fabrication, and test*, vol. **62**, SPIE press (2004).

[29] M. T. Gale, "Replication techniques for diffractive optical elements," *Microelectronic Engineering* **34**(3-4), 321–339 (1997).

[30] C. Zhao, "CGH and stitching technique: powerful combination enabling full-surface figure measurements of certain classes of aspheres," in *Optical Design and Testing VIII*, **10815**, 1081503, International Society for Optics and Photonics (2018).

[31] K. Matsushima, *Introduction to Computer Holography: Creating Computer-Generated Holograms as the Ultimate 3D Image*, Springer Nature (2020).

[32] B. C. Kress, *Optical Architectures for Augmented-, Virtual-, and Mixed-reality Headsets*, SPIE press (2020).

[33] W. Zhao, X. Wang, H. Liu, Z.-f. Lu, and Z.-w. Lu, "Development of space-based diffractive telescopes," *Frontiers of Information Technology & Electronic Engineering*, 1–19 (2020).

[34] A. Zhang, "Multifocal diffractive lens design in ophthalmology," *Applied Optics* **59**(31), 9807–9823 (2020).

[35] H. Ries and J. Muschaweck, "Tailored freeform optical surfaces," *JOSA A* **19**(3), 590–595 (2002).
[36] P. Gimenez-Benitez, J. C. Miñano, J. Blen, R. M. Arroyo, J. Chaves, O. Dross, M. Hernández, and W. Falicoff, "Simultaneous multiple surface optical design method in three dimensions," *Optical Engineering* **43**(7), 1489–1503 (2004).
[37] F. R. Fournier, W. J. Cassarly, and J. P. Rolland, "Fast freeform reflector generation using source-target maps," *Optics Express* **18**(5), 5295–5304 (2010).
[38] M. Karlen, C. Spangler, and J. R. Benya, *Lighting design basics*, John Wiley & Sons (2017).
[39] R. Wu, Z. Feng, Z. Zheng, R. Liang, P. Benítez, J. C. Miñano, and F. Duerr, "Design of freeform illumination optics," *Laser & Photonics Reviews* **12**(7), 1700310 (2018).

Aizhong Zhang graduated with a Ph.D. from the Institute of Optics at the University of Rochester in 2017. He cofounded the company Lightrino, LLC in 2017, focusing on ophthalmic instruments development. He also worked as a part-time optical engineer in Optimax Systems Inc. from 2018 to 2020. He is the inventor of ten U.S. patents or patent applications, and he is a member of SPIE and OSA.

Richard N. Youngworth is a Fellow of SPIE and OSA. He is the founder and chief engineer of Riyo LLC, an optical design and engineering firm. He obtained his Ph.D. in optics at the University of Rochester. Dr. Youngworth has authored and delivered numerous short courses, papers, talks, and lectures on optical design and engineering. He is active in optical standards development and in professional society work.

Chapter 2
Optical Materials

Ralf Jedamzik and Uwe Petzold
Schott AG – Advanced Optics, Hattenbergstrasse 10, 55122 Mainz, Germany

Frank Nürnberg and Bodo Kühn
Heraeus Quarzglas GmbH & Co KG – Optics, Quarzstrasse 8, 63450 Hanau, Germany

Gordon von der Gönna
Hellma Materials GmbH, Moritz-von-Rohr-Straße 1, 07745 Jena, Germany

2.1 Introduction

Photonics plays an important role driving innovation and enabling technology for many markets. The application of photonics spreads across very different areas: from optical data communication to imaging, lighting, manufacturing, life sciences, health care, security and safety. Photonics offers new and unique solutions where today's conventional technologies are approaching their limits: speed, capacity and accuracy. The impact of photonics in our daily lives is remarkable. It becomes visible by looking at some current challenges:

- Augmented and virtual reality for consumers and even more relevant for industrial applications.
- Assistance systems in automotive as rear/side-view cameras, LED/laser lighting, lidar, night vision support, head-up displays, etc.
- Industry 4.0 with its connected individual production lines and substitution of humans in mechanical and electric production lines by robots requires exact three-dimensional imaging, object recognition, barcode scanning, quality inspection, distance control, and absolute optical measurements.
- Continuous development of displays with higher resolutions or trends like "internet of things" need lithographic production lines with

accurate resolution that need interferometric measurement and recording devices with precise imaging objectives.
- Enlarged usage of laser technology for cutting, engraving, or welding.
- Use of cameras for security and defense applications or in drones.
- Three-dimensional printing requires three-dimensional imaging.
- Continuously increasing performance and power of laser sytems.

All these applications rely to a certain extent on elements out of the optical material, which guide, transport or modify light. Each optical element has its dedicated function, so that a very special optical material needs to be chosen. In a portfolio of more than 200 existing optical materials, every single glass/crystal has its purpose and complements all the others.

A selection of optical materials is shown in Figures 2.1 and 2.2 sorted along their most important properties: "refractive index" and "transmission". The following subsections will provide an overview of the production processes and the unique properties of optical glasses, quartz glass and optical crystals. Detailed information is provided via the references herein.

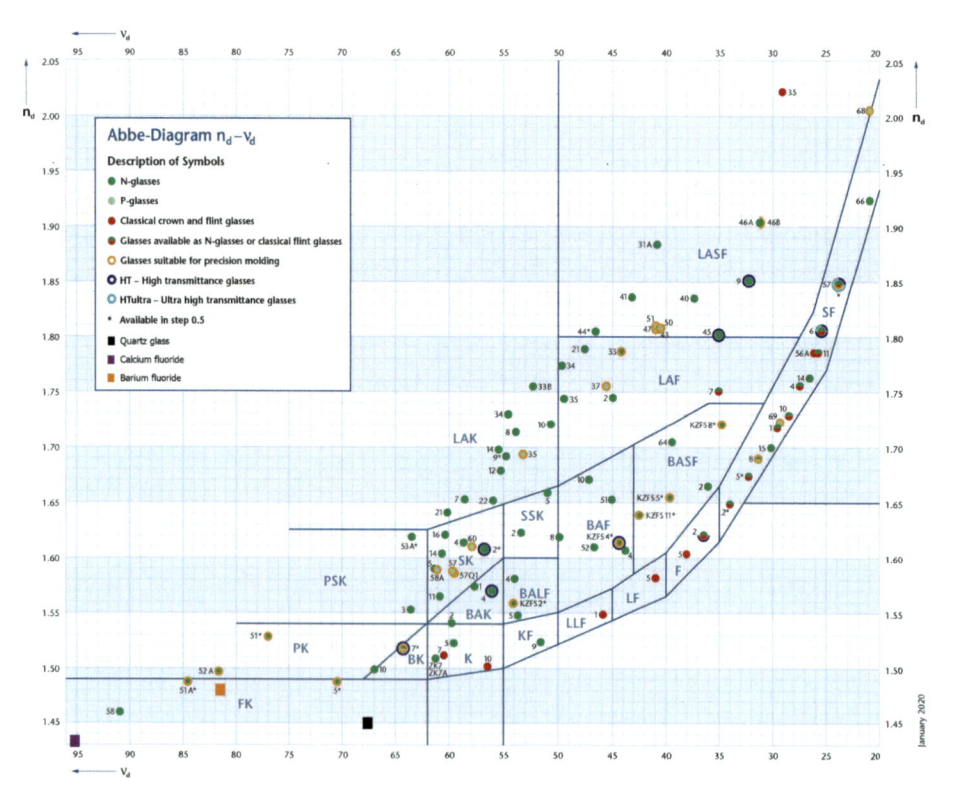

Figure 2.1 The Abbe diagram, displaying the optical position of the various glass types. Source: SCHOTT.

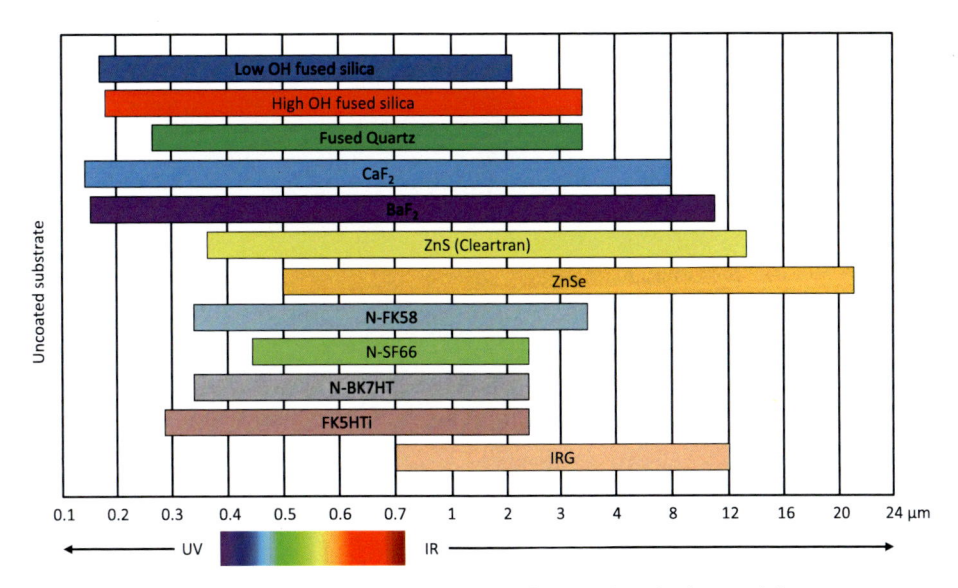

Figure 2.2 Transparency range of several optical materials.

2.2 Optical Glass Production

Prior to 1880, optical components were made from simple crown and flint glasses. The crown glasses were soda-lime silicates with low refractive indices and moderate Abbe numbers. The flint glasses were lead silicates with relatively low Abbe numbers [1].

In 1884, Otto Schott, Ernst Abbe and Carl and Roderich Zeiss founded the SCHOTT & Associates Glass Technology laboratory in Jena, Germany. Otto Schott's target was to develop new glasses on optical positions defined by Ernst Abbe that would tremendously improve the image quality of Carl Zeiss microscopes and optical instruments. By 1884, two dozen flint and crown glasses were available for optical system designs.

Even today, the photonics industry relies heavily on optical glasses to realize the tightest optical design requirements of modern photonics applications. The difference between today and the past is that the designer can choose from more than 120 optical glasses with differences in optical position, transmittance and very specific dispersion requirements. Optical glass is the raw material for the fabrication of spherical lenses, aspheres, prisms, beam splitters, optical fibers, axicons, and other optical components. It is a hugely versatile material: It can be hotformed into various shape to lenses or even fibers. It maintains its requirements even in harsh environments like very high and low temperature or chemical environments. Optical glass is mechanically stable and scratch resistant. The transparency is very high, even in large thicknesses. There are many advantages of this traditional material that, for example, optical plastics cannot cope with.

Optical engineers use components out of various optical glasses to optimize their designs concerning resolution, aberrations, stray light, etc. Such

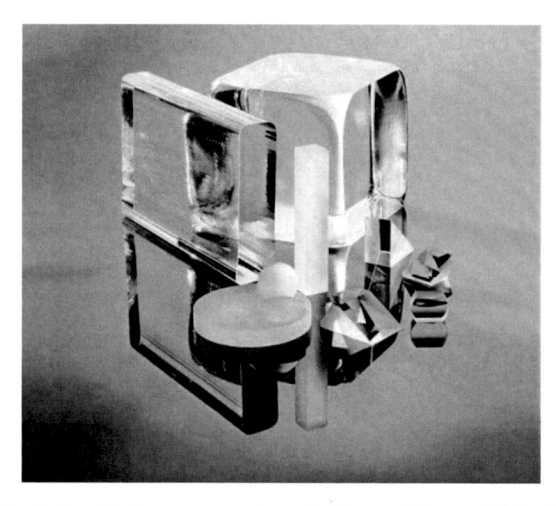

Figure 2.3 Fabricated optical components out of raw strips and blocks of optical glass.

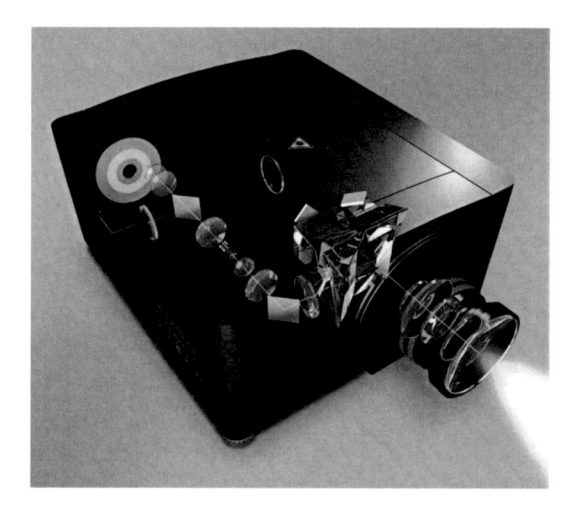

Figure 2.4 Optical components in digital projection. In addition to high-quality optical glasses in prisms for color separation, such applications also use lenses, homogenizers, dichroic mirrors and innovative ceramic materials for light conversion.

applications take advantage of the essential combination of material features of optical glass like high light transmission, large variety but also precise light deflection (index of refraction and dispersion), high uniformity of light deflection, and sufficient environmental resistance.

2.2.1 What is optical glass?

Glass has many unique properties [2]. The most obvious feature of optical glass for a human eye is the high transparency, as indicated in Figure 2.5. Compared to window glass, the optical path in optical glass is

Optical Materials 15

Light transmittance of glass

Figure 2.5 Comparison of the light transmission through window glass and optical glass.

more than 30 times longer for achieving the same transmission, which is a huge difference.

In order to achieve much higher grades of transmission than window glass, more stringent requirements are needed for the purity of the ingredients, haze level, and number of bubbles and inclusions. Figure 2.6 shows the internal transmittance over the visible spectrum and the near-infrared regime.

In contrast to the transmission, the optical position of an optical glass is not obvious for a human eye. Hereby, the refractive index at a specific

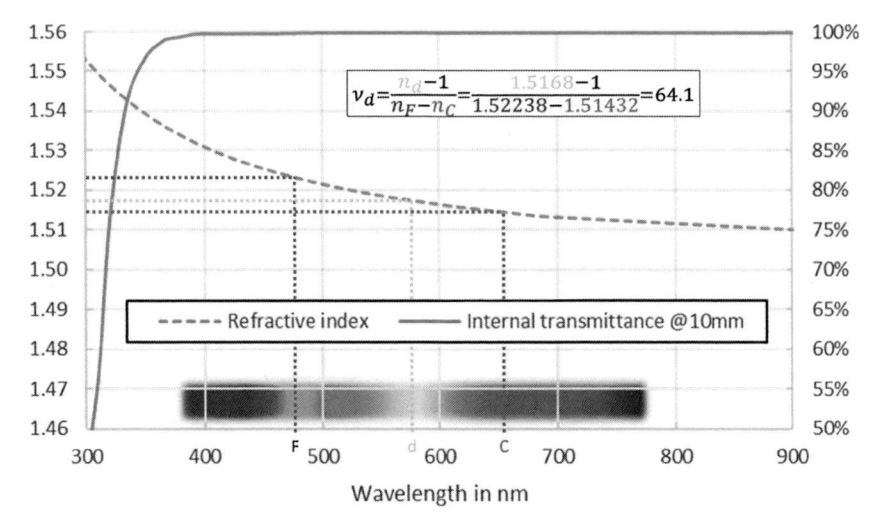

$$v_d = \frac{n_d - 1}{n_F - n_C} = \frac{1.5168 - 1}{1.52238 - 1.51432} = 64.1$$

Figure 2.6 Spectral development of the internal transmittance and refractive index of an optical glass (e.g., SCHOTT N-BK7®).

wavelength and the Abbe number describing the dispersion define the optical position. The graphic in Figure 2.6 shows the decrease of the refractive index starting from the ultraviolet over the visible spectrum to the near-infrared regime. Dispersion is the name of this spectral refractive index development. The Abbe number ν_d as defined in the upper box of Figure 2.6 is a measure for the dispersion. The value n_d at the d-line (587.5618 nm) is typically the reference for the refractive index. These two values n_d and ν_d define the optical position and exhibit the main distinctive features of optical glass types.

This dispersion is one of the main reasons why we need optical glasses and highly sophisticated lens systems for photonic products. If a single lens focuses a blue light ray (e.g., the F-line at 486.1327 nm) and a red light ray (e.g., the C-line at 656.2725 nm), both rays experience different deflections due to the varying refractive index. Therefore, the focus position of both colors differs. If an optical designer combines a flint and a crown glass lens in a proper way, the designer achieves that the focus of the blue and the red ray overlaps. This doublet is an achromatic system. Unfortunately, the focus position of other colors still varies. Therefore, a further chromatic correction and other aberrations require a complex multi-lens design, as depicted in Figure 2.7. Such lens system design relies on a broad portfolio of optical glasses that spread widely in their optical position.

Figure 2.1 shows a n_d-ν_d diagram, also called an Abbe diagram. This diagram maps the different optical glass types by using the refractive index and Abbe number as coordinates. The left part of the diagram corresponding to high Abbe numbers contains the crown glasses indicated with the letter "K" in the end. The right part of the diagram corresponding to low Abbe numbers contains the flint glasses as indicated with the letter "F" in the end. Besides the

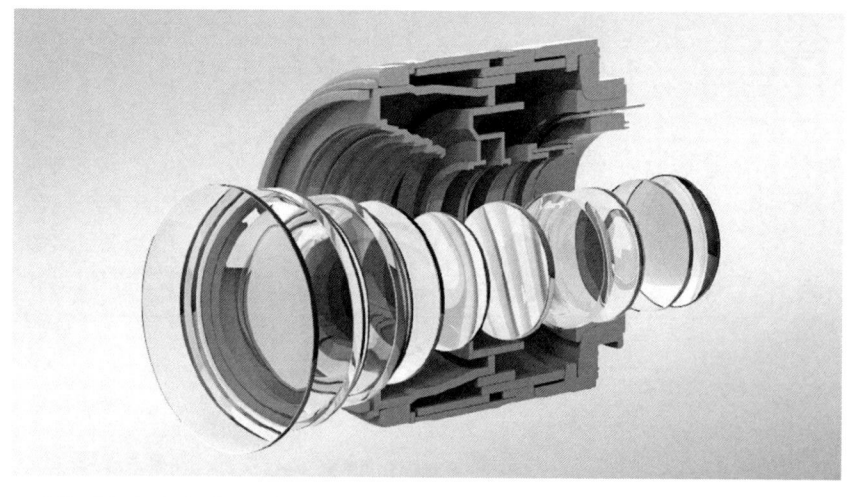

Figure 2.7 Exploded view of a lens system consisting of 10 different lenses made of different optical glass types. This depicts the complexity of such setups.

rough differentiation between crowns and flints, the map shows further areas of similar chemical composition, e.g., the region of barium flints with the label "BAF" or the area of lanthanum crowns with the label "LAK". According to their position in the diagram, the glasses get their labels, e.g., F2 is located in the "F" regime. The number at the very end of the glass-type label has no further information and counts the developments in the relevant area (seldom followed by a letter indicating a new version). The prefix "N-" indicates that no lead and arsenic are contained in the glass. Then, the prefix "P-" types are special glasses that are environment friendly as well but additionally show a low transformation temperature to enable precise molding process. Finally, the suffix "HT" or "HTultra" defines special versions of the glass type with high or even ultrahigh transmittance. All explanations are valid for the optical glass manufacturer SCHOTT as a reference but are in general transferable to other manufacturers as well.

2.2.2 Raw materials

Optical glass is molten from pre-mixed raw material batches with ingredients of high purity and defined grain size. Raw material batches contain main ingredients called network formers; depending on the glass type, these are silicates, borates, aluminates, phosphates or even lead. Although compositions of optical glasses are usually given in the form of the oxides used, in reality, many elements are introduced as, e.g., carbonates or hydrocarbonates. The reason for this is that many oxides have high melting temperatures that would require high energies, endangering the refractory materials used. Network modifiers, for example, (earth-)alkalines and rare earth elements, further reduce the melting temperature and adjust the required viscosity and glass properties.

Each ingredient weight is adjusted to accuracies of single grams in an automated process. The different compounds need to be mixed properly to ensure batch uniformity. The batches are mixed in specific bag sizes (typically several 100 kg for the large tanks), and the reproducibility of the mixture from bag to bag needs to be very high. The ingredients of optical glass cover a wide range in the periodic system of the elements. SCHOTT N-BK7®, one of the most common optical glass types, contains 9 different oxides based on the material safety datasheet [3]. The weight fraction ranges from up to 70% for silica as a main component down to much less than 1% on components like calcium oxide, antimony trioxide or titanium oxide. N-FK51A, a low-dispersion glass, lists fluorine between 20–30% as a main component in the material safety datasheet [3], together with 10–20% phosphorous, aluminum, barium and strontium oxide. N-LASF31A, a high refractive index glass, lists 10 oxides with lanthanum oxide of 30–40% as a main component and several other exotic and expensive main components like 10–20% of gadolinium and tantalum oxide. This shows a glimpse of the complexity of optical glass composition. The different chemical compositions strongly influence the viscosity of the glass during melting. SCHOTT N-BK7® is a high-viscosity glass, whereas N-LASF31A is a low-viscosity glass. The viscosity influences the hot forming process and striae quality in the glass (striae are short-range index variations in the glass). Lanthanum-containing glasses are often more sensitive to crystallization and can therefore not be produced in any thickness. The composition influences not only the optical position but also the internal transmittance. Heavy elements, like lead in SF57, result in a high refractive index. The lead in SF57 is also required to achieve very good transmittance at the blue spectral range. Exchanging lead with titanium maintains the optical position but reduces the achievable maximum transmittance at the blue. Some glass compositions react with the surrounding atmosphere at melting temperature. This reaction leads to changes in the glass surface and can lead to internal quality issues.

2.2.3 Melting and coarse annealing

The two main steps of optical glass production are melting and fine annealing. The melting process itself takes place in melting tanks and is divided into three steps: melting, refining and mixing.

In the melting step, the pre-mixed raw material batches are filled into the melting tank. The melting tanks interior is composed of refractory wall material. The type of refractory material selected depends on the glass that is molten. Typical tank walls are composed from, for example, quartz or ceramic building bricks or palisades. The material used depends on the reactivity of the glass composition. Some melts are chemically very aggressive and require being in contact only with chemically inert materials like platinum. The chemical attack from the melt leads to erosion of the wall material. Therefore, the tank material has a limited lifetime and needs to be replaced after a certain time. Erosion from the tank wall material can also influence the chemical composition melt and therefore the properties of the optical glass.

The temperature of the raw material is increased to the melting temperature range by direct heating using gas burners or indirect electrical heating. Melting temperatures depend on the material composition and typically range from 800°C to 1400°C. The reactivity of the melt depends very much on the chemical composition. Some optical glasses, like the mainly silica containing glass SCHOTT N-BK7®, are easier to melt, shown in the center of the Abbe diagram. Other non-silica-containing glass types like phosphate glasses or fluoro-phosphate glasses are more sensitive and can react with the surrounding atmosphere, changing the composition on the glass surface. Depending on the viscosity of the glass, such reactions can have an influence on the glass quality and homogeneity. The melting process itself roughly determines the optical properties of the optical glass to the third digit after the decimal point of the refractive index. With an additional heat treatment

Figure 2.8 View inside the melting chamber with some still solid raw material on the surface of the melted material.

process, fine annealing, the optical position can be adjusted to the fifth digit, and homogeneities up to the sixth digit can be achieved.

Gas bubbles are generated during the melting process due to several physical and chemical reasons. They are the result of chemical reactions taking place during the melting, entrapped air in the batch, furnace atmosphere gases, decomposition gases, and gases from refractory glass melt interaction, moisture release, and contaminations in the melt. Physical parameters that influence the bubble content are, for example, the grain size of the batch constituent, melting temperature and melting duration.

After melting the raw materials, the glass melt may contain a very large number density of bubbles in the order of 10^5 per kg. To promote the removal of bubbles from the glass melt refining agents are added. These refining agents generate finer bubbles that are distributed all over the melt and have high ascension rates. Dissolved gases diffuse into the fining bubbles leading to an increase of their size and further ascension. The typical refining agent before the year 2000 was arsenic oxide. Today, antimony oxide is the leading refining agent for optical glass at SCHOTT. Other refining agents are, for example, sodium sulfate or tin oxide.

This process leads to stripping of the gases in the melt, supported by an increased temperature. Increasing the temperature reduces the viscosity and increases the bubble diameter, promoting the ascension of the bubbles. This production step is called fining or refining. In the refining step, the internal quality of optical glass in terms of bubbles is mainly controlled.

The last and final step before hot-forming the required production formats is the mixing process. In the mixing process, the number of striae is reduced and the melt is homogenized. Striae are short-range compositional variations that lead to density variations and therefore local refractive index changes. Striae can also result from temperature gradients induced glass flow and density variations. Several aspects lead to homogenization of the melt. Heat convection moves the liquid glass along temperature gradients. Rising bubbles also lead to a certain mixing process. The dominant process in the mixing phase is the physical stirring of the melt. The stirrer is typically made from platinum parts to reduce the chemical interaction with the melt. The shape of the stirrer influences the mixing effect. Today's modern FEM (finite element method) tools allow the development of mixing configurations by simulating the trajectories of glass melt volumes depending on temperature and time. Such simulations are complemented by physical laboratory scale experiments to find an optimum for the setup in production.

In a discontinuous melting tank or pot furnace, the three main steps (melting, refining, mixing) take place in a single pot, separated in time. The raw materials are placed into the pot and melted. Then the temperature is increased for the refining and subsequently the melt is mechanically mixed and poured into the mold.

In a continuous melting tank, the melting area, the refining area and the mixing area are separated from each other. The melting tank is continuously filled with raw material that is subsequently molten. The molten glass flows into the refining chamber through an ascending pipe. In the refining area the temperature is raised, so that the bubbles can rise to the surface. The refined glass melt flows into the mixing crucible to minimize the striae content.

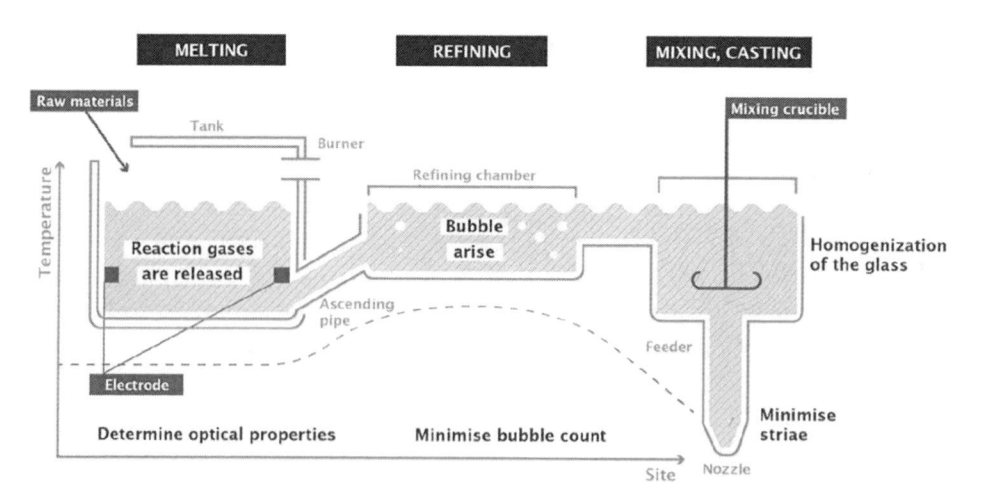

Figure 2.9 Sketch of a melting tank for optical glasses including the spatial temperature profile. The overall time consumption from the raw material melting to the casting takes several hours.

(a) (b)

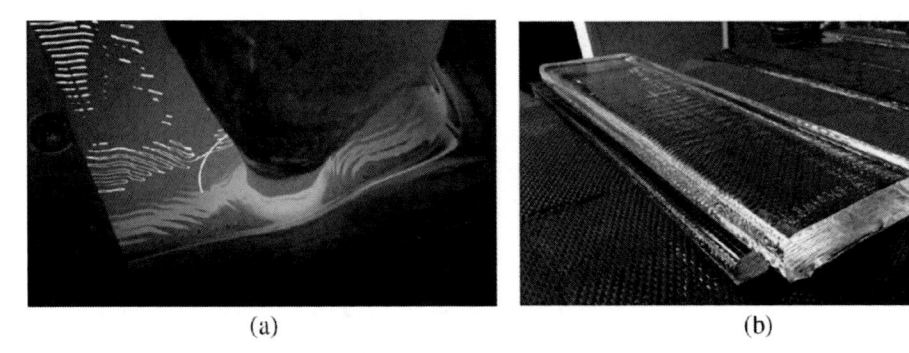

Figure 2.10 Hot forming (a) process to produce an endless strip of width 160 mm and height 40 mm (b). The still liquid glass is glowing due to the black body radiation.

Subsequently the glass flows through a feeder and the glass parts are hot formed.

Typical production formats are continuous strips of glass with typical cross section of e.g., 160 mm × 40 mm thickness. For specific applications, much larger cross sections are achievable. Other formats are block glass that is, for example, produced in typical dimensions 200 × 200 × 190 mm or individual circular molds larger than 1000 mm in diameter and 200 mm in thickness are possible from large tank sizes. Optical glass tanks are usually significantly smaller than glass tanks for bottles and window productions. Figure 2.10 shows the hot forming process of a strip production. The glass is not yet frozen and still glowing red due to black body radiation.

As glass has a rather low heat conductivity (≈ 1 W/(m × K)), a fast cooling process results in a high value of stress. The outer part is already frozen, but the inner part of a strip is still liquid. So, the volume change during the freezing of the inner part cannot be compensated by the already solid outer part. If this stress exceeds a certain threshold, some cracks or breakage occurs. With increasing thickness of a strip, this risk of damage rises. Therefore, a controlled first cooling process, called coarse annealing, is necessary to minimize the risk of glass breakage. In a continuous production process, the hot glass parts are placed on a conveyor belt of an annealing lehr with a length of several meters. At the beginning of the continuous annealing lehr, the temperature is adapted to the hot glass; at the end of the lehr, the temperature of the glass is less than a hundred degrees. This cooling process takes several hours, depending on glass type and format.

After the coarse annealing in the lehr, the glass manufacturer breaks or saws the glass strip into manageable length depending on the final application.

The most important control parameter of the glass quality is the continuous monitoring of the refractive index and Abbe number over time. Typically, glass samples are taken every two hours. These glass samples are reference annealed in a specific procedure (at SCHOTT based on 2°C/h), and the refractive index and Abbe number are measured. Additional samples are taken for inspection of the internal properties like the bubble and inclusion content and striae in regular intervals. The results are monitored continuously.

After coarse annealing, the glass needs to go through a fine annealing process to adjust the refractive index to the required optical tolerance. Cold processing steps like sawing, cutting, grinding, lapping, and polishing convert the fine annealed raw glass into the required optical components.

Another processing route is the fabrication of pressed blanks before fine annealing. Pressed blanks (also called pressings) are hot-formed, semi-finished lenses or prisms for the economic mass production of optical components. In large quantities, pressings are more cost effective than conservative cutting and grinding components from fine annealed raw glass formats. The near-net-shape appearance of pressings reduces the amount of additional machining in the finishing procedure of the customer. Pressings are generated from the hot pressing of cubicle-shaped raw glass parts with defined volume. During hot pressing, the temperature of the glass is increased to a viscosity working range of 1×10^4 dPas and 1×10^6 dPas. Subsequently, the glass part is pressed into the required shape with a dedicated tool at a defined force. The pressing is then first cooled down to room temperature rather quickly in a coarse annealing furnace. In a subsequent production process, pressings are fine-annealed to adjust the refractive index and reduce the stress birefringence to the required levels of the specification. The homogeneity of the pressings is defined by the index homogeneity of the selected raw glass batch and the homogeneity of the temperature distribution during fine annealing. Pressings are usually small glass parts typically 10 to 100 mm in size but can be also produced to dimensions up to 320 mm.

Some applications require very large glass blanks, like atmospheric dispersion correctors for large astronomical telescopes, scientific windows, and Fizeau plates in large interferometers, for example. The maximum

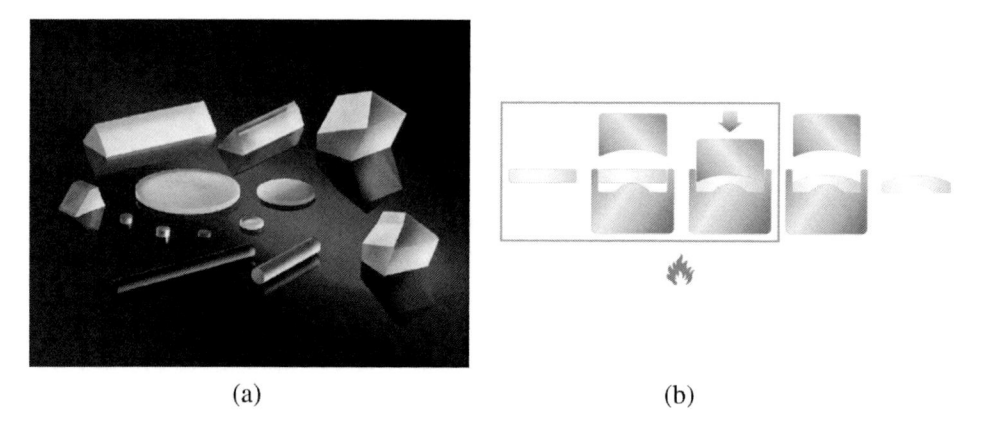

(a)   (b)

Figure 2.11 Pressings of different format and size (a) and schematic pressing process (b).

Figure 2.12 950-mm-diameter SCHOTT N-BK7® prepared for oil-on-plate homogeneity measurement on a Fizeau Interferometer [5].

achievable size of high homogeneous optical glasses is restricted by their viscosity and tendency for crystallization. The tendency for crystallization is a matter of temperature and time. For crystallization, several conditions are required. Crystals grow on nuclei of a few nanometers or on external seeds (called heterogeneous nucleation from, e.g., glass-metal contacts, impurities). Homogeneous nuclei form out of the melt in a glass composition specific temperature region. Both nucleation and crystallization are functions of time and temperature. The temperature range for nucleation is lower than the temperature range of crystallization with a composition dependent overlap. First nuclei occur during the cooling of the melt after casting. Crystals can form if the hot forming temperature of a glass composition is close to the nucleation and crystallization temperature range. The longer the part needs to be kept at high temperatures, the higher the risk of crystallization. Therefore, some optical glass can be produced to sizes up to 1000 mm in diameter, while other glasses are restricted to formats below 200 mm [4].

Classical borosilicate crown, barium crown, dense crown, fluorine crown glasses, and lead silicate flint glasses have the highest chance to be produced in large dimensions. Many other glass types, especially those with outstanding optical properties like very low dispersion or special partial dispersion, cannot be made in large sizes due to the crystallization tendency of the glass.

2.2.4 Fine annealing

The composition of the glass defines the first three digits of the refractive index, but even simple optical designs require more precise refraction values.

The velocity of the annealing process has an influence on the internal structure of the glass matrix and thus on the optical properties. Therefore, an additional fine annealing procedure defines the final optical features [6]. This is necessary because applications like interferometric measurements are sensitive to variations in the sixth digit of the index of refraction. While the melting of the optical glass takes on an order of magnitude of 1 day, the fine annealing procedure could take from several days (small dimensions) up to several months (large dimensions).

Actually, the annealing rate has a significant impact on the optical position of the glass. Tight control of the chemical composition is mandatory to hit the target values of the refractive index and the Abbe number. The order of magnitude of the accuracy required is 10^{-4} to 10^{-6} in the refractive index, depending on the application. By keeping the chemical composition constant, the glass manufacturer can control the refractive index within an accuracy of 10^{-3} to 10^{-4}. The annealing velocity influences the internal glass structure and thus the optical features. The fine adjustment of the refractive index takes place in the so-called fine annealing. Therefore, ovens heat up each piece of glass again. At a target temperature around the glass-type specific transformation temperature, the stress inside the glass relaxes. By cooling the glass with a constant rate, the glass manufacturer can control the refractive index with the required precision of 10^{-4} and 10^{-6}. Basically, slower annealing rates yield higher refractive indices. In practice, the following formula has proven itself:

$$n_d(h_x) = n_d(h_0) + m_{nd} \cdot log(h_x/h_0) \tag{2.1}$$

where h_0 is the original annealing rate, h_x is the new annealing rate and m_{nd} is the annealing coefficient for the refractive index depending on the glass type.

An analogous formula applies to the Abbe number:

$$\nu_d(h_x) = \nu_d(h_0) + m_{\nu d} \cdot log(h_x/h_0) \tag{2.2}$$

where $m_{\nu d}$ is the annealing coefficient for the Abbe number depending on the glass type.

The annealing coefficient $m_{\nu nd}$ can be calculated with sufficient accuracy with the following equation:

$$m_{\nu d} = \frac{m_{nd} - n_d(h_0) \cdot m_{nF-nC}}{n_F - n_C} \tag{2.3}$$

where the coefficient m_{nF-nC} has to be determined experimentally.

Figure 2.13 shows the change of refractive index and Abbe number as a function of the annealing rate for different optical glasses relative to the catalog optical position. The catalog position is referenced to 2°C/h annealing for comparison. It can be seen that the refractive index decreases with

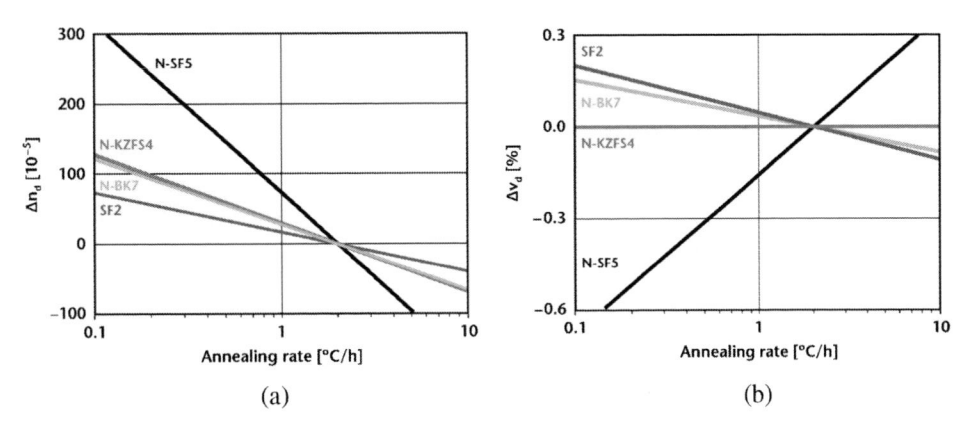

Figure 2.13 Change of refractive index (a) and Abbe number (b) with annealing rate for different glass types.

increasing annealing rate, and the amount of change strongly depends on the glass type.

In general, the Abbe number also increases with decreasing annealing rate. High-index lead-free glass types such as N-SF5 show anomalous behavior. Anomalous behavior means that the Abbe number increases with increasing annealing rate.

The annealing rate can be used to adjust the refractive index and Abbe number to the desired tolerance range. In practice, the annealing rate influences the refractive index and the Abbe number simultaneously. Figure 2.14 shows a diagram of the Abbe number versus the refractive index for SCHOTT N-BK7®. The rectangular boxes indicate the tolerance limits for the refractive index and the Abbe number. The center of the frames is defined by the nominal catalog value. After melting the optical glass is cooled down at a high annealing rate. To control the refractive index during the melting process samples are taken directly from the melt after each casting. These samples are cooled down very fast together with a reference sample of the same glass. The reference sample has a known refractive index at an annealing rate of 2°C/h. By measuring the change in refractive index of the reference sample the refractive index of the sample can be measured with moderate accuracy in the range of $\pm 10^{-4}$. The annealing rate dependence of the Abbe number and refractive index of each glass is represented by a line in the diagram with a slope that is characteristic for the glass type.

For a given melt, the position of the line in the diagram is given by the initial refractive index/Abbe number measurement for a cooling rate of 2°C/h as a fix-point together with the glass typical slope.

The refractive index and Abbe number for a given glass part can be adjusted by a fine annealing step along this characteristic line. Glass for cold processing has to be fine annealed to reduce internal stresses. During this fine

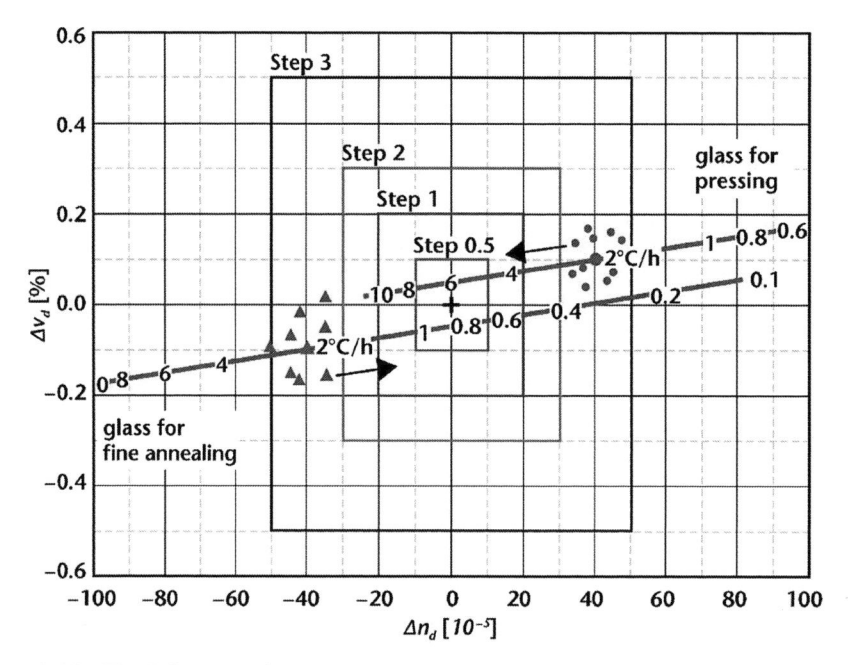

Figure 2.14 The influence of the annealing rate on the refractive index and Abbe number of SCHOTT N-BK7® for different initial refractive indices.

annealing, the annealing rate is in general lower than 2°C/h. The initial refractive index has to be adjusted during melting in such a way that the desired tolerances can be reached during fine annealing. For example, the initial refractive index of SCHOTT N-BK7® is, in general, lower than the target value.

Glass for hot processing (i.e., reheat) pressing is subjected to much more rapid annealing. Reheat pressing is used to generate pressed blanks or prisms that are near net shape hot formed parts with round cross-sections or angled prismatic shape. The heat treatment processes used by the customer in general use annealing rates much higher than 2°C/h. For example, for SCHOTT N-BK7® pressings, the initial refractive index needs to be higher than the target value.

Fine annealing influences not only the optical position but also the refractive index homogeneity and stress birefringence. Slow fine annealing rates are beneficial to generate high homogeneity and low stress birefringence for large optical glass parts.

2.2.5 Refractive index measurement

The refractive index is the most important property of optical glass. Therefore, refractive index measurement is a key characterization method for process control. The requirements for refractive-index control demand a

Optical Materials 27

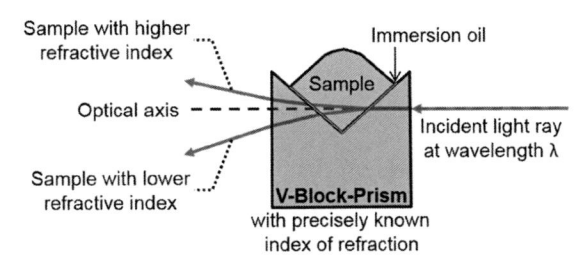

Figure 2.15 V-block refractometer measurement principle [6].

fast and accurate measurement method for production control and a very high accurate measurement method for the tightest requirements [6,7].

Therefore, two different measurement setups are common for refractive index measurements: the v-block refractometer and the spectral goniometer. Figure 2.15 shows the principle of the v-block measurement. The samples are shaped in a nearly square shape. One sample is about 20 × 20 × 5 mm in size. The sample will be placed in a v-shaped block prism. The refractive index of this prism is known very precisely. The refraction of an incoming light beam depends on the refractive index difference between the sample and the v-block prism. SCHOTT utilizes a setup that can take up to 10 samples glued together into one v-block stack. Therefore, many samples can be measured in a very short time. The relative measurement accuracy is very high, and therefore differences in refractive index within one v-block stack can be measured very accurately. The standard measurement temperature is 22°C at SCHOTT but can vary between other glass vendors from 20°C to 25°C. The accuracy of the v-block refractometer is typically $2-3 \times 10^{-5}$.

The spectral goniometric method is based on the measurement of the angle of minimum refraction in a prism-shaped sample; see Figure 2.16. This is the most accurate absolute refractive index measurement method. The measurement is based on the minimum angle of refraction principle. The samples are prism shaped with typical dimensions of about

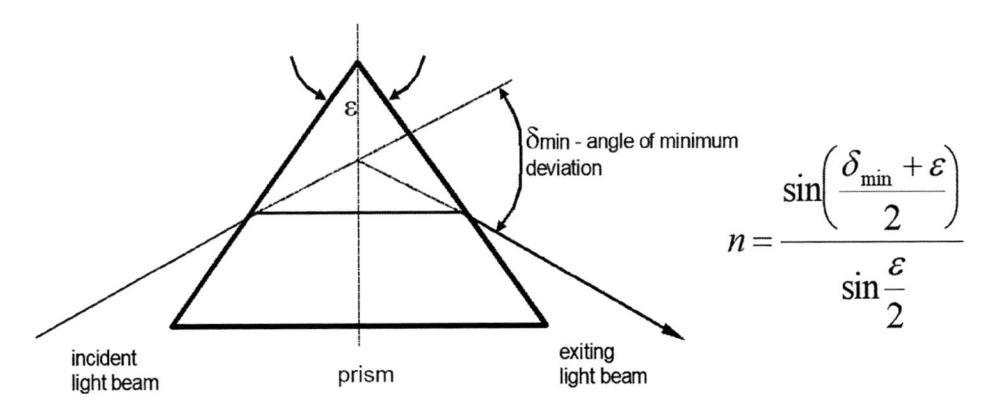

$$n = \frac{\sin\left(\frac{\delta_{min} + \varepsilon}{2}\right)}{\sin\frac{\varepsilon}{2}}$$

Figure 2.16 Spectral goniometer principle.

35 mm × 35 mm × 25 mm. The typical accuracy of the spectral goniometer method is 1×10^{-5}. With SCHOTT's automated spectral goniometer, the Ultraviolet to infrared Refractive Index measurement System (URIS), the refractive index of optical glasses can be measured to an accuracy of $\pm 0.4 \times 10^{-5}$. For single wavelengths, an accuracy of $\pm 1 \times 10^{-6}$ can be achieved. The measurement accuracy for the dispersion ($n_F - n_C$) is $\pm 2 \times 10^{-6}$. These measurement accuracies can be achieved independent of the glass type and over the complete wavelength range from 185 nm to 2325 nm. The standard measurement temperature is 22°C. The temperature can be varied between 18 to 28°C on request. The standard measurement atmosphere is air. On special request, nitrogen is also possible.

2.2.6 Refractive index homogeneity measurement

The refractive index variation over the glass component aperture influences the transmitting wavefront of a part. For specific high-end applications like i-line lithography of measurement technology, the refractive index homogeneity is very important. Typically, the refractive index homogeneity of optical glass is better than 10 ppm for individual glass parts of standard sizes. For i-line lithography (at 365-nm wavelength), the homogeneity requirements are up to 1-ppm refractive index variation or even better on an aperture of typically up to 300 mm.

The refractive index homogeneity cannot be measured absolutely. It is measured interferometrically [8] with specific methods to remove the influence of the surface on the wavefront: either by using additional glass plates fixed with immersion oil to the sample and combining measurements of the sandwiched sample and the glass plate alone, or by using a wedged-shaped sample with 4 individual measurements (front, back, transmitting and empty cavity); see Figure 2.17.

Homogeneity measurement can be performed on parts larger than 1 m (e.g., applications for laser fusion or large parts for atmospheric dispersion correction). The largest interferometers are in the range of a 600-mm aperture. Larger-aperture glass parts are measured using a process of stitching individually measured subapertures. Depending on the interferometer in use, a wavefront measurement accuracy of up to λ/200 can be achieved.

The results can be displayed in color-coded index maps and can be used for analysis in optical design. Striae are categorized according to their peak-to-valley refractive index variations in several grades. These grades can be found in ISO12123: 2018 [9] or ISO10110-18:2018 [10].

2.2.7 Refractive index homogeneity: striae

Striae are measured using the shadowgraph method [12]; see Figure 2.18(a). In this method, a divergent light beam from a strong mercury lamp is directed on a plano parallel polished glass sample. The light traveling through the sample

Optical Materials 29

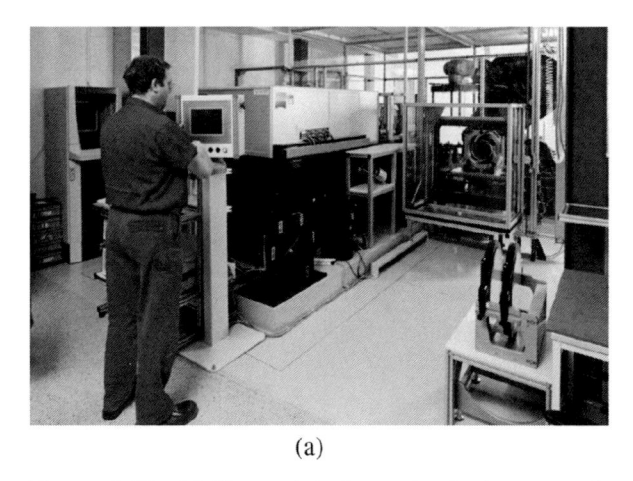

(a) (b)

Figure 2.17 (a) Fizeau interferometer for homogeneity measurement and (b) refractive index homogeneity color distribution with 4.6-ppm maximum index variation (homogeneity) on 166 × 185 mm aperture [11].

is refracted by the striae in the glass sample, leading to dark areas on the screen.

This method is very sensitive. The evaluation is done by comparing the striae level to a reference sample. Striae in optical raw glass are defined in terms of wavefront deviation for a 50-mm path length. Striae appearance is very sensitive to the thickness. Usually, striae visibility is reduced with reduced sample thickness. Striae visibility is also very dependent on the inspection orientation; see Figure 2.18(b). Usually, striae are not a problem in applications, since the glass thickness in many applications is much smaller than 50 mm (more in the range of 10-mm thickness) [13].

The standard ISO 10110-18:2018 contains two methods for specifying striae. The first method specifies the density of striae. Since it refers to finished optical components, it is only applicable to optical glass in its original form of supply to a limited extent. It assigns the striae density to classes 1–4 based on the areal percentage of the test region that they obscure. Thus, it only considers striae that deform a plane wave front by more than 30 nm.

The fifth class specifies glass that is extremely free of striae. It also includes striae below 30-nm wave front distortion and advises the user to make arrangements with the glass manufacturer.

2.2.8 Transmittance measurement

Besides the refractive index and its dependence on the wavelength, the internal transmittance of optical glass [14] is probably the second most important property. The transmittance of an optical glass is inversely proportional to its spectral absorption. The absorption bands of a glass are closely related to its

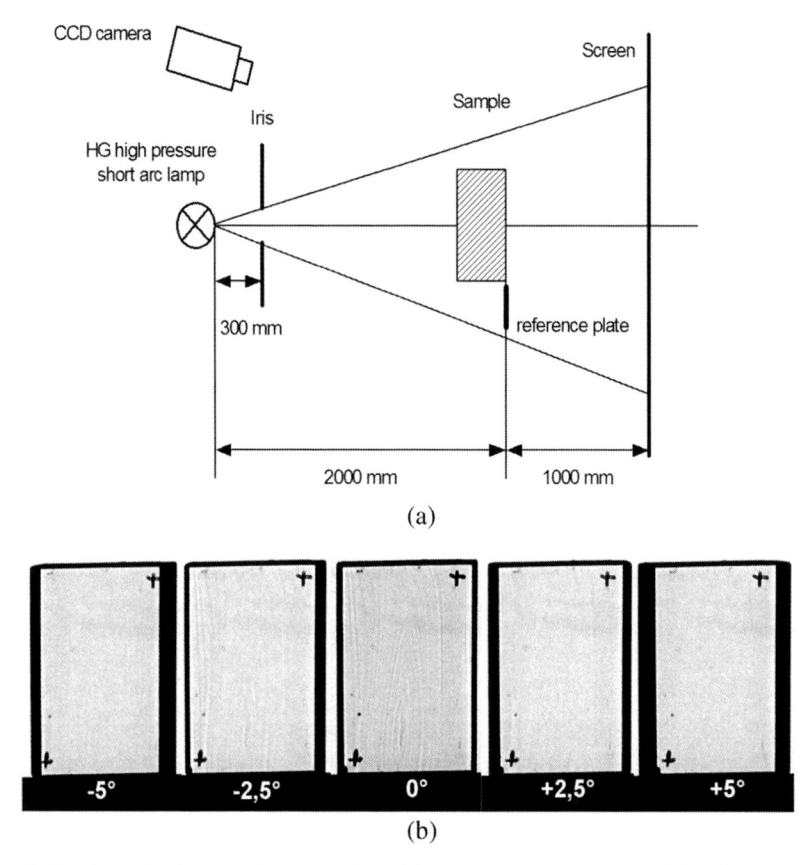

Figure 2.18 (a) Shadowgraph setup for striae measurement and (b) influence on the viewing angle on striae.

dispersion behavior. The dispersion is a measure of the change of the refractive index with wavelength.

The UV transmittance characteristic (see Figure 2.19) is mostly influenced by heavier elements in the glass composition (e.g., lead, barium, niobium, titanium, lanthanum), melting technology and/or residual impurities. These heavier elements are necessary to achieve a high refractive index but decrease the transmittance in the blue region. Therefore, high-refractive-index glasses in bigger thicknesses often show a yellowish color. Nevertheless, there are differences in the characteristic of the UV transmittance edge, depending on which heavy elements are used in the glass composition. For example, SF glasses containing lead exhibit a better transmittance in the blue spectral region compared to N-SF glass types where lead was substituted by titanium or niobium. The melting process also influences the transmittance characteristics of a glass, for example, platinum parts in the melting tank could be the source for platinum impurities in the glass, leading to a weaker UV transmittance.

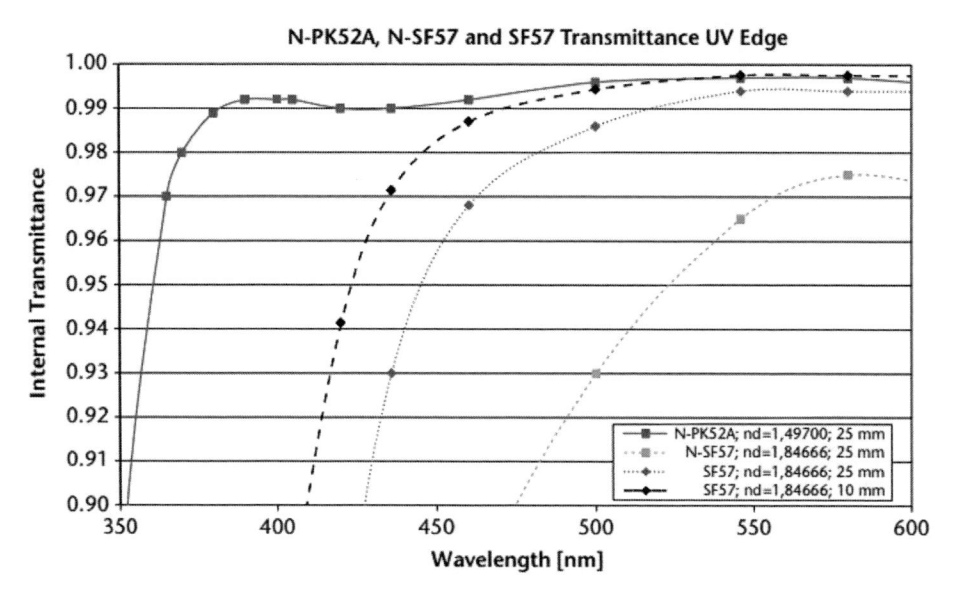

Figure 2.19 UV transmittance of different optical glasses.

Modern melting techniques aiming for the reduction of platinum contact with the melt therefore lead to better transmittance characteristics.

One way to describe the blue edge of the transmittance curve is to use the color code. The color code is a description of the position and slope of the UV transmittance edge. It lists the wavelength λ_{80} and λ_5, at which the transmittance (including reflection losses) is 0.8 and 0.05 at 10-mm thickness. The values are rounded to 10 nm and are noted by eliminating the first digit. Color code 33/30 means for example, λ_{80} = 330 nm and λ_5 = 300 nm. For high-index glass types with n_d > 1.83, the data of the color codes refer to the transmittance values 0.70 and 0.05 (λ_{70} and λ_5) because of the high reflection loss of this glass.

Electromagnetic radiation influences the transmittance of a glass depending on glass type and the wavelength of radiation. The influence of visible and UV radiation on glass is called solarization. The UV radiation generates color centers in the glass, leading to a reduced transmittance. The solarization behavior of optical glass can be investigated by irradiation with a xenon or mercury lamp, or with UV lasers typically below 380 nm. Glasses with low UV transmittance, e.g., with a high lead content (F and SF types), normally have small solarization effects. Several crown glasses with a higher ultraviolet transmittance change their UV-transmittance edge: PSK, BK, K, ZK, BAK, SK and LaK. The steepness of the transmittance edge becomes smaller. This effect can be reversed at higher temperatures (the higher the temperature is, the faster the effect will be reversed).

The transmittance of optical glasses is in general high in the visible and at wavelengths up to 1970 nm. The infrared transmittance is influenced by the

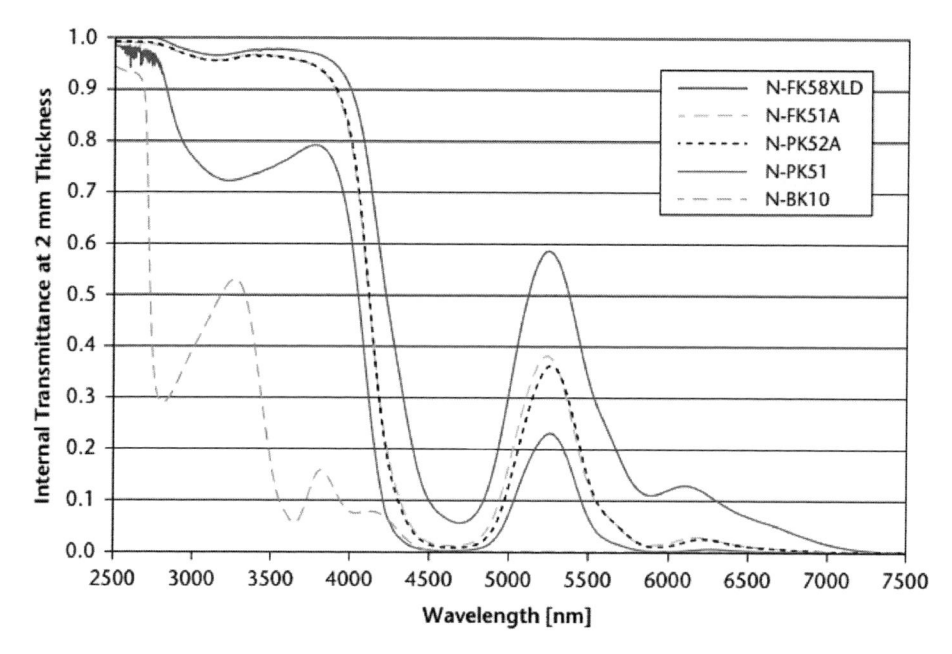

Figure 2.20 IR transmittance of different optical glasses.

OH content in the glass. The first small absorption can usually be found at 1450 nm. The main OH absorption band is sensitive to the atomic surrounding and typically occurs between 2.9 μm and 4.2 μm.

The short-wave infrared (SWIR) spectral band is located between the visible and the thermal infrared. Based on the existing data, all SCHOTT optical glasses can be used in the 0.9 to 1.7-μm SWIR spectral range. Regarding a definition of the SWIR range up to 3 to 4 μm, most of the optical glasses have a low transmittance above 2 μm. Nevertheless, fluorophosphate optical glasses offer excellent transmittance even up to 4 μm [15].

The transmittance of optical glass is measured using double-beam spectral photometers. Typical standard setups enable measurements within a wavelength region from 250 nm up to 2500 nm. The measurement accuracy over the complete spectrum is about ±0.5%. Within 400 nm to 700 nm, the accuracy is ±0.3%. The wavelength can be measured with an accuracy of ±0.2 nm and ±0.8 nm. Standard measurement sample thickness is 25 mm (20 mm x 15 mm x 25 mm). With different setups, it is possible to increase the transmittance measurement accuracy in the UV and visible range (200–850 nm) to ±0.08% and in the near-infrared range (850–2500 nm) to ±0.3% (with ±0.02 nm in the UV/VIS and ±0.08 nm in the NIR). Additionally, it is possible to expand the wavelength range from 2500 nm to 16600 nm. Within this range, the measurement accuracy is ±1%.

2.2.9 Stress birefringence measurement

Stress birefringence [16] is the translation of the thermomechanical stresses that are frozen in the glass due to the cooling of hot-formed and fine annealed glass to room temperature. Such thermomechanical stresses in terms of MPa are translated by means of the glass dependent stress optical coefficients to the so-called stress birefringence that gives optical designers a measure of how the stress inside the glass affects applications with sensitive polarized light requirements.

The stress birefringence amount therefore strongly depends on the glass type, the glass size and the temperature history of the glass part.

ISO 10110-18:2018 defines the typical required maximum stress levels for applications. Photography requires levels less than 10 nm/cm, precision optics less than 5 nm/cm and polarization optics or interference instruments less than 2 nm/cm.

For stress birefringence measurement, automatic stress mapping systems are usually used like the STRAINMATIC M4 from the company ILIS in Erlangen, Germany (see Figure 2.21). This system is loosely based on the de Senarmont and Friedel experimental setup for stress birefringence measurement [16].

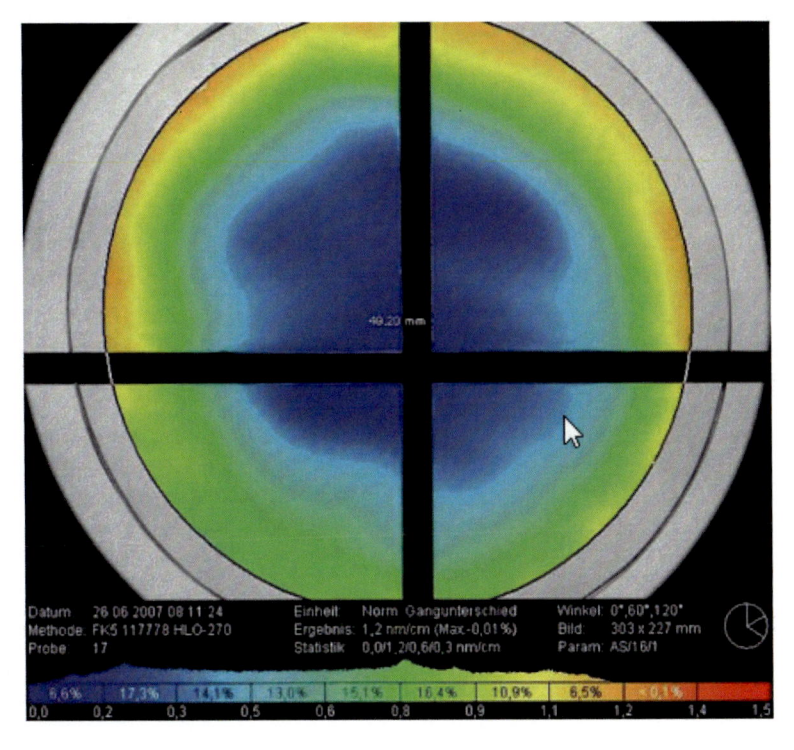

Figure 2.21 Stress birefringence map of a 270-mm-diameter FK5 i-line glass generated with the STRAINMATIC M4 from ILIS, Erlangen, Germany.

The stress birefringence is usually symmetric to the part geometry and higher near the edges. Usually, the measurement value is generated at a position about 5% from the edge.

2.2.10 Bubble and inclusion measurement

Optical glass in general is nearly free of bubbles and inclusions compared to other technical glasses due to sophisticated production processes, which are optimized for low bubble content. Nevertheless, to specify an optical glass for a desired optical component, it is helpful to know the background on the generation of bubbles and inclusions and their impact on the application. This technical information [17] gives an overview on the topics involved in the selection of the right specification for bubbles and inclusions.

The inclusion quality will be assessed by visual inspection. To visualize the bubbles and inclusions, the following measurement setup is used in general. The glass is placed on a black background and illuminated from the side. The glass is viewed from above by looking through it toward the black background. The bubbles and inclusions become visible as bright spots. This arrangement is very sensitive for the quantification of bubbles and inclusions. To determine the sizes, either comparison standards or microscopes are used. The evaluation includes all bubbles and inclusions with dimensions ≥ 0.03 mm.

The bubble content is expressed by the total cross section in mm^2 in a glass volume of 100 ccm, calculated from the sum of the detected cross sections of bubbles. Additionally the maximum allowable quantity per 100 ccm is defined. The evaluation considers all bubbles and inclusions ≥ 0.03 mm. Inclusions in glass, such as stones or crystals, are treated as bubbles that have the same cross section. Typical maximum allowable total cross sections range from 0.03 to 0.006 mm^2 per 100 ccm. The maximum quantity ranges from 10 per 100 ccm down to 2 per 100 ccm (bubbles larger than 0.3 mm).

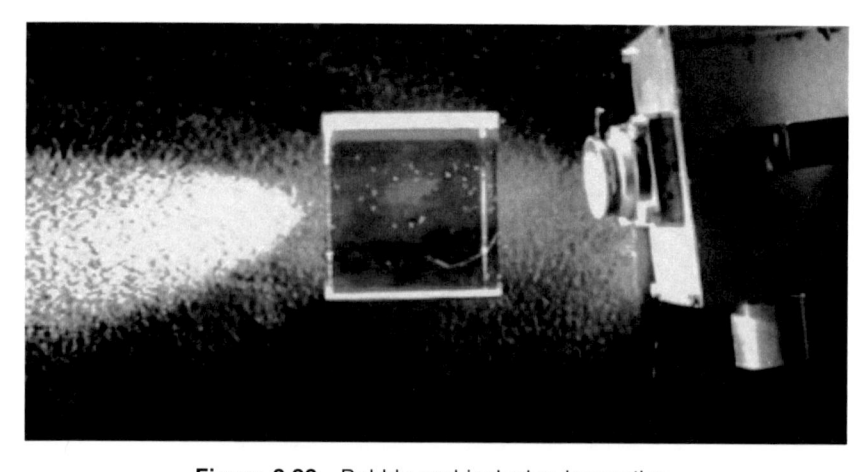

Figure 2.22 Bubble and inclusion inspection.

2.2.11 Other properties

Due to limited space, this book chapter could not cover all relevant properties of optical glasses. Some additional properties are listed here but not elaborated (no guarantee of completeness):

- Temperature dependence of the refractive index [18,19].
- Chemical properties [20]: Attack from acids, alkaline, humidty and stain sensitivity; some glasses are more sentitive than other glasses.
- Mechanical properties [21]: Young's modulus, Poisson number and CTE; some glasses are very thermal shock sensitive.
- Polishability: Some glasses (in the lower left of the Abbe diagram) are harder to polish than others [22].
- Bending strength of optical glasses glued in designs with thermal expansion differences [23].
- Precision moldability of optical glasses (requiring low glass temperature) [24].
- Fluorescence properties of optical glasses in sensitive microscopy applications [25,26].
- Bulk laser damage of optical glass [27].
- Solarization of optical glass with pulsed laser radiation [28].
- Radiation resistance of optical glass for space applications [29].
- Dispersion of optical glass in the near infrared [30].

2.3 Fused Silica Production

Silicon dioxide (SiO_2) is the simplest chemical composition of a glass and has been commercially available for more than 100 years. Nevertheless, it's still an active field of research and improvement. Quartz glass (or silica glass or vitreous silica) can be differentiated into two subgroups:

- Fused quartz, also known as natural fused silica, is made from mined raw materials.
- Fused silica, also known as synthetic fused silica, is made from pure, industrially made, chemical compounds.

2.3.1 Properties of fused silica

Several unique optical, mechanical and thermal properties have made quartz glass an indispensable material in the fabrication of high-tech products [31]. Among these are

- high chemical purity and resistance,

- high softening temperature and thermal resistance,
- low thermal expansion with high resistance to thermal shocks,
- high transparency from the ultraviolet to the infrared spectral range, and
- high radiation resistance.

Chemical composition

At first glance, quartz glass appears very simple both chemically and structurally since it is made from a single oxide component (silicon dioxide – SiO$_2$). Silica, as it is also known, is found throughout the earth's crust. However, only a very small fraction is sufficiently pure (>99.98% SiO$_2$) to be suitable as raw material for quartz glass. Sand at the beach is also mostly SiO$_2$ but isn't suitable for fusing a high-purity quartz glass.

Even at very low levels, contaminants have subtle yet significant effects. Purity is mostly determined by the raw material (see Figure 2.23), the manufacturing method and subsequent handling procedures. Special precautions must be taken at all stages of manufacture to maintain high purity. Additionally, purification steps to further improve the quality of the quartz sand as raw material are possible.

The most common impurities are metals (such as Al, Na and Fe among others), water (present as OH groups) and chlorine. These contaminants not only affect the viscosity, optical absorption and electrical properties of the quartz glass. They can also influence the properties of material processed in contact with the quartz glass during the final use application. Synthetic fused silica from Heraeus, for example, contains total metallic contamination below 1 ppm. For fused quartz, the amount is approximately 20 ppm and consists primarily of Al$_2$O$_3$ with much smaller amounts of alkalis, Fe$_2$O$_3$, TiO$_2$, MgO and ZrO$_2$.

Figure 2.23 (a) Raw Pegmatite is a very special rock type including different mineral assemblages (picture: prima91 - stock-adobe.com) [32]; (b) Natural quartz crystal (picture: Xavier - stock-adobe.com) [32]; (c) Pegmatitic quartz sand after different purification steps.

For light to be transmitted over many km, the used quartz glass needs to have a very low absorption, and it has to be of exceptional purity. Lots of glasses are impermeable to ultraviolet (UV) radiation, but quartz glass is transparent for wavelengths <350 nm. Quartz glass also exists in doped varieties with different coloration to change properties like transmission or absorption for a specific wavelength (filter glass). Opaque materials like Heraeus HOD 500 are also used as a heat barrier for diffuse scattering of infrared (IR) radiation [33].

Structure

In the quartz glass structure, all atoms are bonded with at least two others, see Figure 2.24. In combination with the strength of the silicon oxygen (Si-O) chemical bond, this is the cause of the high temperature stability and chemical resistance of quartz glass. But the structure is also rather open with wide spaces (interstices) between the structural units. This accounts for the higher gas permeability and much lower thermal expansion coefficient of quartz glass relative to most other materials.

In addition to metallic impurities, fused quartz and fused silica also contain water present as OH units. The OH content influences the physical properties like attenuation and viscosity. General, higher OH content mean lower use temperature but which is still much higher than for common optical glasses. Electrically fused quartz has the a very low hydroxyl content (<1–30 ppm) since it is produced in vacuum or a dry atmosphere. Hydroxyl incorporated by this production method is not fixed in the glass structure [34]. It can go up or down depending on the thermal treatment and amount of moisture to which the quartz glass is exposed at elevated temperature. Flame-fused quartz has significantly more hydroxyl (150–250 ppm) since fusion occurs in a hydrogen/oxygen flame. Here, the OH content cannot be changed after build up.

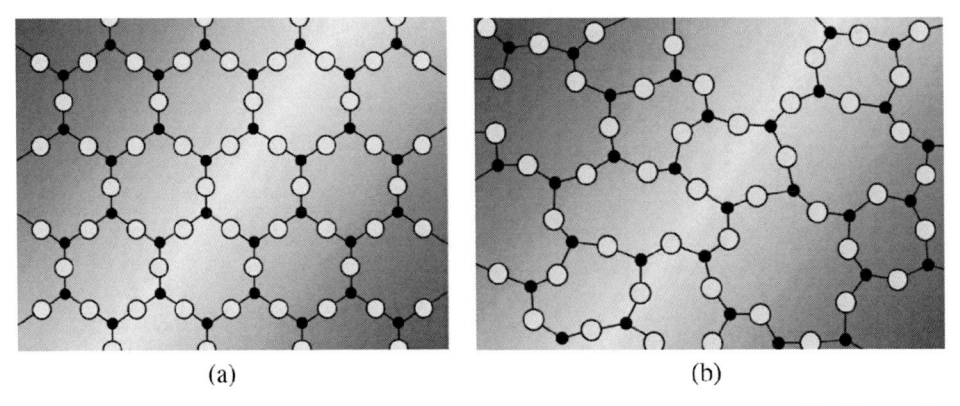

Figure 2.24 (a) Crystalline SiO_2 structure before the fusing process; (b) Glassy (or amorph) SiO_2 structure after the fusing process.

Synthetic fused silica produced by flame hydrolysis of silicon tetrachloride can have high (up to 1300 ppm) or very low (<1 ppm) hydroxyl content depending on production conditions and whether a hot chlorination step is employed to remove it. The hydroxyl level can be so high because the silica particles that result from the hydrolysis reaction are extremely small and therefore have a tremendous surface area capable of absorbing moisture present in the flame.

The main attributes of electrically fused materials are the low hydroxyl content and reduced devitrification rates. The low hydroxyl content increases infrared transparency and viscosity. The higher viscosity results in an increased maximum use temperature as well as helping to inhibit devitrification. Devitrification is also restrained by the neutral/reducing atmosphere used during melting. This causes the material to be slightly oxygen deficient, which helps to restrain devitrification.

The high resistance of quartz glass to elements and compounds is another advantage for high-end applications. Fused quartz is outstandingly resistant to water, salt solutions and acids. It is only attacked by hydrofluoric and phosphoric acid. Metals which are free from oxide, except for alkalis and alkaline earths, do not react with fused quartz or fused silica.

Quartz glass is sensitive to all alkali and alkaline-earth compounds because even slight traces of them hasten devitrification at high temperatures. It is always advisable to remove fingerprints, which contain traces of alkalis, from quartz glass with alcohol before heating [35].

Thermal Properties

One of the most attractive features of quartz glass is its very low coefficient of thermal expansion (CTE); see Figure 2.25. The average CTE

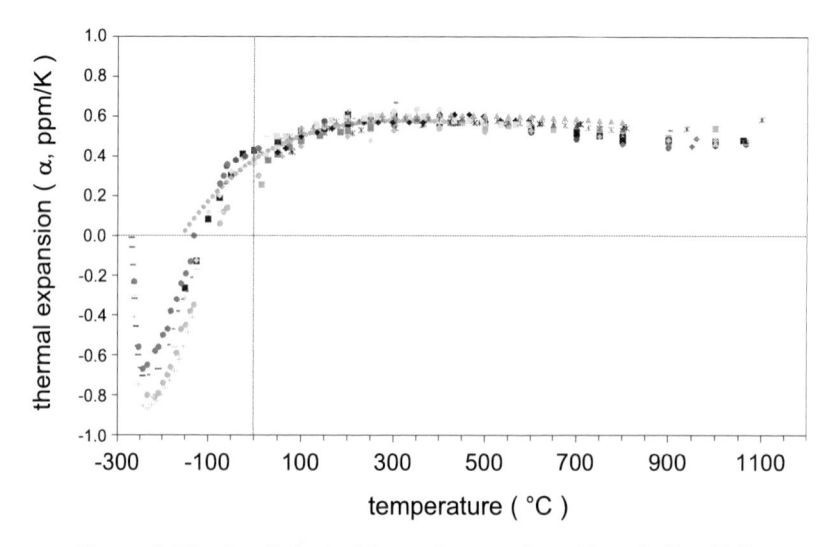

Figure 2.25 Coefficient of thermal expansion of fused silica [36].

value for quartz glass is about $5 \times 10^{-7}/°C$, is many times lower than that of other optical materials. To put this in perspective, imagine if 1-m^3 blocks of stainless steel, borosilicate glass and quartz glass were placed in a furnace and heated by 500 °C. The volume of the stainless-steel block would increase by more than 28 liters and that of the borosilicate block by 5 liters. The quartz block would expand by less than one liter. Such low expansion makes it possible for the material to withstand very severe thermal shock.

It is possible to rapidly quench thin articles of quartz glass from over 1000 °C by plunging them into cold water without breakage. However, it is important to realize that the thermal shock resistance depends on factors other than CTE such as surface condition (which defines strength) and geometry. The various types of fused silica and fused quartz have nearly identical CTEs and thus can be joined with no added risk of thermally induced breakage [36].

Mechanical Properties

The theoretical tensile strength of silica glass is greater than 1 million psi. Unfortunately, the strength observed in practice is always far below this value. The reason is that the practical strength of glass is extrinsically determined rather than being solely a result of chemistry and atomic structure, e.g., an intrinsic property like density. It is the surface quality in combination with design considerations and chemical effects of the atmosphere (water vapor in particular) that ultimately control the strength and reliability of a finished piece of quartz glass. Because of stress concentration on surface flaws, failure almost always occurs in tension rather than compression.

This could also be stated as the probability that the piece will experience a mechanical stress greater than the strength of any existing flaws. As a result of this dependence on probability, reliability decreases as the size of the glass article increases. Similarly, if the number of pieces in service increases, so does the chance of experiencing a failure.

Surface condition is very important. For example, machined surfaces tend to be weaker than fire-polished ones. Also, older surfaces are usually weaker than younger ones due to exposure to dust, moisture or general wear and tear. These factors must be considered thoroughly when comparing the strengths of different "brands" of quartz glass. This is because these tests often turn out to be just comparisons of surface quality resulting from sample preparation, small differences which easily overwhelm any differences in intrinsic strength [37].

Optical Properties

The optical properties of fused silica and fused quartz offer striking opportunities for research and industry. The broad transparent transmission range covers the complete visible spectrum and extends far into the infrared and ultraviolet regions; see Figures 2.26 and 2.27. Transmission is mainly be influenced by material purity. Additionally, transmission ranges can be customized by adding doping materials.

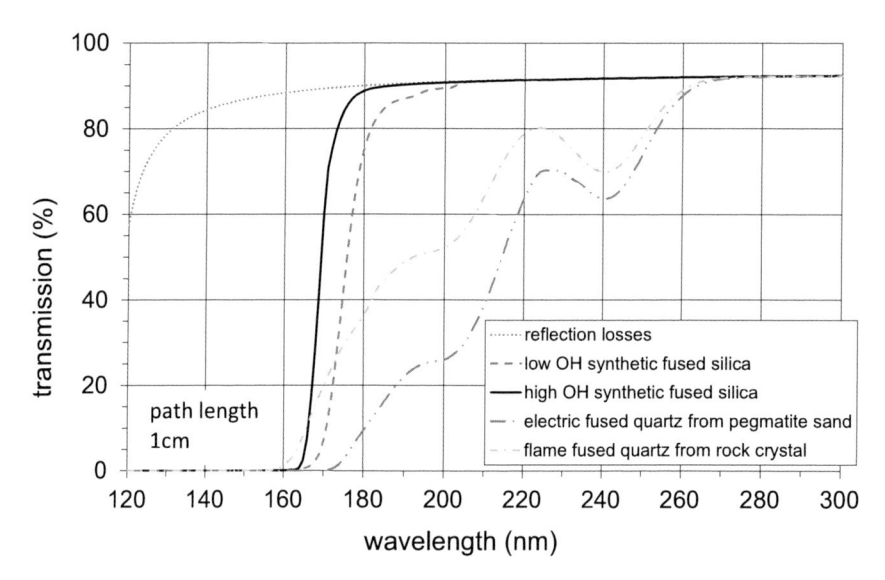

Figure 2.26 Typical quartz glass transmission in the ultraviolet for a 1-cm-thick sample.

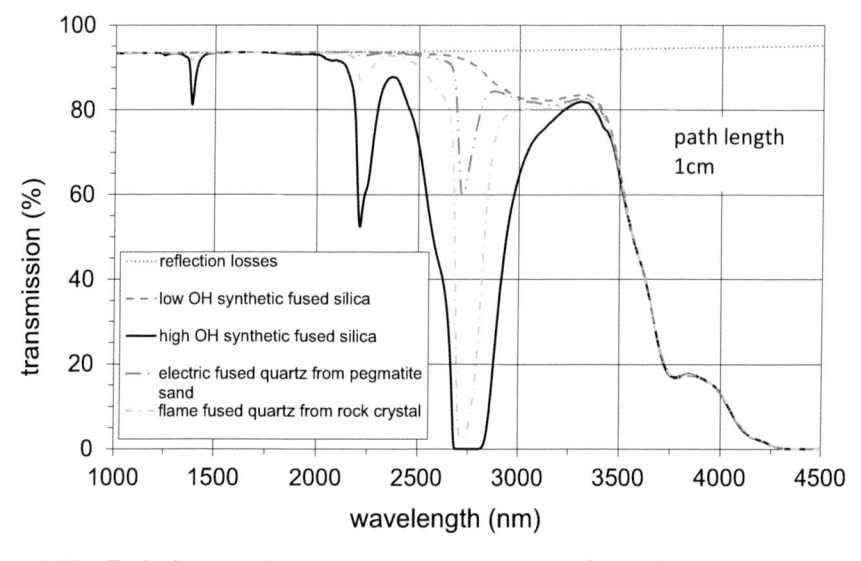

Figure 2.27 Typical quartz glass transmission in the near-infrared for a 1-cm-thick sample.

The intrinsic UV and IR absorption edges in silica glass are located at roughly 0.180-and 3.5-μm wavelengths, respectively. The intrinsic UV absorption edge results from the onset of electronic transitions with Si-O network at the point where photon energy exceeds the network bandgap energy. The intrinsic IR edge arises due to lattice (multi-phonon) vibrations of the Si-O network.

Various overtones of the fundamental SiO4 tetrahedron vibrational modes are the first to be observed. These intrinsic absorption edges are then further modified by the presence of impurities. Metallic impurities shift the UV edge to higher wavelengths. Water (OH) introduces absorption bands just below the IR edge. The strongest of these is the fundamental O-H stretching band at 2.73 μm [38].

Electrical Properties

Quartz glass is a good electrical insulator, retaining high resistivity at elevated temperatures and excellent high-frequency characteristics; see Figure 2.28. The large band gap inherent in the electronic structure of the silicon-oxygen bond results in electrical conduction being limited to the current carried by mobile ionic impurities. Since the level of these impurities is very low, the electrical resistivity is correspondingly high [39].

Since ionic conduction is related to the diffusion coefficient of the ionic carriers, the resistivity also has a strong exponential temperature dependence. Unlike typical conductors such as metals, the resistivity decreases with increasing temperature.

The dielectric constant of quartz glass has a value of about 4 which is significantly lower than that of other glasses. This value changes little over a wide range of frequencies. The reason for the low dielectric constant is, once again, the lack of highly charged mobile ions, but it also results from the stiffness of the silicon oxygen network which imparts a very low polarizability to the structure [40].

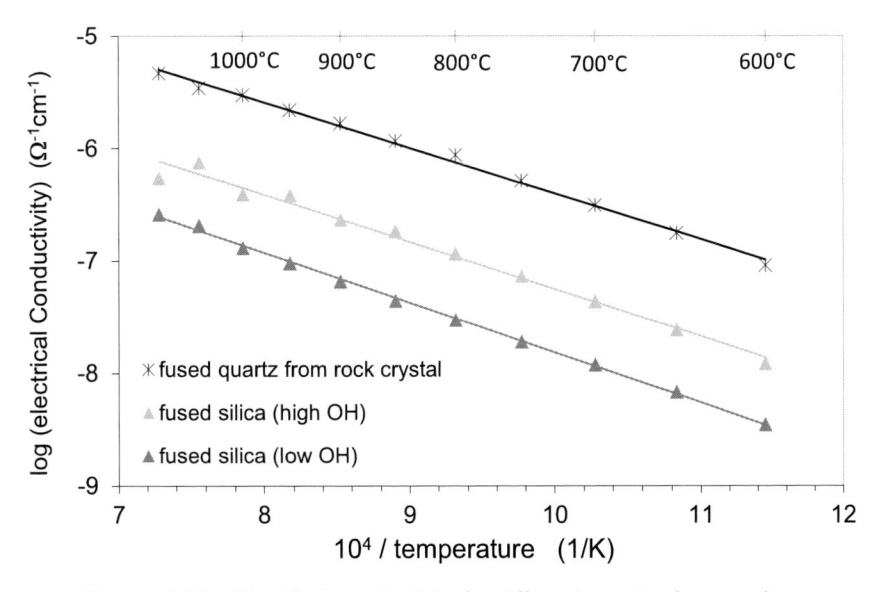

Figure 2.28 Electrical conductivity for different quartz glass grades.

2.3.2 Making of fused quartz and fused silica

There are two distinctive ways to produce quartz glass. One starts with high purity quartz or other silicon-dioxide-containing minerals that are fused using various heat sources. The other starts with gaseous silicon containing chemicals (e.g., $SiCl_4$) that are burned in the presence of oxygen to form silicon dioxide.

2.3.2.1 Fused quartz

Electric fusion

Electric fusion is the most used melting process for manufacturing quartz glass. Two methods of electric fusion can be used:

- Continuous fusion: In the continuous method, quartz sand is poured into the top of a vertical melter that consists of a refractory metal crucible surrounded by electric heating elements; see Figure 2.29(a). The interior is maintained in a neutral or slightly reducing atmosphere that keeps the silica from reacting with the refractory metal. The melted material exits the bottom orifice of the crucible, which is shaped to produce rods, tubes, plates or other products of various dimensions.

- Batch or boule fusion: In the batch fusion method, a large quantity of raw material is placed inside a refractory lined vacuum chamber which also contains heating elements. Although this method has historically been used to produce large single boules of material, it can also be adapted to produce much smaller, near-net shapes.

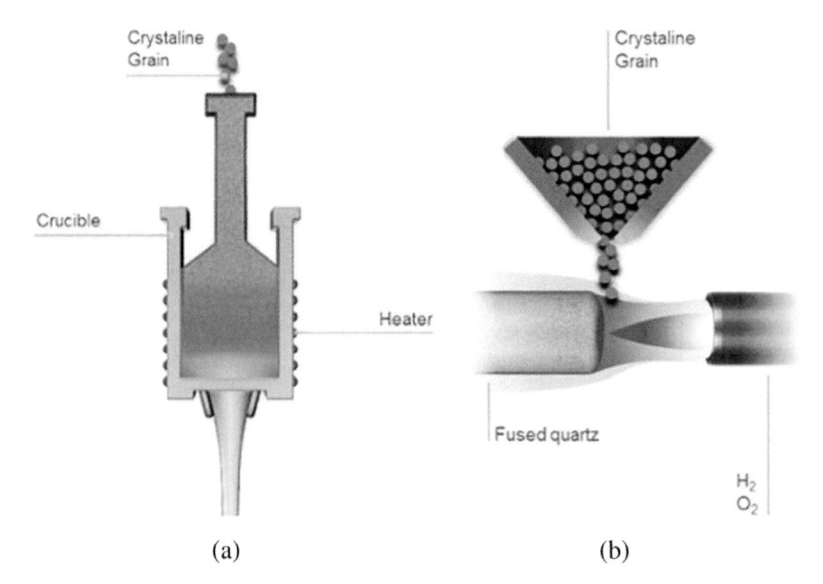

Figure 2.29 (a) Continious electric fusion of fused quartz; (b) flame-fused quartz manufacturing.

Flame fusion

Historically, the first method of producing fused quartz was by small-scale fusion of quartz crystals in a flame. Heraeus chemist Richard Küch first began fusing quartz rock crystal in a hydrogen/oxygen (H_2/O_2) flame more than 100 years ago. Heraeus has been producing quartz glass on an industrial scale with this process ever since.

Today, flame-fused quartz is manufactured on a large scale by a continuous process in which highly refined quartz sand is fed through a high-temperature flame and deposited on the surface of a melt contained in a tank lined with refractory material; see Figure 2.29(b). The viscous melt is withdrawn slowly through a die in the bottom of this tank, and it solidifies in a shape determined by the die. In this way, it is possible to produce an ingot of transparent fused quartz of the desired cross-section (round, rectangular or hollow), which is cut off at intervals and removed for further processing.

2.3.2.2 Fused silica

In this process, the silicon containing precursors (e.g., $SiCl_4$) are burned in the presence of oxygen to form nanoparticles of silicon dioxide, also called soot. Because the precursors are specifically produced and refined, they are available in exceptionally high purity and the resulting fused silica has a very low metallic impurity content.

As the production process involves vapors of chemicals (silicon-containing precursors), it is called chemical vapor deposition (CVD). There are two sets of process families, one where the deposited nanoparticles are directly melted to a condensed fused silica layer, and one, where the soot is accumulated and in a secondary process step condensed to transparent fused silica (this process is called vitrification).

One-step fused silica production

Direct quartz (DQ): This method is widely used to produce fused silica for optical applications. The silicon dioxide nanoparticles generated by the combustion of a silicon containing precursor are directly and transparently fused to a quartz glass ingot which is redrawn in the same speed as the length of the ingot increases due to deposition. OH contents of more than 400 ppm are typical.

Chemical vapor deposition (CVD): To produce optical fiber core rods, deposition of fused silica with a defined refractive index is done inside of fused silica tubes. The chemicals are brought into the tube by a carrier gas. The reaction to form soot is triggered by a heat source. There are different heat sources employed, and they differentiate the flavors of the CVD process. The heat source is either a flame (MCVD), a furnace (FCVD) or a plasma (PCVD). All gases that have not reacted are treated in a scrubber.

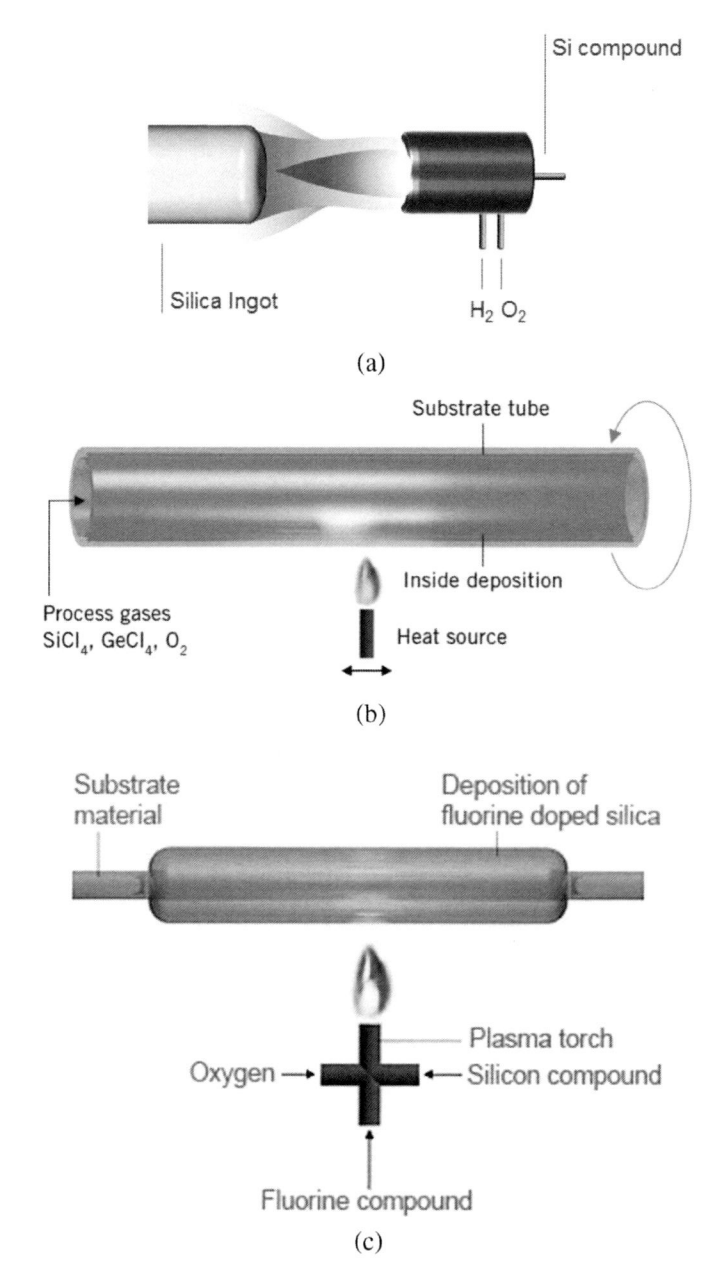

Figure 2.30 (a) Direct quartz (DQ); (b) chemical vapor deposition (CVD); (c) plasma outside deposition (POD).

Plasma outside deposition (POD): For this process, a plasma heat source is used to "burn" chemical precursors and deposit a thin glass layer on a rotating target. The target can be a tube or a solid rod, but it is not necessarily round. Because of the high temperature of the plasma, this process is best suited to produce fluorine-doped fused silica. The achievable maximum content of fluorine in silica is a function of deposition temperature. There is a limit to the maximum fluorine content, because fluorine also etches away fused silica. The plasma outside deposition is typically employed to produce highly-fluorine-doped silica, which has a lower refractive index than undoped fused silica. This difference in refractive index is needed for optical fibers.

Two-step fused silica production

Outside vapor deposition (OVD): In this process, soot is deposited on a rotating bait rod or on the end of a rod that is pulled upwards (vapor axial deposition; VAD). The soot accumulates and forms a porous body with a density that is less than 25% of that of fused silica. This porous body is then vitrified in a consecutive step to transparent fused silica; see Figure 2.31. Because of its high surface, it is easy to dope the porous soot body. In the fiber optic industry, hydrogen is substituted by chlorine in a dehydration step before the soot body is vitrified.

2.3.3 Homogenization of quartz glass

While the processes described above may yield in a quite reasonable peak-to-valley refractive index homogeneity [41], the residual local inhomogeneities do not necessarily correspond to the shape of the optical element in mind. However, low-order spherical aberrations like the power term can be corrected quite easily.

To improve the local homogeneity of quartz glass further, homogenization processes can be used. Unfortunately, the techniques developed for

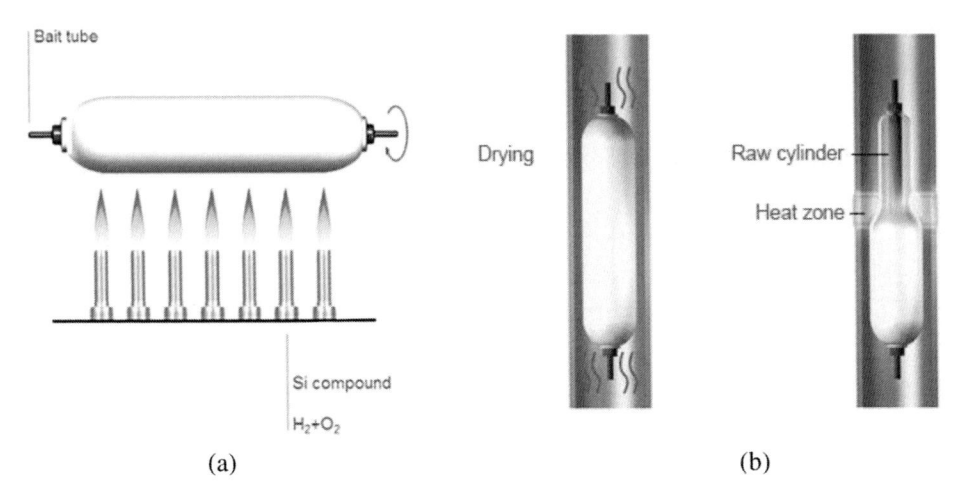

Figure 2.31 Outside vapor deposition (OVD): (a) porous soot body generation; (b) dehydration and vitrification of the soot body.

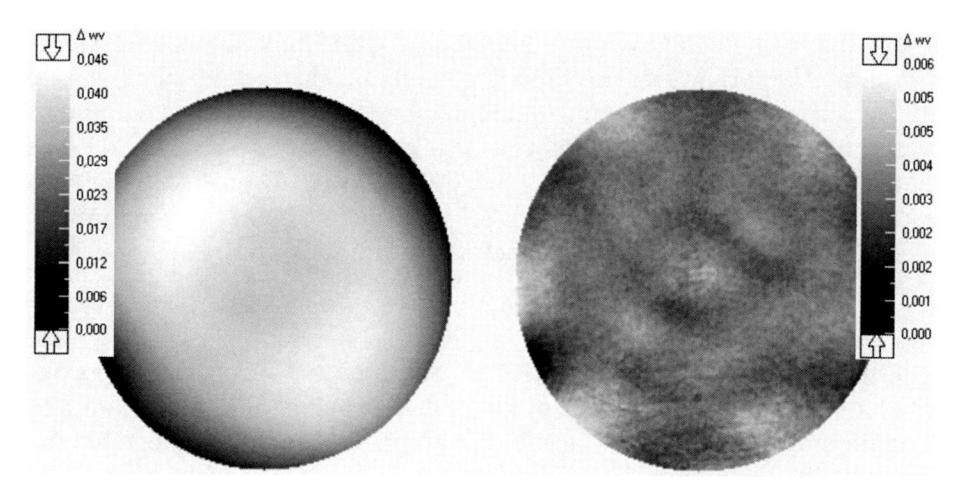

Figure 2.32 Optical homogeneity: (left) offset and tilt removed; (right) all 36 Zernike terms are removed.

optical glass do not work because quartz glass does not feature a liquid phase. Nevertheless, tool-free methods have been developed and used for decades now. However, Heraeus Quarzglas is the only company worldwide using a post-production homogenization process to further optimize the optical homogeneity. It is possible to produce quartz glass to the requested specification without the need to select the material. Particularly for larger sizes, it may prove difficult to find the desired properties by selection.

The homogenized quartz glass features a low residual inhomogeneity after removal of all 36 Zernike terms of optical aberration and in its 3D variant shows no striae which are nearly inevitable after raw material deposition [42].

2.3.4 Shaping and forming of quartz glass

You can process fused quartz and fused silica as any other glass. The processes can be grouped into mechanical processing and hot forming.

2.3.4.1 Mechanical processing

Due to its hardness, quartz glass requires diamond tools to process it mechanically. Because it is fragile, there is a limit to the force it can withstand before cracking, and thus the feed rate during processing needs to be chosen carefully. The general processes include

- Cutting: Typically, band and wire saws as well as chop saws are used to cut quartz glass. The cut loss and the surface quality depend on the machine and the feed speed. It is also possible to cut quartz glass using CO_6 lasers or a water jet machine. While a laser produces flame-glazed surfaces, a water jet produces a rougher surface. If the thickness of the quartz glass is too large for a single cut with a laser, it is possible to do multiple consecutive cuts. After a laser cut, it is necessary to anneal the glass piece to relieve thermally induced stress and keep the part from shattering. Both a laser and a water jet allow the manufacturing of quartz glass plates of simple or complex contours.

- Drilling: Holes in or small rods of quartz glass can be produced with diamond drills. It is important to ensure proper cooling to prevent the tools from failing prematurely. For thin pieces of quartz glass, e.g., in plates, it may be easiest to use a laser to drill holes.
- Grinding
- Milling

2.3.4.2 Hot forming

A lot of experience is necessary to do hot forming of quartz glass. Due to the high melting point and steep viscosity, the temperature range in which it can be processed is rather narrow. If the temperature is too low, quartz glass is still solid; if it is too high, it gets very soft, almost liquid, and it evaporates to form soot dust. Additionally, hot processing quartz glass requires an annealing step (or multiple intermediate annealing steps) to relieve thermal stress. If not done correctly, fracturing of the glass can happen any time.

- Welding: Two components of quartz glass can be joined by welding. It is important to heat the two components and keep the temperature high enough to avoid a critical level of thermal stress. A small rod of quartz glass (a welding rod) is melted locally in order to fill the gap between the two glass components. Once the melt seam is done, or for large parts in multiple sections, the part needs to be annealed to relieve thermal stress.
- Glass blowing
- Elongation / compressing
- Collapsing: Typically, a tube is locally heated to the softening temperature; a partial vacuum is applied to the inside, and the tube reduces in diameter. The process is frequently employed in fiber optic production to collapse the cladding onto the core rod. In its extreme form, the process can yield rods, from a tube. This is typically the case in CVD processing.
- Slumping: A state-of-the-art process to generate big optical blanks is hot slumping. The raw material could go through several steps to increase the size up to 1.5 m in diameter; see Figure 2.33. The limit of this process is given by the furnace size and the ability of fused silica to flow into the empty part of the mold provided.

Figure 2.33 1.5-m-diameter fused silica blank produced via hot forming.

2.3.5 Doping quartz glass

Although fused silica has excellent optical properties, sometimes the addition of dopants is required.

Because of its high purity, small impurities can change the properties of fused silica quite dramatically. If the impurities are added on purpose, it is called doping. The reason to dope is primarily to change either the transmission characteristics or the mechanical properties.

- Titanium for applications requiring UVB light and UVC radiation blockage.
- Cerium for UV blockage below 380 nm.
- Ytterbium, erbium, neodymium, holmium for laser light converters or fiber lasers.
- Fluorine or germanium for modification of refractive index.
- H_2 / D_2 for radiation resistance.
- Samarium or copper as optical filter.
- Aluminum for higher operation temperature.
- Boron for localized strain.

If quartz glass is used as an edge filter in the UV, the operation temperature is important. With increasing temperature, the band edge shifts to longer wavelengths [43].

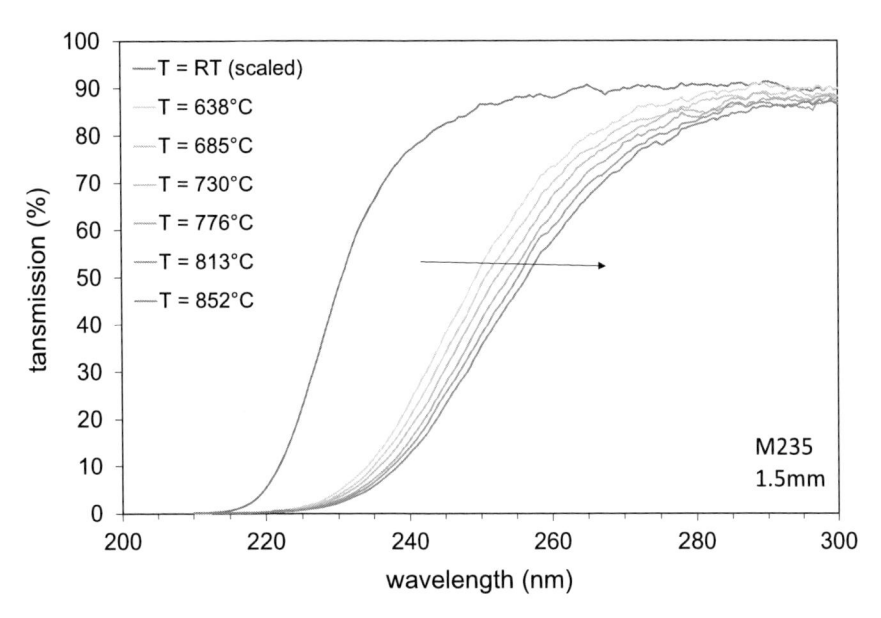

Figure 2.34 Temperature-dependent band edge shift of M235.

2.3.6 Quartz glass for applications in the near-infrared

The wavelength region from 800 nm to 3000 nm is defined as the near-infrared (NIR) spectrum. For all optical applications, it is important that the optical components transmit light with as little distortion as possible. For laser applications, an additional requirement is very low absorption [44]. Absorption causes local heating of the optical component. This heat can cause stress and a change in refractive index, resulting in focus shift. Absorption may cause not only poor performance but also damage to and destruction of the optical component. There are two primary sources for absorption in the near IR: metallic impurities and OH groups. For synthetic fused silica, the metallic impurities are typically in the ppb range and can be neglected [45].

For the larger part of the NIR spectrum, the absorption is caused by OH molecules. These OH groups are present in the SiO_2 glass matrix due to the production process. Among other parameters, fused silica grades can be differentiated by their OH content. There are grades that have a high, medium or low OH content. The position of the vibrational or rotational excitation of the OH molecules in fused silica has been extensively analyzed in optical fibers, where it is the cause of signal attenuation [46]. In Figure 2.35, there are two lines: one where the influence of the OH content is large (yellow line), and one where it is small (red line). For those lines, we will compare the performance of various fused silica grades. To convert attenuation into absorption, the attenuation value in dB/km must be multiplied by 2.3 to get absorption in ppm/cm.

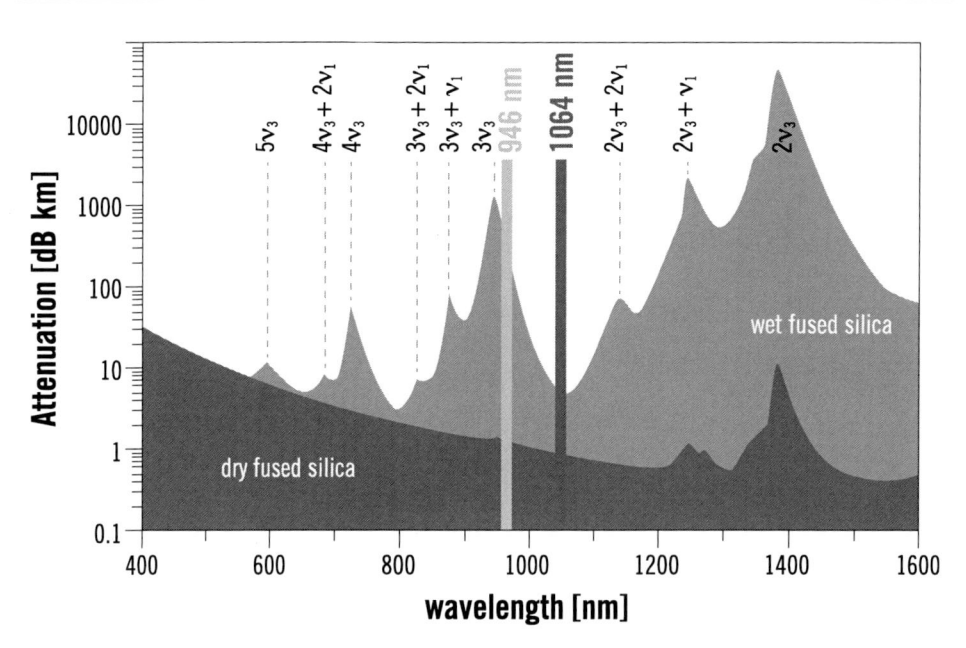

Figure 2.35 Vibrational or rotational excitation of the OH molecules in fused silica [45].

Absorption measurements on samples of Heraeus quartz glass grades (Infrasil® 30X (electric fused quartz), Suprasil® 300X and Suprasil® 31X (both synthetic fused silica) were done for the strong OH absorption case (946 nm) and the weaker OH absorption case (1064 nm) [47].

The absorption at 946 nm is dominated by the OH content. For very-low-OH content, the absorption due to metallic impurities comes into effect. For applications at 1064 nm the OH content plays a minor role compared to the metallic impurities. All synthetic grades, even the high-OH grades, show a low absorption; however, there are applications where even such low absorptions must be minimized [42].

Besides absorption, bulk scatter also contributes to intensity loss, especially if a long optical path length like in optical fibers is involved. Depending on wavelength, bulk scatter can be an important loss channel [48] and scales with n^8/λ^4.

2.3.7 Laser-induced damage threshold of quartz glass

The development of today's laser systems is reaching the limit of optical components: coatings come off, surface damage due to polish quality and insufficient cleaning or bulk damage. Several properties of the laser light source are driving the damage, e.g., fluence, energy, intensity, pulse duration. Multiple measurements can be found in the *SPIE Laser Damage* proceedings, e.g., of 2019 [49] or in the years before. It is being attempted to map the full portfolio of different parameter sets, but it is not fully understood up to now.

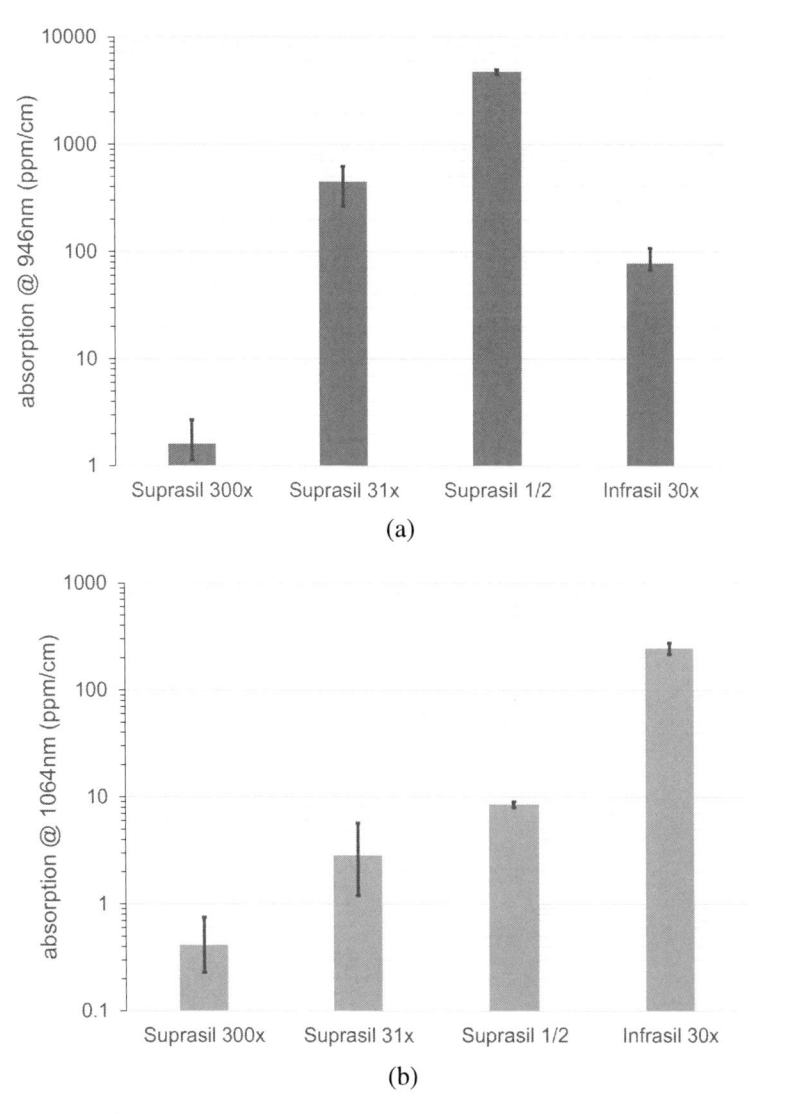

Figure 2.36 Quartz glass absorption coefficients: (a) at 946 nm; (b) at 1064 nm.

Basically, there are a lot of laser-induced damage threshold (LIDT) values for individual sets of parameters as mentioned above. The focus of the studies is on the coating and the surface of the optical component, partially on the bulk material itself. Past observations during laser operation have not revealed the bulk material as the main driver for LIDT. However, since the laser operating performance is reaching tens of kilowatts, raw material manufacturers try to understand the contribution of the bulk material to the overall LIDT.

- **Optical homogeneity / striae**: the conservation of the wavefront is crucial for a lot of transmissive optical components. Strong gradients could lead to disturbance and generation of spikes in the intensity distribution of the light, leading to values above the LIDT. Optical homogeneity and striae measurements enable high-performance quartz glass grades.

- **Inclusions**: Quartz glass might be interspersed with inclusions depending on production process, necessary oversize or handling (tools). Inclusions are typically dissolved in the fused silica matrix during further hot forming. Remaining inclusions >60 μm can be detected by a trained naked eye. Inclusions down to 5 μm can be measured by an automated bubble and inclusion inspection system (ABII); see Figure 2.37.

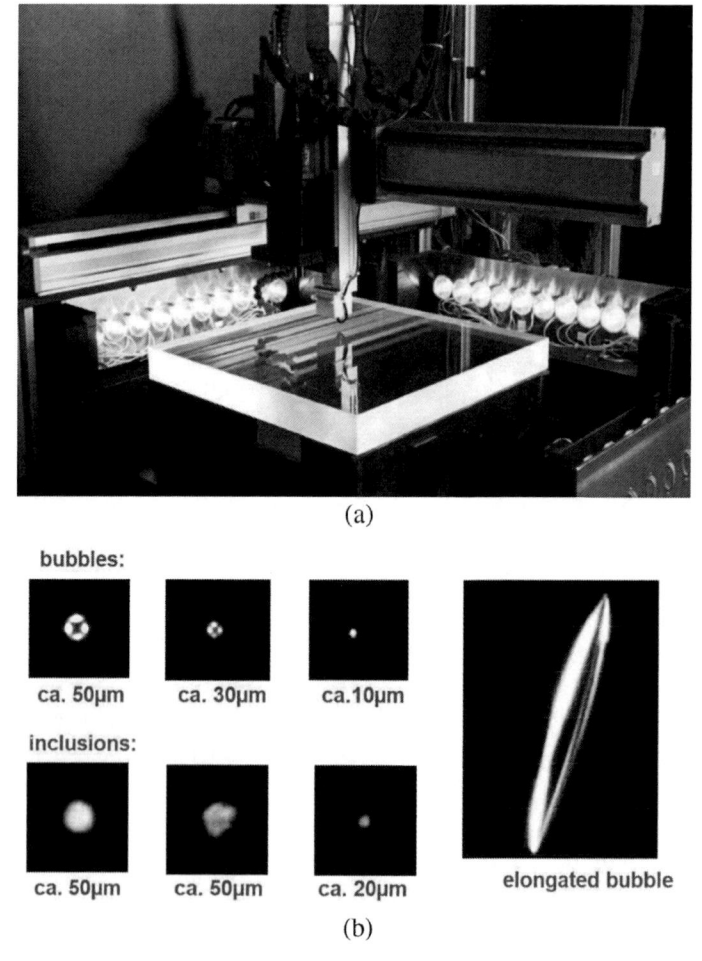

Figure 2.37 (a) Automated bubble and inclusion inspection system, resolution 5 μm; (b) examples and differences of inclusions and bubbles.

- **Bubbles**: The nature of glass production does not prevent gas bubbles from becoming incorporated into the material. You can tune the process parameters and the growth speed, so that a minimum number of bubbles occur. Compared to regular glass, optical quartz glass is nearly free from bubbles. Glass manufactures group their materials in bubble classes according to ISO 58927 or in total cross-sections of all bubbles (TBCS), ISO 10110. A standard bubbles specification for an optical glass is class 0, where no bubbles <80 μm are counted. However, single bubbles with a diameter of several tens of micrometers will affect the performance of high-power high-energy operating laser systems. Bubbles as optical elements modulate the wavefront and generate intensity spikes in the material that reach the LIDT. In addition, the distorted wavefront can lead to material flaking at the back side of the optical component. The optical element is damaged, and the damage growth can even affect subsequent elements. Heraeus has a unique homogenization process (described in Section 2.3.3) that reduces the bubble content significantly so that quartz glass can be manufactured with no bubbles >10 μm in diameter and measured with an ABII; see Figure 2.37.

2.3.8 Applications

Due to its unique properties, quartz glass is used for a wide variety of applications:

- Extended range of transmission from UV to NIR, e.g., analytics, 3D printing, UV annealing and curing, UV disinfection, semiconductor metrology, view ports.
- Low NIR absorption, e.g., laser material processing, fiber lasers, laser components for long-pulse, short-pulse and cw systems, astronomy, fusion energy, medical.
- Purity: chemistry, semiconductor, pharma, spectroscopy or aerospace.
- Radiation hardness: UV applications, space exploration [50], accelerator technology, fundamental research.
- Thermal stability: high- and low-temperature applications, e.g., in furnace view ports, semiconductor production, cryogenic applications, space.
- Low coefficient of thermal expansion, e.g., astronomy [51] or metrology.
- Hardness: optical finishers love to work with quartz glass.

2.4 Crystalline Materials Production

2.4.1 History

The use of crystals for optics can be traced back to the dawn of history, but no one can define an exact start. In the British Museum in London, one will find the so called "Nimrud Lens", a convex lens made of rock crystal dating back to around 750 BCE in modern-day Iraq. Whether it was used for optical purposes is not known. Pliny the Elder tells us in his *Natural History* the Roman emperor Nero wore spectacles made of emerald:

Nero princeps gladiatorum pugnas spectabat in smaragdo. – Pliny, *Natural History*, 37.16.

The princeps Nero viewed the combats of the gladiators in a smaragdus. Whether these green gemstones were only a sun shade or carved into lenses to aid his nearsightedness will remain unclear.

Through the ages, one can find lenses made of crystal – mostly rock crystal – but it is glass that has dominated optics.

In the 1880s, Ernst Abbe, professor for physics at Jena University, laid the foundation for modern, theory-based optics. This not only led to scientific glass making and the foundation of the Schott company in Jena in 1884 but also to a new assessment of crystals and their optical properties. One of the materials of interest was calcium fluoride (CaF_2), a crystal with low dispersion and a high Abbe number which was definitely beyond the range achievable with optical glasses at that time. In 1886, Abbe started journeys to Switzerland where he "harvested" CaF_2 from nature to be used as color-correction lenses in microscopes of the Jena-based Zeiss company.

In the 1930s, the Swiss deposits were depleted of optical-grade material. At the same time, technology was ready to provide all prerequisites needed for the artificial growth of crystals. The technology developed then is basically the same until today. CaF_2 for optics might have been the first crystal grown on industrial scale.

2.4.2 The crystalline state

The optical materials dealt with here have a defined chemical composition, but there is a difference between glass and crystals. Glass as an under-cooled liquid has no medium-range or long-range order of the atoms and ions, distances and angles between the components of a crystal are fixed in a lattice; see Figure 2.24.

There is one easy consequence: In a glass, it doesn't matter which direction the light is propagating. The average structure is the same, i.e., it is isotropic. In a well-ordered lattice, the number and distances of components are different for different directions. The material is *anisotropic*. Since this anisotropic behavior is sometimes connected with anisotropic optical properties, it is important to agree on a convention describing the lattice.

Optical Materials

Commonly used are the Miller indices, based on the works of crystallographer William Hallowes Miller (1801–1880).

Since a crystal is an ever-repeating set of atoms (ions) and distances, it is sufficient to describe the elementary cell, the smallest volume of the lattice-containing component and angle relation. For a cubic crystal like calcium fluoride or sodium chloride, a 2 × 2 × 2 cube with positive ions sitting in the corners is sufficient. One corner is the origin of coordinates, directions are defined by the number of lengths of elementary cells from this point in x, y and z direction. The notation in Miller's system is [h,k,l] in square brackets; see Figure 2.38. Perpendicular on each direction there is a crystal plane with notation (h,k,l) in round brackets. Such directions and planes are more than just an atomic description. They are connected with macroscopic properties like stress birefringence and cleavability. This is why knowledge about this notation is important for optical designers as well as for production engineers.

2.4.3 Crystal growth

One of the best-known crystal-generation processes is growth from an oversaturated solution. This is how rock salt is formed, being the remains of a prehistoric ocean. Industrial growth of CaF_2 and other members of the fluoride family works a different way. The starting material is the crystal composition already but comes as a powder or granulate. It is melted, and the liquid is then cooled down in a time regime allowing the crystal to be built atomic or ionic layer by layer. It is a 100% physical process with no chemical reaction taking place. Since solidification takes place at one specific temperature – the melting point – control of temperature and temperature gradient is key to growth success. Two main ways are pursued here.

A seed of the desired crystal is dipped into a liquid of the same composition and then slowly pulled out. At the interface between the liquid

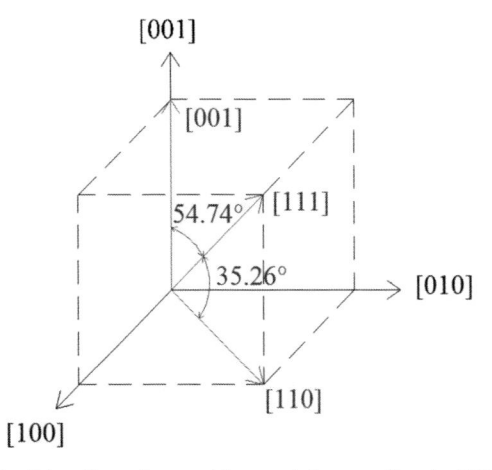

Figure 2.38 Directions in a cubic crystal according to Miller indices.

and the cooler ambient environment, the crystal grows onto the seed material. By adjusting the pull velocity, the diameter of the produced crystal can be varied. This so-called Czochralski process is the method of choice for some semiconductor crystals like silicon and germanium or oxides like sapphire (Al_2O_3).

In a different approach, the liquid remains in its reservoir. A temperature gradient has a cooler bottom and a warmer top. By lowering the temperature, the melting point moves through the volume from bottom to top, allowing the crystal to grow. At the end, the crystal has the shape of the reservoir and can be removed; see Figure 2.39. This Bridgman-Stockbarger process is the method of choice for fluorides like CaF_2, BaF_2 and MgF_2.

2.4.4 Optical ceramics

Besides glass and crystals there is a third class of material used in optics – ceramics. Strictly speaking, ceramics should be seen as a special form of crystalline material. Whereas the crystals we discussed in the previous paragraphs are produced to have an endlessly repeating lattice over the whole volume – called mono crystalline, a ceramic consists of crystallites (grains) in the micrometer size range, which are connected to their neighbors with no preferred crystal orientation. This makes ceramics crystalline and anisotropic on the microscopic scale but isotropic on the macroscopic scale.

Common ceramics for optical purposes are zinc sulfide (ZnS) and zinc selenide (ZnSe). They are widely used in optics working in the infrared (IR) region.

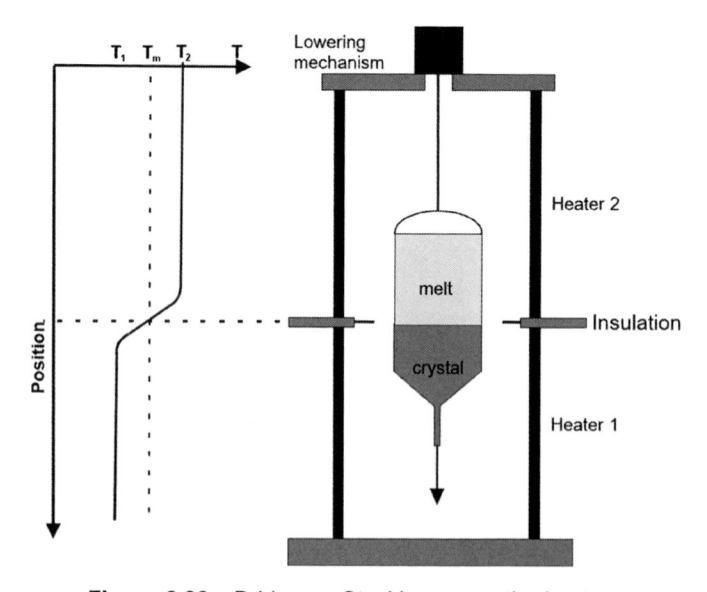

Figure 2.39 Bridgman-Stockbarger method setup.

While most classic ceramics are made by sintering an existing powder mixture using elevated temperature and sometimes pressure, most of the ZnS and ZnSe material is produced on a different route.

Chemical vapor deposition (CVD) is a process including a chemical reaction, which is different from the crystal growth process described above; see Figure 2.40. For zinc-based ceramics, one starts with metallic Zn and gaseous compounds, either H_2S or H_2Se. In an environment of reduced atmospheric pressure and elevated temperatures, the Zn evaporates, enabling a gas phase reaction and a subsequent deposition on a cool target within the reaction chamber. Management of flow and concentration of reactant gases is key to obtain a uniform stoichiometric product during the whole production cycle.

Other properties directly influenced by the CVD parameters are grain size distribution, bulk density and scattering. The production target is a uniform distribution of all properties throughout the harvested material. The CVD process yields either plates or pre-shaped geometries, depending on the shape of the deposition target. Sizes of plates produced can range up to 50" in diagonal.

2.4.5 Post-growth heat treatment

After unloading the crystalline materials from their growth or deposition vessels, a first quality assessment is carried out. For some products, the

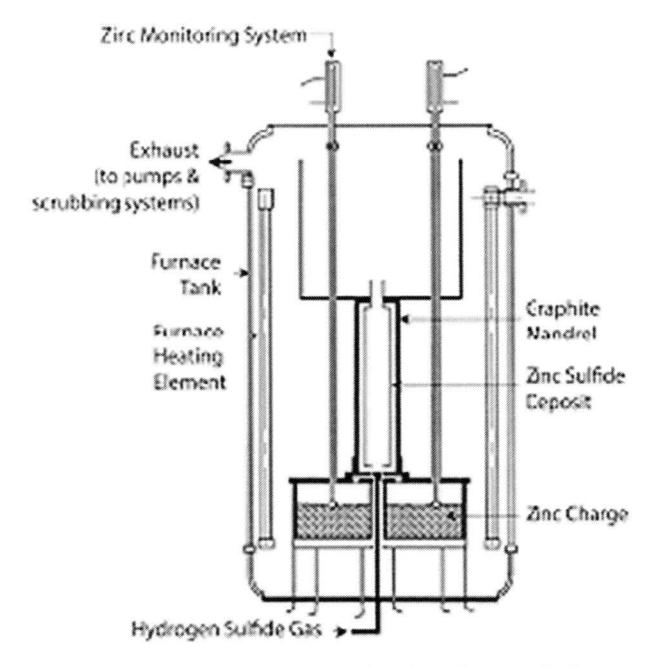

Figure 2.40 Schematics of a CVD furnace [52].

obtained quality is already sufficient, and the material can be used to make optical components. But there are features that can be improved by an additional heat treatment, e.g., stress in fluoride crystals and transmission in ZnS.

2.4.5.1 Crystal annealing

While growing, crystals incorporate defects of different lateral dimensions in their lattice. Point defects are un-occupied places in the lattice or interstitial atoms or ions. Even with a 99.999% purity, each volume of 1 cm^3 contains 10^{18} impurity sites. Point defects cannot be directly visualized, but they affect material properties like transmission and laser irradiation resistance. Since they are unavoidable their effect has to be minimized. Line defects are additional or missing planes within a certain volume of the crystal lattice. The macroscopic effect of a non-aligned crystal lattice is stress which often looks similar to striae in optical glasses. Other than in glasses, these "striae" are no change in composition and density of the material. And last but not least, an overall stress can occur due to too-fast and non-isotropic cooling after the growth process.

These defects of an as-grown crystal are addressed by a subsequent annealing step. Mono-crystalline and in most cases pre-oriented blocks of crystalline material are heated to a temperature well below melting point but will allow the point defects to move through the lattice occupying their site of minimized energy. Some of the planar defects can move, too and the overall stress level can be reduced with the right cooling regime down to room temperature. Effects on stress in a CaF_2 crystal disc are shown in Figure 2.41.

For the non-isotropic crystal, annealing results are dependent from crystal orientation. Targeting for a certain stress level may require different annealing temperatures and durations. Applying the same "recipe" to different

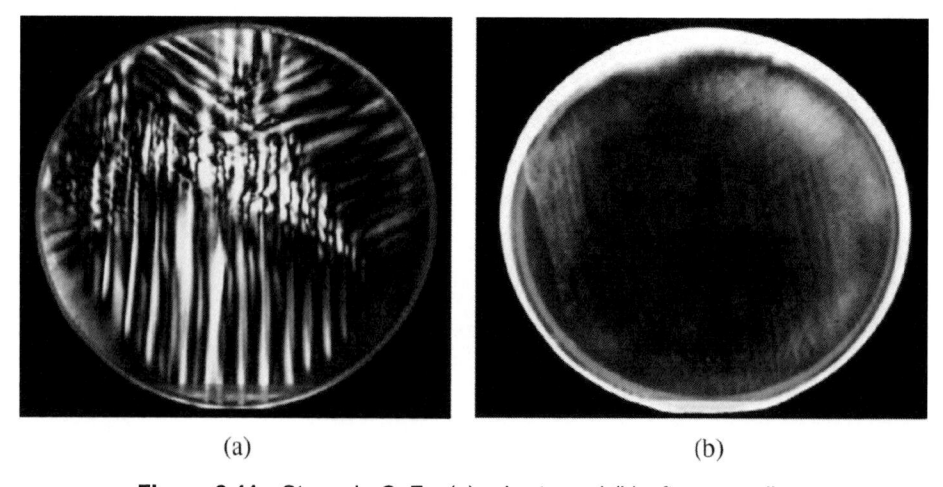

(a)   (b)

Figure 2.41 Stress in CaF_2: (a) prior to and (b) after annealing.

Optical Materials 59

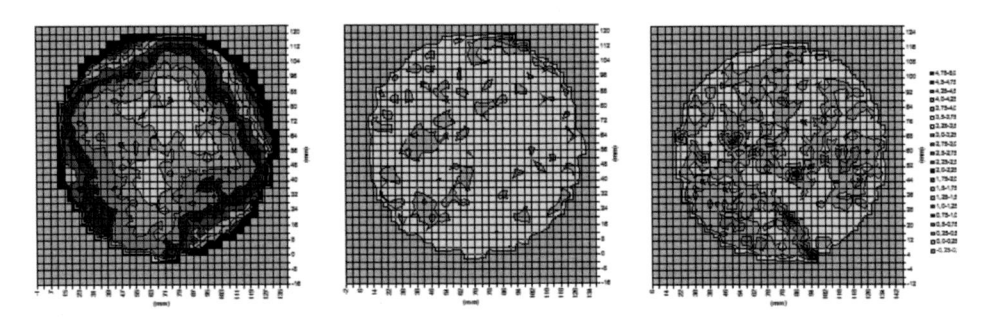

Figure 2.42 Stress in CaF$_2$ along different crystal axis: (left) (100) orientation, stress = 8.5; (middle) (111) orientation, stress = 1; (right) (110) orientation, stress = 1.5.

orientations results in huge differences in stress. Figure 2.42 shows three CaF$_2$ blanks, cut out of the same crystal in different orientations and annealed simultaneously. The stress is normalized to the (111) orientation.

2.4.5.2 Hot isostatic pressing of ceramics

Comparable to the monolithic crystals for optics, ceramics can be used in the "as deposited" state or after a further treatment, depending on the properties requested by the application. For the CVD-produced ZnS ceramics there is a tremendous change in properties. Whereas the material directly obtained from deposition is yellow and opaque, it appears almost colorless after hot isostatic pressing (HIP), and it is almost identical to the transmission of grown ZnS mono-crystals. This transformation from an infrared (Forward Looking InfraRed - FLIR) to a multispectral (clear-) grade and its effect on transmission is shown in Figure 2.43.

The change in properties is mostly attributed to a change in grain size. Elevated temperature and pressure allow the structure to recombine, resulting in grains with more than 100× the volume of the initial grain size before HIPping. The process does not change the crystal structure, and so the density remains constant. As a result, scattering on grain boundaries is significantly reduced. Furthermore, interstitial remains of the sulfuric components can either recombine or leave the material via diffusion processes.

2.4.6 Properties and qualities

2.4.6.1 Deep UV materials

Classic optical glasses are transparent in the spectral range accessible to the human eye with some extent to the near-UV (UV-A) and IR. Glasses with a sufficient transmission at mercury i-line (365 nm) are very rare and have to be produced from raw material with an improved impurity level. Fused silica is transparent down to wavelengths in the 160 nm range. For even shorter wavelengths, the only suitable transparent materials are the fluoride crystals

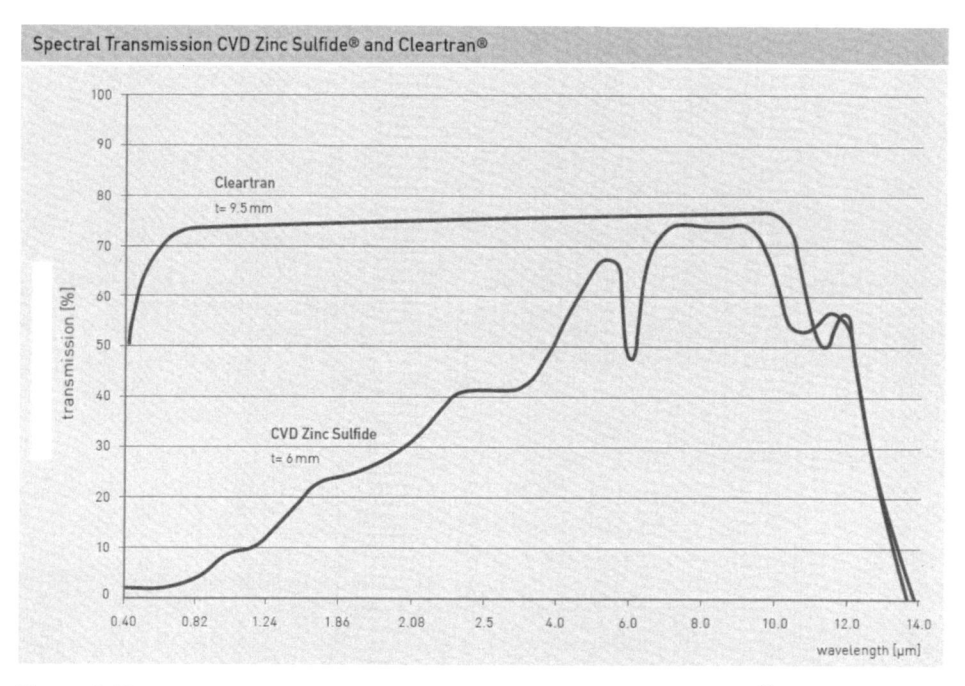

Figure 2.43 Transmission before (CVD zinc sulfide) and after (Cleartran®) the HIP process [53].

CaF$_2$ and MgF$_2$, the latter being birefringent over its entire transmissive range.

In the first decade of our millennium, 193-nm lithography became the workhorse of modern chip manufacturing. The only two optical materials with a sufficient set of properties to support this technology are fused silica and CaF$_2$.

2.4.6.2 Laser damage in the DUV

When discussing damage in optics, most papers are referring to a macroscopic destruction of a surface or even the bulk material. Such processes are observed under strong irradiation especially by high-power or short-pulsed lasers. The conditions of the element's surface like coating, roughness and sub-surface damage are important contributors to the resistance against this kind of damage and shall not be discussed in this chapter.

The laser damage discussed here is a deterioration of transmission under irradiation without macroscopic destruction. Potential sites for the start of such processes are defects in any lattice. This might be un-occupied sites, free dangling bonds or impurities. Radiation with a high enough energy will always free electrons from their site, giving them the opportunity to move through the lattice. After a certain lifetime and in a perfect structure, they will recombine with their now un-occupied site. The real crystal with its defects can trap such electrons, creating absorption centers of a much longer lifetime. Tiny effects add up over time and lead to measurable transmission degradation. In a state-of-the-art 193-nm-lithography scanner, a lens has to work properly for several years under almost 24/7 irradiation of an excimer laser in the kHz regime. Therefore, meticulous control of impurities and the establishment of laser test routines are key for manufacturers of DUV optical materials.

2.4.6.3 VIS materials

The range of visible light is the home ground of optical glasses. Still, there are some features that make crystals a desirable material. As already mentioned, calcium fluoride was found to be a material with low refractive index and very low dispersion. The Abbe number of 94 puts the material to the bottom left corner of the Abbe diagram almost un-occupied by any glass type.

Another property making the fluorides attractive to optical designers is the special refractive index change with temperature. Whereas most common glasses show an increase of the index with increasing temperature, it is the opposite for CaF_2 and BaF_2.

Combining materials with positive and negative dn/dT makes the lens performance robust against temperature changes, by which one can create an athermal design.

2.4.6.4 IR materials

Optical glasses transparent in the visible range are made of oxides with silicon oxide SiO_2 being the network former in the most cases. The glasses contain silicate- (SiO_2) and "water" (-OH) ions in their network. OH absorbs the IR radiation in the 2.5 to 3.0 μm range, and the SiO_2 network itself absorbs all

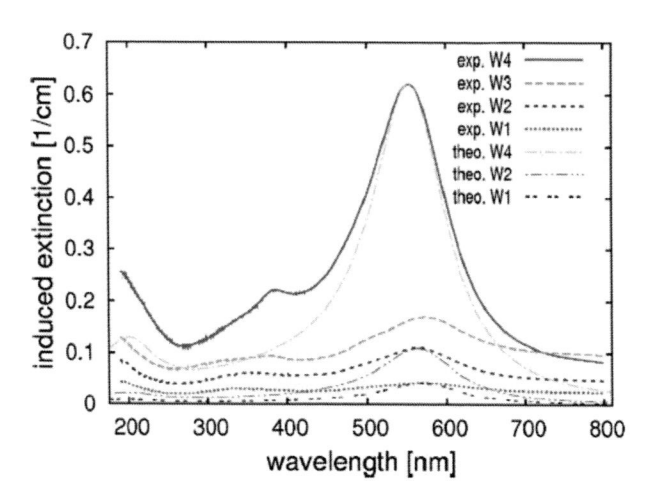

Figure 2.44 Induced extinction in CaF_2 of increasing laser durability (from W4 to W1) after 193 nm irradiation.

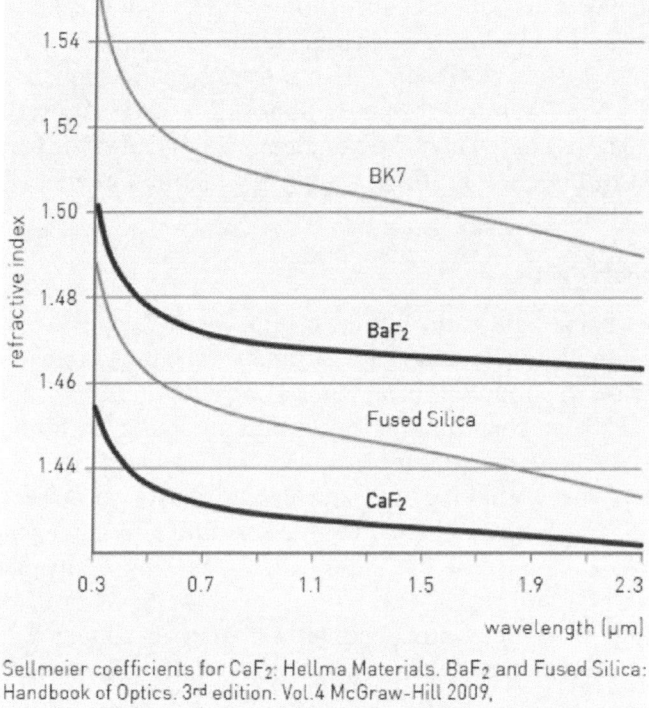

Figure 2.45 Refractive index of several optical materials in the UV to NIR range.

radiation with wavelengths longer than 4 μm. This restricts the use of optical glasses to the spectral range below these values. There are semiconductors (germanium, silicon) and metallic glasses (chalcogenides) which are transparent in these IR wavelengths.

In contrast to silicate-based glasses, those based on the chalcogens S, Se, or Te, are nominally transparent out to 10 μm or more. In general,

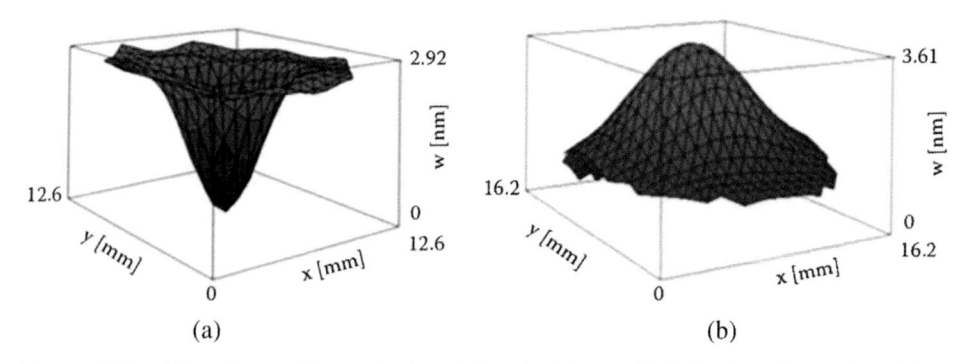

Figure 2.46 Wavefront deformation in (a) fused silica and (b) CaF$_2$ under excimer laser irradiation due to thermal lensing [54].

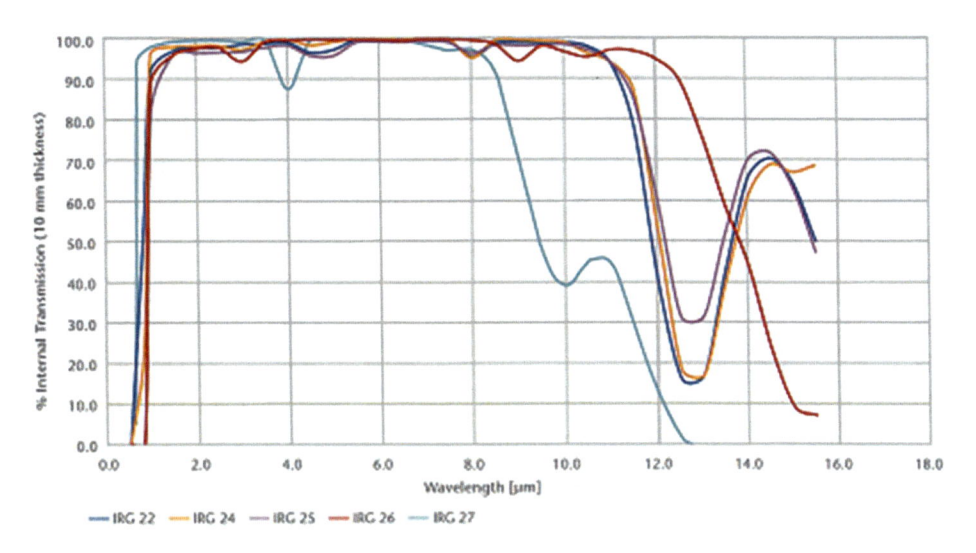

Figure 2.47 Transmittance curves of SCHOTT IRG materials IRG 22 to IRG 27, 10 mm thickness.

transparency ranges are limited at the short-wavelength, high-energy side by the electronic band-gap of the material and at the long-wavelength, low-energy side by the interaction of photons with lattice vibrations. These limitations are intrinsic features, resulting directly from the glass bulk composition. Examples of extrinsic features include absorption due to Ge-O just below 13 µm for Ge-containing glasses and at 4 µm due to S-H absorption. In principle, these two absorptions can be reduced via increased raw material cleaning procedures and modified processing, but such activities can be time- and cost-intensive.

Infrared chalcogenide glasses (IRG) offer excellent transmission in the SWIR (short-wave IR), MWIR (mid-wave IR), and LWIR (long-wave IR) range; see Figure 2.47. Physical properties such as low dn/dT and low dispersion enable optical designers to engineer color-correcting optical systems without thermal defocusing. The IRG family of chalcogenide glasses is optimized for pairing with the family of IR glasses and other IR materials to support high-performance optical designs. These glasses encompass the common IR transmission bands 3–5 µm and 8–12 µm, but they can transmit as low as 0.7 µm.

Their obvious disadvantage is the absorption in the visible range. For applications that require transmission in the VIS and IR range at the same time, crystals and ceramics are the material of choice.

2.5 Optical Material Trends

As described in the introduction, various industries have promising applications pushing optical materials. This leads to a variety of trends for optical materials.

- Augmented reality requires high refractive indices in order to enlarge the field of view of the glasses. This should be paired ideally with a low density to reduce the weight of the device.
- Laser processing devices need glass types that withstand higher and higher laser radiation.
- Space applications require an equal refractive index distribution, so high homogeneity within a component and optionally, depending on the exact mission profile, a resistance against ionizing radiation.
- Automotive applications depend on athermal designs consisting of optical glasses with a dedicated coefficient of thermal expansion and thermal change of the refractive index.
- Machine vision applications need compact lenses and optical systems that provide high image resolution, with good contrast. They need specific glass type combinations to enable high monochromatic and chromatic aberration correction.
- Short-wave infrared (SWIR) applications can visualize features that normally would not be detectable with visible light alone, such as rotten fruits in fruit sorting, fakes in paintings, content levels in opaque bottles, etc. All these machine vision applications employ specific optics that ideally transmit in the visible spectral range and also in the SWIR wavelength range as broad-band optics. Optical designs require materials that transmit in the visible and the SWIR range, sometimes even up to 4 μm.
- Lenses and optical systems in outdoor civilian and military high-end security and surveillance applications typically demand robustness, compactness, high resolution and excellent performance in any weather and temperature conditions. Additionally, these systems operate at dim light situations as well so that high-transmission glasses are relevant.
- Professional movie cameras are used for cinematography, mobile broadcasting and videography. Constantly growing resolution (up to 8K currently) requires complex high-resolution lens designs with larger apertures and excellent color correction. With large apertures, the homogeneity requirement increases as well.
- Even though most optics are a few centimeters in dimension, astronomical or other science applications require large blanks, even beyond one meter.
- Demanding applications like microscopes rely on the tightest tolerances of the optical position to achieve the Abbe limit and special glass types that enable superior correction of chromatic aberrations.

Fused silica and fused quartz, known as optical glasses with a very high purity, are indispensable nowadays in industry and science: from enabling totally new laser-based material processing routes in the automobile industry up to outstanding scientific finding about gravitational waves ending in a Nobel prize. Applications become more and more challenging, specifications need to be tigher and reliability in quality is essential. The growing trend of the last few years improving NIR laser performance from year to year in power, fluence, pulse duration and beam profile quality drives optical materials to their edge. Fused silica with the lowest absorption values in the NIR is qualified for kW cw or PW short-pulse systems. Enviromental safety and health gain importance: water and air disinfection in the UV wavelength range is a standard process in everyday life. Chips for mobile phones need to be smaller and more powerful, requiring even better optical material to make them. Laser surgury and medical laser treatment are getting more and more popular. Fused silica is a realiable material to serve as a potential substrate for all these growing trends and new challenges in the photonics world.

Crystalline materials witnessed the birth of optics. The intricate ways of making them allowed their use in niche applications only for a long time. The improvement of process technology made production technologies affordable and predictable. From a few tons per year, perspective crystals and ceramics are still in a niche compared to optical glass, but the number of applications is increasing. Trends supporting this growth are

- Development in multi-wavelength metrology for semiconductor fabrication equipment
- Moore's law of shrinking chip features at a constant price
- Sensing in the VIS/IR range for autonomous driving
- Demand for highly transparent wide-band materials for ground-based and airborne astronomy
- With the growing number of applications, the understanding of non-isotropic materials and the knowledge in machining them to the highest precision is growing, too. Crystals are not an exotic material anymore but are valued for their properties contributing to the progress in photonics.

References

[1] H. Bach and N. Neuroth, *The properties of optical glass*, Springer Science & Business Media (1998).
[2] U. Petzold, "Optical glass: A high-tech base material as key enabler for photonics," *Advances in Glass Science and Technology*, 79 (2018).

[3] SCHOTT, "Material safety datasheets of SCHOTT optical glass," (2020). https://www.schott.com/advanced_optics/english/knowledge-center/certificates-guidelines.html.

[4] R. Jedamzik, U. Petzold, V. Dietrich, V. Wittmer, and O. Rexius, "Large optical glass blanks for the elt generation," in *Advances in Optical and Mechanical Technologies for Telescopes and Instrumentation II*, **9912**, 99123E, International Society for Optics and Photonics (2016). https://www.schott.com/d/advanced_optics/21cea63c-d62f-42b7-9161-8b7ef2d7eda5/1.3/schott_tie-25_striae_in_optical_glass_eng.pdf.

[5] P. Hartmann, *Optical Glass*, SPIE Press Bellingham, WA (2014).

[6] SCHOTT, "*Technical Information No. 29: Refractive index and dispersion,*" (2020). https://www.schott.com/advanced_optics/english/knowledgecenter/technical-articles-and-tools/tie.html/.

[7] U. Petzold, R. Jedamzik, P. Hartmann, and S. Reichel, "V-block refractometer for monitoring the production of optical glasses," in *Optical Systems Design 2015: Optical Fabrication, Testing, and Metrology V*, **9628**, 962811, International Society for Optics and Photonics (2015).

[8] SCHOTT, "Technical Information No. 26: Homogeneity of optical glass," (2020). https://www.schott.com/advanced_optics/english/knowledgecenter/technical-articles-and-tools/tie.html/.

[9] International Organization for Standardization, "ISO 12123:2018: Optic and photonics - Specification of raw optical glass," (2018).

[10] International Organization for Standardization, "ISO 10110-18:2018: Optics and photonics — Preparation of drawings for optical elements and systems — Part 18: Stress birefringence, bubbles and inclusions, homogeneity, and striae," (2018).

[11] R. Jedamzik and U. Petzold, "Optical glass: refractive index homogeneity from small to large parts-an overview," in *Optical Components and Materials XVI*, **10914**, 109140V, International Society for Optics and Photonics (2019).

[12] SCHOTT, "Technical Information No. 25: Striae in optical glass," (2020). https://www.schott.com/d/advanced_optics/21cea63c-d62f-42b7-9161-8b7ef2d7eda5/1.3/schott_tie-25_striae_in_optical_glass_eng.pdf.

[13] S. Reichel, P. Hartmann, U. Petzold, S. Gärtner, and H. Gross, "Effects of striae inside optical glasses on optical systems," in *Optical Design and Engineering VII*, **10690**, 106900L, International Society for Optics and Photonics (2018).

[14] SCHOTT, "Technical Information No. 35: Transmittance of optical glass," (2020). https://www.schott.com/advanced_optics/english/knowledge-center/technical-articles-and-tools/tie.html/.

[15] R. Jedamzik, U. Petzold, and G. Weber, "From vis to swir: a challenge for optical glass and ir materials," in *Optical Components and Materials*

XV, **10528**, 105280P, International Society for Optics and Photonics (2018).

[16] SCHOTT, "Technical Information No. 27: Stress in optical glass," (2020). https://www.schott.com/advanced_optics/english/knowledge-center/technical-articles-and-tools/tie.html/.

[17] SCHOTT, "Technical Information No. 28: Bubbles and inclusions in optical glass," (2020). https://www.schott.com/advanced_optics/english/knowledge-center/technical-articles-and-tools/tie.html/.

[18] M. Englert, P. Hartmann, and S. Reichel, "Optical glass: refractive index change with wavelength and temperature," in *Optical Modelling and Design III*, **9131**, 91310H, International Society for Optics and Photonics (2014).

[19] SCHOTT, "Technical Information No. 19: Temperature coefficient of refractive index," (2020). https://www.schott.com/advanced_optics/english/knowledge-center/technical-articles-and-tools/tie.html/.

[20] SCHOTT, "Technical Information No. 30: Chemical properties of optical glass," (2020). https://www.schott.com/advanced_optics/english/knowledge-center/technical-articles-and-tools/tie.html/.

[21] SCHOTT, "Technical Information No. 31: Mechanical and thermal properties of optical glass," (2020). https://www.schott.com/advanced_optics/english/knowledge-center/technical-articles-and-tools/tie.html/.

[22] R. Jedamzik, H. Yadwad, and V. Dietrich, "Results of a polishing study for schott xld glasses," in *Optical Systems Design 2015: Optical Fabrication, Testing, and Metrology V*, **9628**, 96280T, International Society for Optics and Photonics (2015).

[23] P. Hartmann, "Mechanical strength of optical glasses," in *Optical Fabrication, Testing, and Metrology VI*, **10692**, 1069204, International Society for Optics and Photonics (2018).

[24] SCHOTT, "Technical Information No. 40: Optical glass for precision molding," (2020). https://www.schott.com/advanced_optics/english/knowledge-center/technical-articles-and-tools/tie.html/.

[25] R. Jedamzik, F. Elsmann, A. Engel, U. Petzold, and J. Pleitz, "Introducing the quantum efficiency of fluorescence of schott optical glasses," in *Current Developments in Lens Design and Optical Engineering XVIII*, **10375**, 1037508, International Society for Optics and Photonics (2017).

[26] SCHOTT, "Technical Information No. 36: Fluorescence of optical glass," (2020). https://www.schott.com/advanced_optics/english/knowledge-center/technical-articles-and-tools/tie.html/.

[27] R. Jedamzik and F. Elsmann, "Recent results on bulk laser damage threshold of optical glasses," in *High-Power Laser Materials Processing: Lasers, Beam Delivery, Diagnostics, and Applications II*, **8603**, 860305, International Society for Optics and Photonics (2013).

[28] R. Jedamzik and U. Petzold, "Latest results on solarization of optical glasses with pulsed laser radiation," in *High-Power Laser Materials Processing: Applications, Diagnostics, and Systems VI*, **10097**, 1009703, International Society for Optics and Photonics (2017).

[29] R. Jedamzik and U. Petzold, "Schott optical glass in space," in *Astronomical Optics: Design, Manufacture, and Test of Space and Ground Systems*, **10401**, 104010I, International Society for Optics and Photonics (2017).

[30] P. Hartmann, "Optical glass: dispersion in the near infrared," in *Optical Design and Engineering IV*, **8167**, 816702, International Society for Optics and Photonics (2011).

[31] Heraeus Conamic, "Quartz Glass for Optics - Data and Properties," (2020). https://www.heraeus.com/media/media/hca/doc_hca/products_and_solutions_8/optics/Data_and_Properties_Optics_fused_silica_EN.pdf.

[32] Heraeus, "Usage rights with heraeus, property right with adobe." https://stock.adobe.com.

[33] J. D. Mason, M. T. Cone, M. Donelon, J. C. Wigle, G. D. Noojin, and E. S. Fry, "Robust commercial diffuse reflector for uv-vis-nir applications," *Applied optics* **54**(25), 7542–7545 (2015).

[34] A. Schreiber, B. Kühn, E. Arnold, F. Schilling, and H. Witzke, "Radiation resistance of quartz glass for vuv discharge lamps," *Journal of Physics D: Applied Physics* **38**(17), 3242 (2005).

[35] R. Brückner, "Silicon dioxide," (1997).

[36] *Compiled CTE data from 25 publications.*

[37] C. A. Klein, "Characteristic strength, weibull modulus, and failure probability of fused silica glass," *Optical Engineering* **48**(11), 113401 (2009).

[38] F. Nürnberg, B. Kühn, and K. Rollmann, "Metrology of fused silica," in *Laser-Induced Damage in Optical Materials 2016*, **10014**, 100140F, International Society for Optics and Photonics (2016).

[39] M. Stamminger, D.-K. Lee, M. Nebe, and J. Janek, "Electrical conductivity of quartz glass measured by impedance spectroscopy between 600° c and 1100° c," *Physics and Chemistry of Glasses-European Journal of Glass Science and Technology Part B* **53**(5), 181–185 (2012).

[40] M. Lübben, F. Cüppers, J. Mohr, M. von Witzleben, U. Breuer, R. Waser, C. Neumann, and I. Valov, "Design of defect-chemical properties and device performance in memristive systems," *Science Advances* **6**(19), eaaz9079 (2020).

[41] D. Schoenfeld, B. Kühn, W. Englisch, and R. Takke, "Interferometry at the physical limit: how to measure sub-parts-per-million optical homogeneity in fused silica," *Applied optics* **42**(10), 1814–1819 (2003).

[42] J. Degallaix, C. Michel, B. Sassolas, A. Allocca, G. Cagnoli, L. Balzarini, V. Dolique, R. Flaminio, D. Forest, M. Granata, et al., "Large and extremely low loss: the unique challenges of gravitational wave mirrors," *JOSA A* **36**(11), C85–C94 (2019).

[43] S. Franke, H. Lange, H. Schoepp, and H. Witzke, "Temperature dependence of vuv transmission of synthetic fused silica," *Journal of Physics D: Applied Physics* **39**(14), 3042 (2006).

[44] Heraeus Conamic, "Fused silica for applications in the near infrared (nir)," (2020). https://www.heraeus.com/media/media/hca/doc_hca/products_and_solutions_8/optics/Fused_Silica_for_Applications_in_the_NIR_EN.pdf.

[45] F. Nürnberg, B. Kühn, A. Langner, M. Altwein, G. Schötz, R. Takke, S. Thomas, and J. Vydra, "Bulk damage and absorption in fused silica due to high-power laser applications," in *Laser-Induced Damage in Optical Materials: 2015*, **9632**, 96321R, International Society for Optics and Photonics (2015).

[46] O. Humbach, H. Fabian, U. Grzesik, U. Haken, and W. Heitmann, "Analysis of oh absorption bands in synthetic silica," *Journal of non-crystalline solids* **203**, 19–26 (1996).

[47] *Measurements from Dr. Mühlig, Institut für Photonische Technologien (IPHT), Jena.*

[48] S. Schröder, M. Kamprath, A. Duparré, A. Tünnermann, B. Kühn, and U. Klett, "Bulk scattering properties of synthetic fused silica at 193 nm," *Optics express* **14**(22), 10537–10549 (2006).

[49] D. R. C. Wren Carr, V.E. Gruzdev, and C. Menoni, *Laser-induced Damage in Optical Materials 2019*, vol. **11173**, International Society for Optics and Photonics (2019). https://www.spiedigitallibrary.org/conference-proceedings-of-spie/11173.toc.

[50] G. Lu, "The simplest and most successful experiment onboard apollo 11," *Scientific American* (2019).

[51] J. Criddle, F. Nürnberg, R. Sawyer, P. Bauer, A. Langner, and G. Schötz, "Fused silica challenges in sensitive space applications," in *Advances in Optical and Mechanical Technologies for Telescopes and Instrumentation II*, **9912**, 99120K, International Society for Optics and Photonics (2016).

[52] J. S. McCloy and R. W. Tustison, "Chemical vapor deposited zinc sulfide," *SPIE* (2013).

[53] Hellma, "Data sheet," (2020). https://www.hellma.com/en/crystalline-materials/optical-materials/cvd-zinc-sulfide-and-cleartran/.

[54] D. Ristau, *Laser-induced damage in optical materials*, CRC Press (2014).

Dr. Frank Nürnberg is the Global Head of Sales Optics at Heraeus Quarzglas (Hanau, Germany). He received his Ph.D. on laser-plasma-physics from Technical University Darmstadt (Germany) in collaboration with Lawrence Berkeley National Laboratory (USA). Frank has concentrated his career in sales and technical support of fused silica in optical applications. He started as European Key Account Manager. In 2016, he took over the globally responsibility for sales projects in the scientific field of lasers, astronomy, space and other research area. In July 2021, he was appointed as the Global Head of Sales Optics.

Bodo Kühn, a physicist, graduated from University of Bayreuth in 1995 and joined Heraeus Quarzglas developing fused silica for optical microlithography. Later on he headed the company's laboratories and specialized in material properties of quartz glass. Since 2019 he is working as Senior Expert Analytics for different R&D projects.

Dr. Ralf Jedamzik is a Principal Scientist at the SCHOTT AG. He is one of the leading experts on optical glass properties and ZERODUR® glass ceramics. After completing a degree in physics at the University in Düsseldorf and a doctorate in material science at the Technical University in Darmstadt, he began his professional career in business as quality technology expert at SCHOTT in Mainz, germany in 1999. Since 2013 he is principal scientist and project leader of important ZERODUR® and optical glass development projects. As an Application Manager for optical glass and ZERODUR®, he is a renowned international technical consultant for numerous industrial projects. He has authored more than 50 papers.

Dr. Uwe Petzold, a physicist, graduated from Technical University in Darmstadt 2009 and received a Dr.rer.nat. in Applied Optics there in 2013. After a short period as a post doc scientist within the area of nonlinear microscopy, he started an industrial career at SCHOTT AG, Mainz as an application engineer in 2014. Within this position he was especially dealing with optical measurement devices. Since 2015, he is taking care about SCHOTT's historical core business the optical glasses as a product manager. As a part-time job he is a lecturer at the University of Applied Sciences in Darmstadt. He is a member of SPIE.

Dr. Gordon von der Gönna is Director Sales and Business Development at Hellma Materials GmbH. He graduated as a chemist at the University in Jena and received his Dr.rer.nat. at the university's Institute of Glass Chemistry in 2000. Joining SCHOTT Lithotec in Jena in the same year, he held several positions in R&D and Sales in the company's Fused Silica and Calcium Fluoride division. After shutdown of the business in 2010 he was co-founder of Hellma Materials GmbH, a manufacturer of crystals and ceramics for optics and radiation detection.

Chapter 3
Optical Fabrication

Jessica DeGroote Nelson
Optimax Systems Inc., 6367 Dean Pkwy, Ontario, NY 14519, USA

3.1 Introduction

Optical components, or lenses, have been used for thousands of years. Ancient artifacts that appear to be lens shaped pieces of rock crystal have been dated to approximately 1600 BCE [1]. It was hundreds of years later when we first have documentation about the fabrication processes used to create these optics. Corrective lenses were said to be used by Abbas Ibn Firnas in the 9th century, who had devised a way to produce very clear glass. He shaped and polished his glass into round rocks that he used for magnified viewing. He called these rocks "reading stones" [2]. Circa 1719, Antonie van Leeuwenhoek built a simple microscope with only one lens to examine blood, yeast, insects and many other tiny objects. Leeuwenhoek was the first person to describe bacteria, and he also invented new methods for grinding and polishing microscope lenses that allowed for curvatures providing magnifications of up to 270-mm diameters, the best available lenses at that time [3]. Optical fabrication methods continued to advance over the years, highlighted by Sir Isaac Newton and Frank Twyman. Newton is often cited for his use of optical polishing pitch to polish the optics for his telescope [4]. Twyman also made significant contributions to the field of optical fabrication [1].

For several centuries, optical fabrication methods stayed fairly consistent. In the 1960s, there were two disruptive technologies that significantly changed how optics were manufactured, the laser and the computer [5,6]. The laser and the computer enabled laser-based interferometry and analysis methods and computer numerically controlled (CNC) deterministic grinding and polishing, all of which are critical in the modern optical fabrication shop.

This chapter will review the fundamentals of traditional optical fabrication methods and mechanisms of material removal, and then focus on advancements in modern optics manufacturing.

3.2 Traditional Fabrication Methods

3.2.1 Cast iron lapping

Cast iron lapping, or conventional loose abrasive optical lapping, has been used for years to grind optical surfaces in preparation of polishing. The lapping or grinding process produces cracks in the glass that create chips to promote material removal. The three major goals of grinding are to rapidly remove material, achieve the desired form and prepare the surface for polishing leaving minimal damage. Glass surfaces are typically ground with cast iron tools and loose abrasive (e.g., aluminum oxide) in water creating a slurry. The conventional lapping process is commonly used for spherical or flat optical surfaces. Each spherical radius requires unique and dedicated tooling, and while in use the cast iron tools continuously change and require periodic reconditioning. However, conventional optical lapping processes are still very cost-effective for producing spherical optics and are still widely used today.

Loose abrasive lapping is a brittle removal process, where the abrasive creates cracks in glass when pressed against a moving cast iron tool. The grinding material removal rate is proportional to pressure times velocity; this relationship is also known as Preston's equation [7], as shown in Equation 3.1, where *MRR* is the material removal rate, C_p is the Preston's coefficient, *P* is pressure and *V* is velocity. Grinding rates also increase with abrasive size. For a given set of constant operating parameters, there are two phases in the glass removal process: removal of the surface relief and removal of the cracked layer, which are indicated in Figure 3.1. Removal of the surface relief layer is very fast because the pressure of the tool is concentrated on the glass peaks and pits offer access to slurry to efficiently wet the surface and remove debris. Removal of the cracked layer goes more slowly because the tool and part are in more contact.

$$MRR = C_p PV \tag{3.1}$$

Preston's equation provides an accurate correlation of material removal under a constant set of parameters. T. S. Izumitani took his investigation further and explored correlations between grinding rates between different

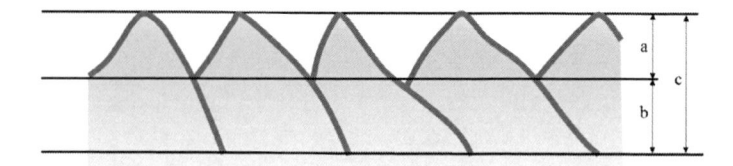

Figure 3.1 Schematic illustration of the cross-section of a ground surface. Surface layer c consists of relief layer a and cracked layer b.

glass types. When comparing glasses, harder glasses will grind slower than softer glasses. Izumitani studied the correlation between glass hardness and grinding rate, and he determined that there is a correlation, but it is not perfect. Glass hardness is a measure of plastic deformation typically measured using a Knoop microindentor. The cross-section of a Knoop indent is shown in Figure 3.2, and the volume outside the indent (bump) equals the volume in the depression. Hardness is a measure of plastic deformation and does not create cracks. Izumitani instead found excellent correlation between the length of a Vicker's indentation crack (see Figure 3.3) and grinding rates, indicating the strong importance of fracture in material removal [8].

Study of material properties and grinding lead Lawn, Evans and Marshall and subsequently Lambropoulos et al. to propose the lateral fracture mechanism for material removal in grinding [10,11]. Figure 3.4 shows this mechanism schematically, highlighting the pseudo-plastic deformation, plastic

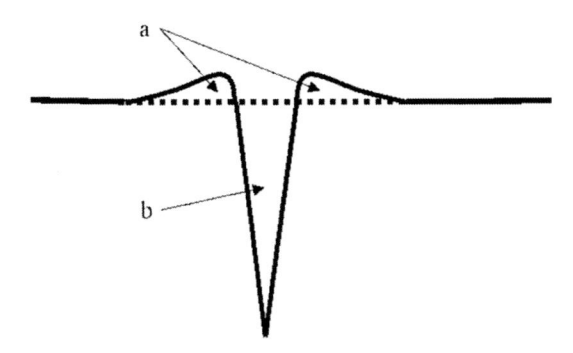

Figure 3.2 Cross-sectional sketch diagram of a Knoop indentation. The volume of the displaced material in the bumps (a) equals the volume in the depression (b) due to plastic deformation.

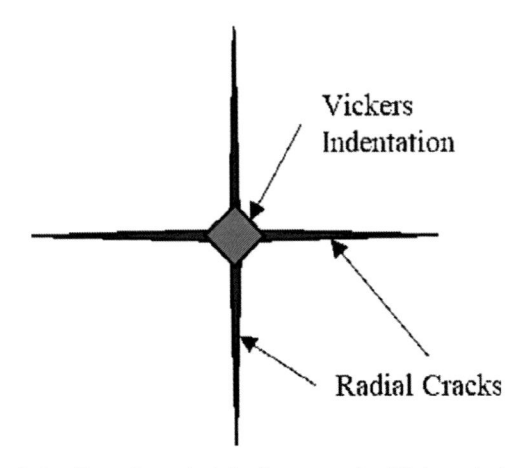

Figure 3.3 Top view sketch diagram of a Vickers indentation.

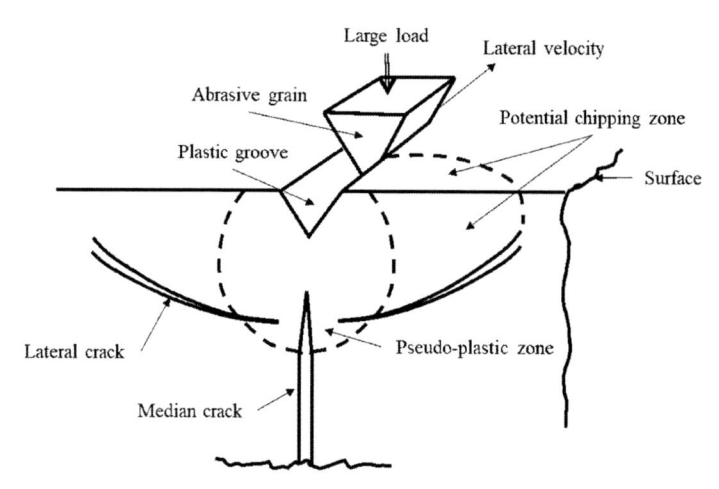

Figure 3.4 Sketch of the lateral fracture mechanism adapted from Reference [9].

deformation and fracture or cracking of the surface [9]. Evans used the Vickers microindent and cracks to calculate the fracture toughness of glass [12]. Lambropoulos et al. extended the work done by Evans, Izumitani and others and found a material figure of merit (FOM) that correlates grinding rates for different materials with their values for the Young's modulus (E), fracture toughness (K_c) and hardness (H_k), as shown in an extended Preston's equation:

$$MRR = C'_p \frac{E^{7/6}}{K_c H_k^{23/12}} PV \qquad (3.2)$$

where C'_p is an augmented Preston's coefficient, MRR is material removal rate, P is pressure and V is velocity [11]. Lambropoulos et al. also studied the effect of material properties with the resulting microroughness based on work done by Buijs and Korpel-Van Houten [13,14] and found that surface roughness scales with the inverse of the square root of Knoop microhardness [11]. Surface roughness measures the visible surface relief layer of the ground surface. The measure of surface roughness of a ground surface is important to estimate the depth of the cracked layer, which will turn into sub-surface damage once the surface is polished, which will be discussed later [15].

A commonly accepted procedure for grinding is to include multiple steps from a larger abrasive to smaller abrasive and progressively achieve a smaller cracked layer. This is often referred to as step grinding. With each step, the optician will remove approximately two times the size of the previous abrasive size. For example, an optician will start off with a 20-μm abrasive, step down to a 9-μm abrasive removing approximately 40 μm of material, as shown in Figure 3.5.

Step grinding, or an equivalent process is vitally important in ensuring that the last trace of damage is removed from the previous abrasive size. If not

Optical Fabrication 75

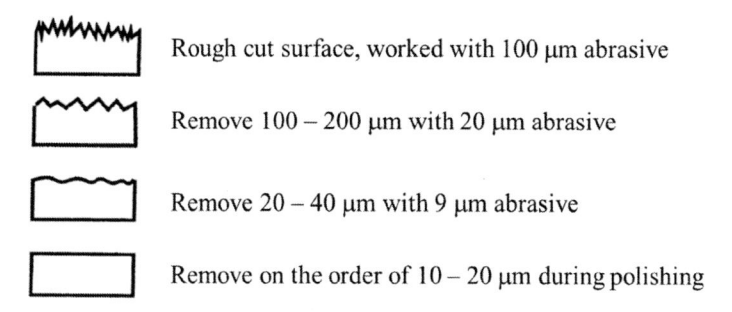

▲▲▲▲▲	Rough cut surface, worked with 100 μm abrasive
▲▲▲	Remove 100 – 200 μm with 20 μm abrasive
︵︵	Remove 20 – 40 μm with 9 μm abrasive
▭	Remove on the order of 10 – 20 μm during polishing

Figure 3.5 Commonly accepted step grinding material removal process used by opticians.

done properly, sub-surface damage (SSD) can result. Sub-surface damage is the residual damage left behind under the polished layer. Many times, the residual cracks or damage close up and are invisible to the naked eye. Sub-surface damage can be a source of scattered light, figure instability and catastrophic failures for high-power laser systems [16]. It is important to remember that just because an optic looks shiny after polish, it does not mean that it is free from SSD. Figure 3.6 shows a cross-section sketch of an optic with SSD highlighting that the SSD lies beneath the top polished layer.

Polishing seldom causes SSD, unless it is done incorrectly (e.g., with excessive load or too hard of polishing lap). Polishing covers up damage caused by grinding, which makes it difficult to recognize and address SSD. Unfortunately, to date there is not a non-destructive measurement for SSD on a polished surface; the best characterization methods are destructive. Typically, SSD is limited through process control that has been developed and tested with destructive methods. A few of the methods include surface etching to enlarge cracks and reveal SSD [17], taper polishing and etching [16] and spot

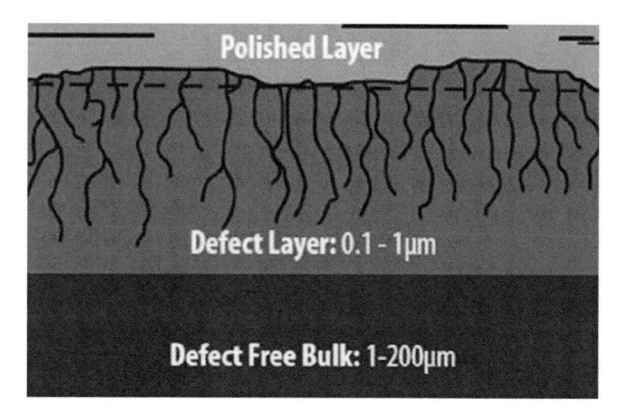

Figure 3.6 Schematic diagram illustrating the sub-surface damage lying beneath the surface of an optic.

polishing using a ball method. The ball method has been referred to as the Itek ball method or Center for Optics Manufacturing (COM) ball method [18,19].

The destructive SSD characterization methods mentioned above can be time consuming. In recent years, several have researched ways to speed up the process utilizing small tool polishers. Small tool polishers will be discussed later in this chapter. Deterministic small tool polishers such as Magnetorheological Finishing (MRF) and the Zeeko Intelligent Robotic Polishers (IRP) have been used to polish spots [15,20,21], troughs [22] or wedges [23] on ground parts in order to identify the last trace of damage created by the grinding step with the aid of microscopes and/or white light interferometry. These methods produce results faster than conventional SSD characterization methods to allow processes to be characterized in real time. A polished surface cannot be analyzed for SSD through non-destructive methods; therefore, it is important to rely on process control to ensure minimized or no SSD.

3.2.2 Conventional or pitch polishing

Conventional or pitch polishing combines a visco-elastic polisher with a polishing agent to remove small amounts of material to achieve three main goals of polishing. The goals of polishing are to converge to final specifications (surface form and mechanical dimensions simultaneously), remove all trace of damage and result in smooth, defect-free surfaces.

Polishing agents, or abrasive particles, are required to achieve a material removal rate and a smoother surface [8]. Cerium oxide is commonly used as a polishing agent for glass polishing. Other abrasives include iron oxide, aluminum oxide, chromium oxide and zirconium oxide. In glass polishing, opticians mix polishing agents with water to form a polishing slurry. Often, this is done empirically. Izumitani experimentally showed the optimal concentration of cerium oxide with water is 10–30 wt.% for BK7 [8].

Pressure is applied to the glass part indirectly by the polisher. Optical polishing pitch is a unique material that is commonly used as a polisher material for polishing spherical glass optics. Pitch is a seemingly rock hard substance that will fracture upon impact, but its visco-elastic properties cause it to slowly flow and deform under pressure at room temperature [24]. In polishing, the visco-elastic properties of pitch are desirable because the pitch will continue to slowly flow to gradually deform to fit the irregular surface of a part and entrap and release grains of polishing compound. There are several types of pitch, including natural wood-based, rosin-based or petroleum-based, [25] as well as synthetic polymer-based pitch [26].

Theories have existed for year surrounding the mechanism of removal in optical pitch polishing of glass [27,28,29,30]. Similar to grinding, Preston's equation of material removal proportional to pressure times velocity also holds true for polishing [7]. Izumitani, formally of the Hoya Corporation, studied the glass removal function extensively, and found that the polishing process involves more than mechanics. He concluded that polishing is a chemo-mechanical process that proceeds as follows: water penetrates the glass surface; hydration (ion exchange) causes formation of a soft, hydrated layer; pressure from a polisher forces abrasive polishing agent particles against glass; the porous, hydrated surface layer is scraped or sheared off the glass surface and into the polishing slurry; the removed glass is carried away as a coating bounded with the abrasive polishing agent particles; and then the cycle is immediately repeated [8,31,32,33]. L. Cook and M. Cumbo both continued to explore the effects of chemistry in polishing, specifically looking at the role of slurry pH in the polishing removal mechanism and resulting surface roughness [34,35,36]. Most recently, significant contributions to better understanding the glass removal process in conventional polishing have come from O. Fähnle who studied polishing parameters and used them to create a "three wagon approach" to process control [37]. T. Suratwala and his team at Lawrence Livermore National Laboratory explored the material science of optical fabrication to create tools such as convergent polishing [38].

3.3 CNC Optics Manufacturing

3.3.1 Spherical CNC generation

Computer numerically controlled (CNC) generation was enabled by computers and relies heavily on bound abrasive tools. A bound or fixed abrasive tool typically consists of a metal or resin matrix with diamonds or other abrasives dispersed within the matrix. One of the major differences between bound abrasive CNC generation and loose abrasive lapping is the wear mechanism. Loose abrasive lapping consists of three body abrasive wear and bound abrasive lapping is two body abrasion. Figure 3.7 illustrates the three- versus two-body abrasion. Two-body abrasion is not exclusive to CNC generation; it can also exist in non-CNC applications such as diamond pellet grinding [39]. Similarly, CNC equipment can also utilize three-body abrasive wear [40].

Conventional grinding processes require dedicated tooling for every unique radius and diameter. Bound abrasive ring tools for CNC generation significantly reduce the number of tools necessary. Multiple radii can be generated with one bound abrasive ring tool by adjusting the head angle. Ring tools can be used to generate flat, spherical, concave or convex surfaces using Equation 3.3:

$$\sin \alpha = D/2R_c \qquad (3.3)$$

where α is the wheel angle, D is the wheel diameter and R_c is the desired radius of curvature. Schematic drawings of both concave and convex configurations are shown in Figure 3.8.

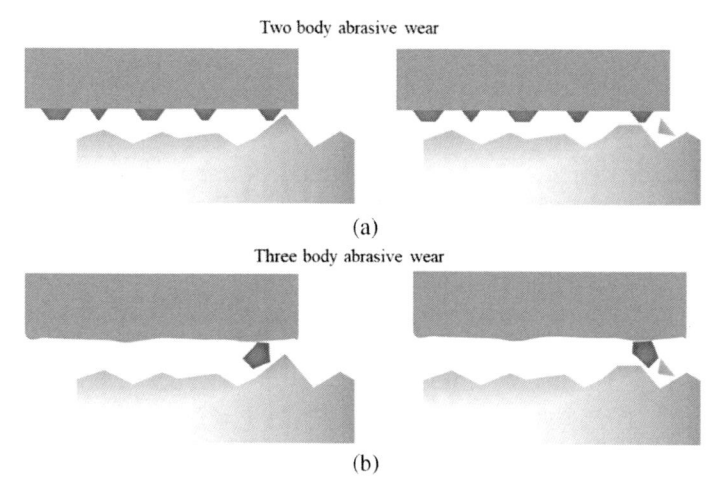

Figure 3.7 Schematic diagram of (a) two-body versus (b) three-body abrasive wear.

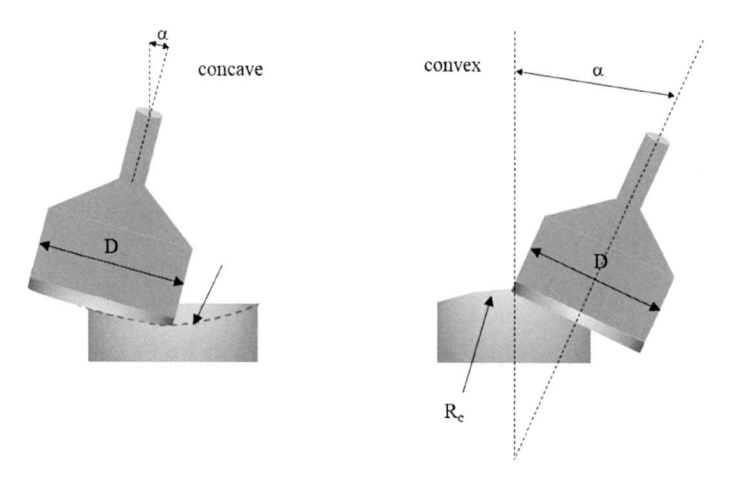

Figure 3.8 Ring tools are used to generate concave and convex spheres.

From 1989–2004, significant progress in CNC and deterministic manufacturing was made at the Center for Optics Manufacturing (COM) at the University of Rochester. COM was a collaborative effort between universities, industry and the US government to further develop, demonstrate and implement optics manufacturing technology to meet the increasing needs of industry and the Department of Defense (DOD) [41,42]. Several industrial companies collaborated with COM, including OptiPro Systems, Inc. and Moore Nanotechnology Systems [43,44]. OptiPro introduced the Opticam (Optics Automation and Management) in 1991 in collaboration with COM and the American Precision Optics Manufacturing Association (APOMA). Opticam technology was developed with the goal to automate the optical manufacturing process and eliminate the need for dedicated tooling, replacing it with semi-universal tooling such as the ring tool geometry mentioned above [41]. CNC generation was a driving factor in significantly reducing lead times for custom optical components and paved the way for companies to specialize in custom optics manufacturing for prototype applications [45].

In parallel to the work being done in the US, Ernst Loh of Wilheilm LOH KG Optikmaschinen was utilizing bound abrasive diamond pellets to generate the spherical surface and then utilize ring tools, both with LOH Diamond Lapping Machines [46]. Loh and others did significant work to advance CNC optical manufacturing methods in Europe and Asia at companies like LOH (now Satisloh), OptoTech, DMG MORI and SCHNEIDER [47,48,49,50,51].

Additional work has been done by several to reduce machine vibration, increase stability and advance computer programming for CNC generation. The goal is to increase the complexity of geometries available as well as the level of resulting precision of the generated surfaces. The introduction of ultrasonic vibration-assistance to the generation process enabled faster material removal rates with less tool wear. This results in more predicable surface form and surface texture [52,53]. Experimental results showed that material removal rates were three times higher with ultrasonic-assistance on than when it was off. Further testing showed that the amount of residual damage was similar between the two methods [54]. As mentioned earlier, the amount of damage induced in each generation step creates potential sub-surface damage in subsequent grinding and polishing steps.

3.3.2 Deterministic small tool polishers

The insertion of computers into the optical manufacturing process, along with significant materials science research, enabled the use of small tool polishers to deterministically figure correct optical surfaces. Deterministic polishing involves a tool that is smaller than the full aperture of the part, locally removing material across the part. Dwell-based algorithms are created to remove material based on initial surface metrology* and the tool influence function. The tool influence function, also referred to as a spot, is calculated prior to each deterministic polishing run. It is calculated by measuring the amount of material removed from the surface of a stationary part by the small tool for a given amount of time. Units for the tool influence function are given in depth per time.

Similar to conventional polishing, the material removal rate for small tool polishers follows Preston's equation (see Equation 3.1 above) and is proportional to pressure multiplied by velocity [55]. The material removal rate can also be adjusted by all of the variables included in Preston's coefficient.

---

*Deterministic polishing is dependent on accurate metrology methods, as discussed in Chapter 4.

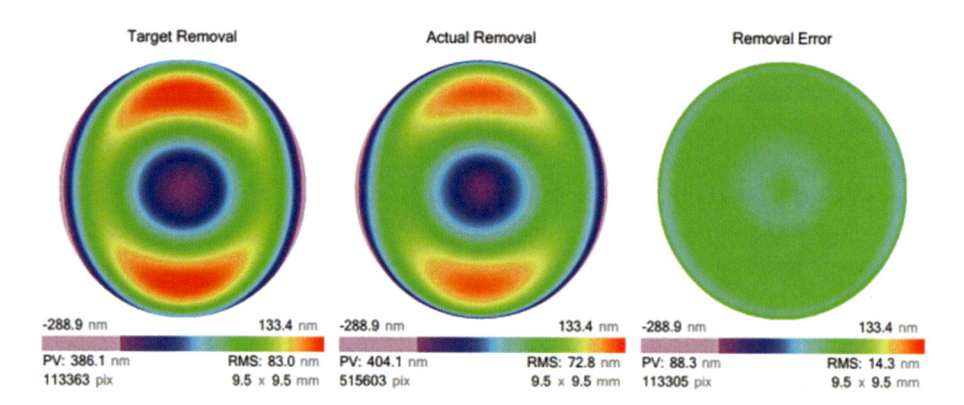

Figure 3.9 A magnetorheological finishing (MRF) example of asymmetric removal maps showing the target removal, actual removal and the removal error (difference between the target and actual) [56].

Typically, in dwell-based small tool polishing, once a spot has been taken, all other variables are held constant, and the only variable to change is the amount of time that the tool dwells on each point of the surface to remove material preferentially in the high points and traverse faster over the lower points to converge on the nominal surface shape. An example is shown in Figure 3.9.

The size of the spot can be adjusted based on the nominal tool size and for compliant small tool polishers; it can also be adjusted by changing the z-offset, or the amount of compression between the tool and the part. There are several trade-offs in choosing a spot size. A larger spot will increase removal rate, which decreases process time, but overall the user will have less control over the surface form. Whereas with a smaller spot the removal rate is less which increases process time but the user will have more control over the surface form, enabling the introduction or correction of smaller features on the surface. Choosing the right spot size is often a balance between the computer software packages predicting the residual error and the experience of the technician [57,58].

Deterministic polishing is performed by optical polishing shops using both commercially available machine tools, as well as custom computer-controlled polishing.

3.3.2.1 Custom computer-controlled polishing (CCP)

Early work with custom computer-controlled polishing (CCP) platforms included converting a precision coordinate measuring machine into a custom CCP platform. The CCP platform utilized composite polishing tools consisting of a soft pitch molded onto a flexible, elastic layer. This flexible tool compensates for local changes in the surface shape. A load sensor incorporated into the tool head maintains constant force per Preston's equation [59,60]. Similar custom CCP platforms were developed at Goodrich Corporation (acquired by United Technologies Corporation and then Collins Aerospace), where Askinazi et al. developed a CCP platform utilizing small tool polishers with varying size diameters to tailor their removal functions [61].

Work done at the Steward Observatory Mirror Lab (SOML) at the University of Arizona combined computer controlled polishing with stressed-lap polishing. Stressed-lap polishing uses stiff spherical polishing tools that are actively continuously controlled and deformed to fit to an aspheric surface profile [62]. Large 8.4-meter off-axis Giant Magellan Telescope (GMT) and Large Synoptic Survey Telescope (LSST) mirrors relied on sub-aperture stressed-lap polishing to meet their final surface figure error of 25 nm rms. The sub-aperture stressed-lap tool used on the 8.4-m mirrors was 1.2 m in diameter, as shown in Figure 3.10. The stressed-lap could be displaced several hundred microns across the diameter in order to compensate for deviations from a spherical surface across the surface [63].

The use of robotic arms [66,67] in optical manufacturing has steadily increased as their availability and ease of use has increased, creating an affordable platform for manufacturing companies to integrate into their processes. Several companies and facilities have used robotic arms to replace or augment manual operations such as tool changing, part mounting/loading in CNC machines, part alignment and automated cleaning. Interchangeable tools have been used with robotic arms to create flexible automated platforms for optical metrology and optical polishing [68,69,70,71,72,73].

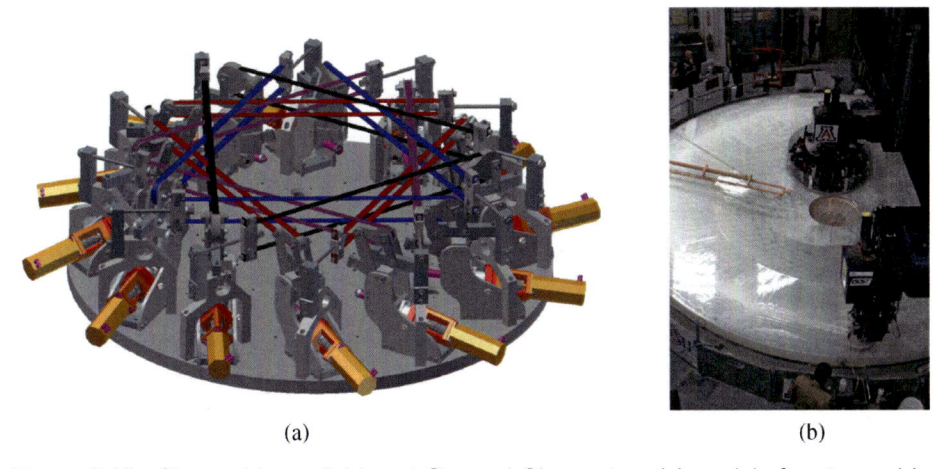

Figure 3.10 Stressed-lap polishing at Steward Observatory (a) model of a stressed lap [64], (b) photograph of a stressed-lap polishing a Large Synoptic Survey Telescope (LSST) mirror [65].

3.3.2.2 Commercially available deterministic polishing platforms

Each commercially available small-tool deterministic polishing platform offers unique features and is often used in some level of combination to complete a finished optic.

Zeeko's bonnet polishing technology uses an air-filled bladder covered with a cloth or polyurethane pad in combination with polishing slurry to remove material. Figure 3.11 is a schematic diagram of Zeeko's bonnet in an off-set angular position, referred to as a 'precessed' position. Zeeko introduced the 'precessions' motion to create a more uniform shaped influence function instead of a "W" shaped function that occurs if the tool were normal to the surface, as illustrated in Figure 3.12. A minimum of four 'precess' positions are required to obtain a Gaussian-like influence function. This Gaussian-like influence function is utilized in the dwell-based removal algorithm included on all Zeeko polishing platforms [57,74,75,76].

OptiPro's UltraForm Finishing (UFF) uses a polyurethane or bound abrasive material cut into a thin belt. This belt is continuously rotating while wrapped around a wheel made of compressive material. Several different wheel durometers and diameters are used to vary the spot size. The rotating belt is used in combination with abrasive slurry or coolant to remove material in a sub-aperture spot. A photograph of the OptiPro UFF is shown in Figure 3.13. The general process flow for UFF polishing consists of taking and measuring the removal function, inputting the initial figure error and desired surface figure into a deterministic algorithm that will use the removal function to optimize a polishing tool path [43,78,79,80].

The aspheric deterministic adaptive polishing technology (ADAPT) by Satis-Loh is a sub-aperture polishing tool made of polyurethane foam core

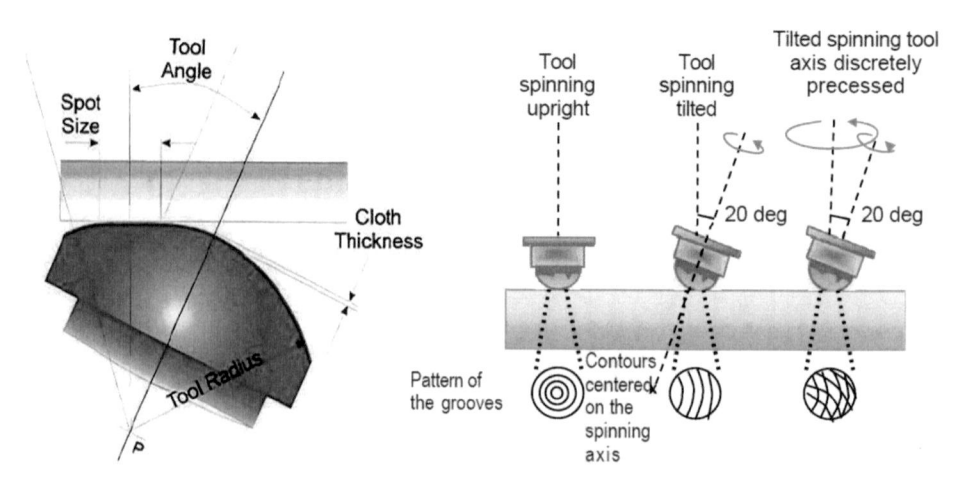

Figure 3.11 Illustration of the Zeeko air-filled pressurized membrane at a 'precessed' angle [77].

Optical Fabrication

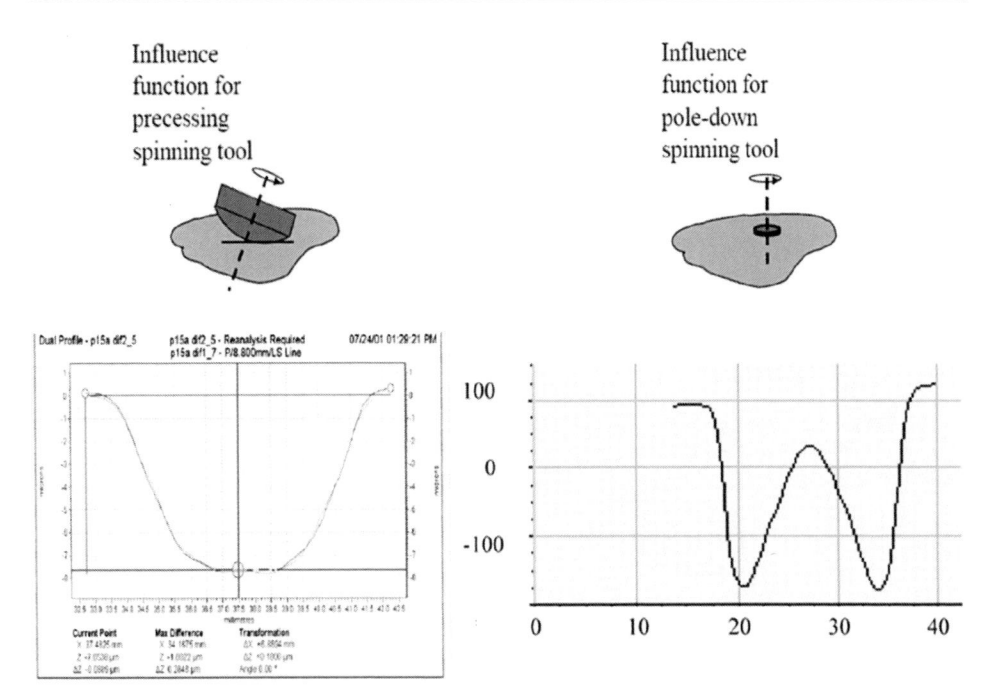

Figure 3.12 Effect of 'precession' motion on influence functions; figure adapted from Reference [57].

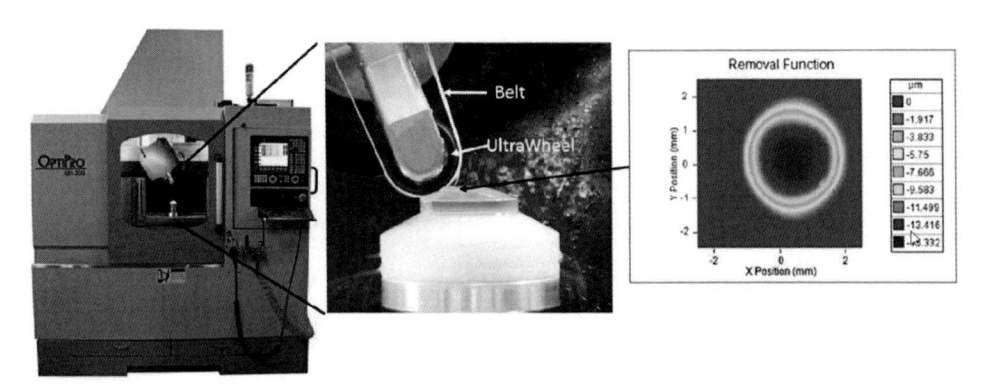

Figure 3.13 Photograph of the UltraForm Finishing system, with a magnified view of the polishing wheel and belt. Included on the right side is a characteristic influence function from the UFF [78].

machined to a specific radius covered with a polishing lap, as shown in Figure 3.14 [81]. The software to control the ADAPT tool calculates the removal function based on finite element analysis (FEA) combining information regarding the lens geometry, tool geometry, specific process information and the tool influence function [47,82].

Illustration of the ADAPT tool
(a)

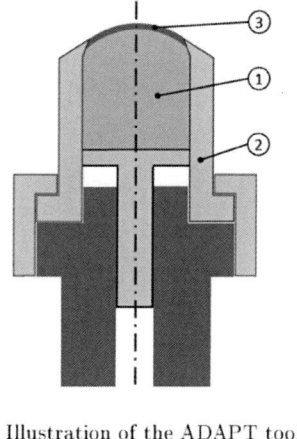

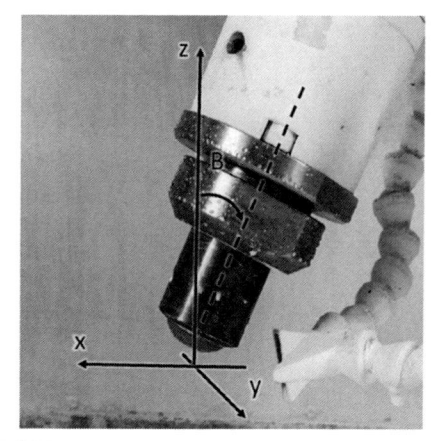

ADAPT tool mounted in OptoTech MCP250 machine
(b)

Figure 3.14 The ADAPT tool by Satisloh [81].

Magnetorheological finishing (MRF) is a deterministic sub-aperture polishing process developed at the Center for Optics Manufacturing (COM) by a group of international collaborators [83] and commercialized by QED Technologies, Inc [84,85]. MRF is based on a magnetorheological (MR) fluid that consists of carbonyl iron (CI), non-magnetic polishing abrasives, water, other carrier fluids and stabilizers. The MR fluid in the absence of a magnetic field has the viscosity of approximately 0.04 to 0.10 Pa·s (at a shear rate of ~800 sec^{-1}). Once introduced to a magnetic field (~2-3 kG), the viscosity increases four orders of magnitude [86].

Figure 3.15 schematically shows an optic depressed into MR fluid flowing left to right on a rotating wheel. The CI particles are shown as red spheres. The non-magnetic abrasives are depicted as blue irregular particles. In the presence of a magnetic field, water and polishing abrasives concentrate at the MR ribbon surface [87,88]. Material removal for MRF has been shown to be dominated by shear stress, as opposed to normal force, which has been

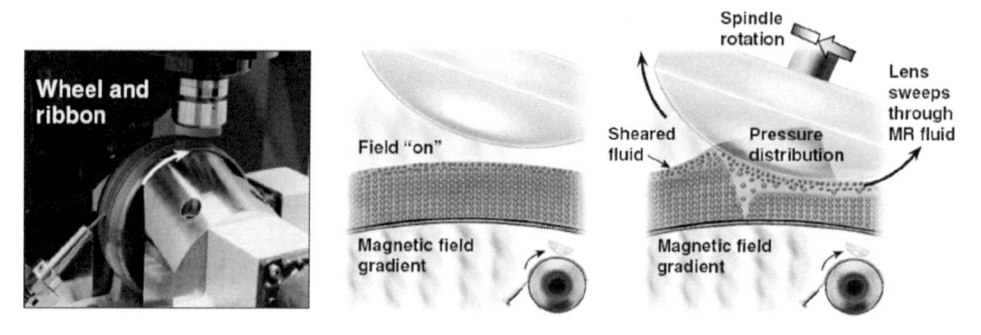

Figure 3.15 Magnetorheological finishing process [91].

hypothesized to be very small and approximately to 1×10^{-4} mN [89]. Particles involved in conventional pitch or pad polishing remove material with normal forces in the range of 5–200 mN [90].

The MRF tool influence function is characterized with MRF spots. An MRF spot is created by lowering a non-rotating optic into the rotating MR fluid ribbon for a known period of time. Material is removed in a characteristic D-shaped spot, as shown in Figure 3.16. When calculating the dwell instructions, four MRF spots are taken and averaged in proprietary software to calculate the MRF machine instructions.

As mentioned above, MRF is dominated by shear force and has relatively low normal force. This has shown to be very beneficial in applications such as ultraviolet (UV) laser damage resistant surfaces for deep-UV lithography and 351-nm laser fusion. Figure 3.17 shows the reduction in visible damage in dark field microscopy images of a conventionally polished part versus a MRF polished part [92].

In early MRF platforms, the magnetic field was concentrated on the apex of the wheel, as depicted in Figure 3.17. Certain geometries, such as steep

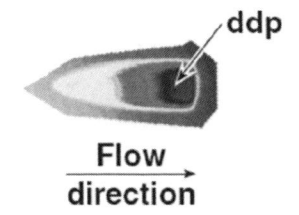

Figure 3.16 MRF removal function, or spot; ddp indicates the depth of deepest penetration.

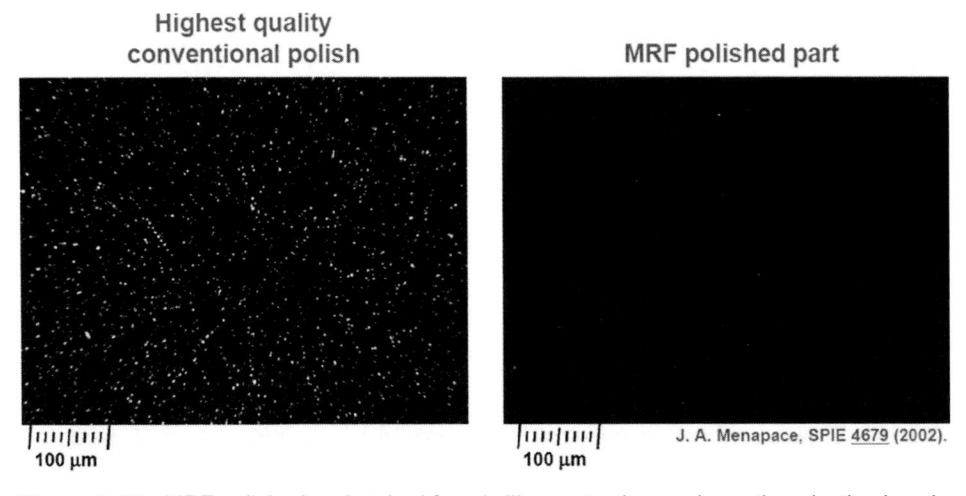

Figure 3.17 MRF polished and etched fused silica parts show a dramatic reduction in subsurface damage [92].

concave surfaces and/or complex geometries, were challenging or not always possible with a 150-mm standard diameter wheel. The introduction of varying wheel sizes and the Q-Flex machine gave rise to the ability to finish more geometries. Alternate wheel sizes offer the ability to vary the size of the MRF spot and allow the user the ability to adapt to the need for higher removal rates or smaller features on the surface [93]. The Q-Flex system is a 5-axis system that can operate in traditional rotational and raster modes and also in freeform mode, where the contact zone and polishing spot is not limited to the apex of the wheel but can utilize any section within a ±40° zone, as indicated in Figure 3.18 [94].

Ion beam figuring (IBF) and plasma jet machining (PJM) are sub-aperture deterministic figuring methods that are abrasive-free. IBF works under vacuum as an ion source accelerates ions, typically argon, to remove material by atomic bombardment and dislodgement. PJM, on the other hand, occurs in atmospheric pressure where plasma jet is incident on the workpiece to remove material. The sketch and photograph in Figure 3.19 show the PJM setup [96].

As mentioned earlier in this chapter, the invention of the laser made a significant impact on the field of optical manufacturing and testing through the use of laser based interferometry. In addition, lasers and laser material processing have also been introduced into the fabrication process chain through laser polishing or smoothing, corrective laser processing, and laser-assisted diamond turning. Laser polishing or smoothing is accomplished by irradiating glass with a continuous wave (CW) laser causing a thin layer of glass to heat up. As mentioned in the materials chapter (see Chapter 2), glass does not have a phase transition temperature, and the viscosity will gradually decrease with increasing temperature. The thin layer of glass is heated until an optimum viscosity is reached to allow the glass to flow and achieve a smoother surface as a result of lower surface tension. Laser polishing results in

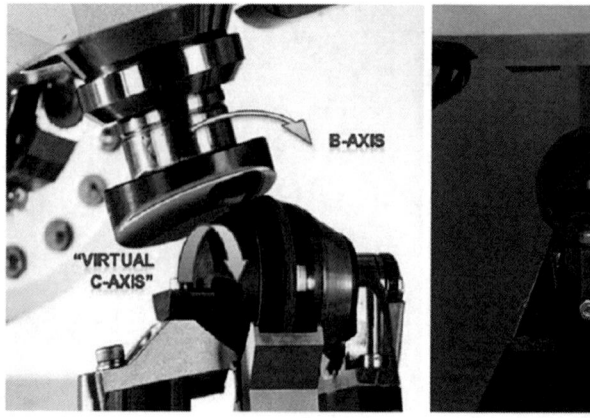

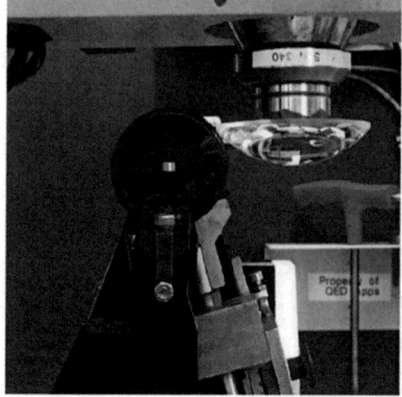

Figure 3.18 Freeform polishing configuration on an MRF Q-Flex platform [95].

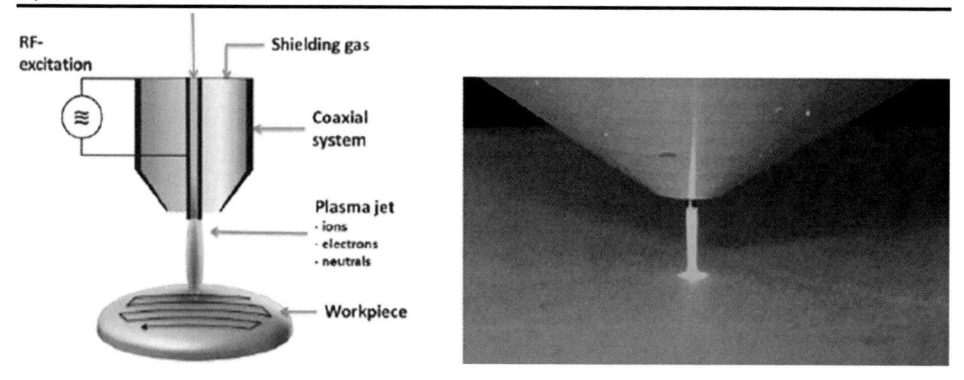

Figure 3.19 Plasma jet machining (PJM) setup and photograph highlighting a raster material removal pattern [96].

a smoother surface but also leaves residual surface waviness and does not correct for surface figure [97]. Additional processing, such as laser-based figure correction, is necessary. In laser figure correction, short-pulse lasers are used to ablate the surface [97]. Work done at Rochester Institute of Technology has recently demonstrated femtosecond-laser-based polishing to deterministically remove material from a germanium substrate while maintaining a low surface roughness value [98].

Additive manufacturing or 3D printing is emerging in optical fabrication. It is commonly used to create custom rapid tooling and models [99]. In addition, advances in additive manufacturing have led to 3D printed optical substrates. Polymer optics have been printed using layer-by-layer inkjet printing [100] and two-photon polymerization [101,102]. Inkjet printing is also used in volumetric index of refraction gradient optic (VIRGO) 3D printing of gradient index (GRIN) polymer optics. The VIRGO process has the capability to print 3D rotationally variant GRIN profiles [103]. Work at Lawrence Livermore National Lab (LLNL) is being done to apply the 3D printing process to glass optics. Dylla-Spears et al. are developing an approach to produce additively manufactured GRIN optics using direct ink writing. Their approach mixes silica nanoparticles with varying concentrations of titanium oxide to vary the index of refraction, as depicted in Figure 3.20. The resulting 3D printed object is in its green body state and requires post-processing heat treatment to fully densify the substrate and polishing to achieve desired surface roughness [104]. Additive manufacturing is an active field of research, and additional advances are expected in the coming years.

3.3.3 Mid-spatial frequency smoothing methods

Mid-spatial frequency (MSF) errors are errors on an optical surface that result in small-angle scatter or blur for an imaging system [105]. MSF errors, also

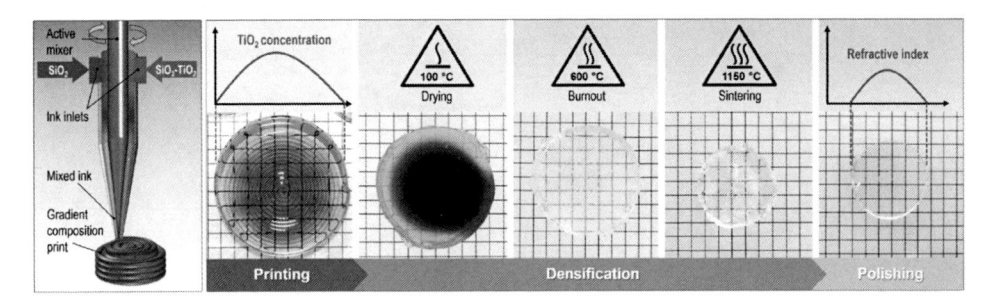

Figure 3.20 Process of additive manufacturing of gradient index (GRIN) silica-titania glass via direct ink writing at LLNL. Figure courtesy of Nikola Dudukovic and Rebecca Dylla-Spears.

referred to as surface ripple, are residual, periodic undulations in the surface profile. MSF became more pronounced with the introduction CNC generation and small tool polishing into the modern optical manufacturing process. Traditional cast iron lapping and full-aperture pitch polishing errors typically manifest in low- and high-spatial-frequency errors (figure error and surface roughness, respectively), but the signature from small tool polishing often falls in the MSF region [106].

Several smoothing methods have been introduced to help control, minimize or prevent MSF errors. Enhanced small-tool dwell map algorithms have been introduced to minimize the creation of MSF errors, including pseudo-random tool paths [107], predictive algorithms to control MSF errors [108] and parametric smoothing models [109]. The most popular method to minimize MSF errors is semi- or full-aperture smoothing methods. The University of Arizona developed a rigid-conformal (RC) tool that incorporates a non-Newtonian fluid layer between a solid back plate and polishing interface [110]. Non-Newtonian fluid is visco-elastic that acts solid-like with an impulse stress (<~0.1 seconds) and liquid-like with a long induced stress (~17 seconds); an example of a commercially available non-Newtonian fluid is Silly PuttyTM. Figure 3.21 shows the results of progressive smoothing of a 250-mm-diameter Pyrex work piece with 12-mm ripple period using an RC smoothing lap [110].

VIBE finishing, which was developed to rapidly polish aspheres [112] with a conformal full-aperture lap, is also used for smoothing mid-spatial frequency errors. The conformal full-aperture lap exhibits long range flexibility and a short range stiffness that allows the lap to maintain a proper fit to the low spatial frequency form while maintaining a "smooth" lap surface in the mid-spatial frequency region in order to smooth the MSF errors [113]. Alternative methods to minimize MSF errors include avoiding inducing them in the first place, such as stressed mirror polishing. In stressed mirror polishing, the mirror blank is under a calculated amount of stress and polished like a sphere with a full- or semi-full-aperture lap. When the stress is removed, the resulting mirror is in the prescribed aspheric form [114,115,116].

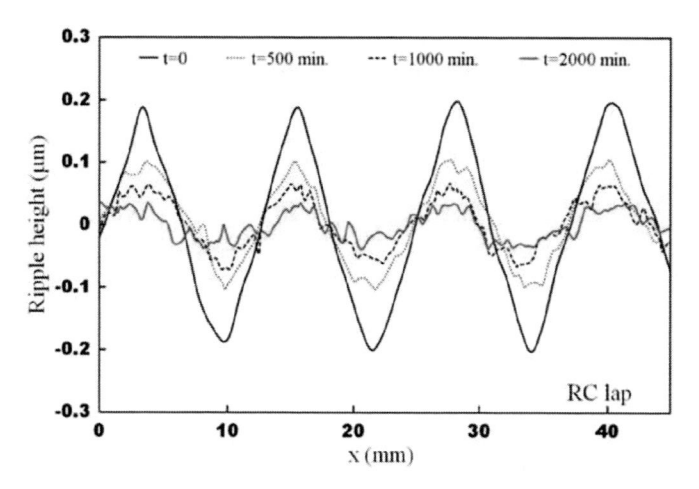

Figure 3.21 Progressive smoothing of a Pyrex substrate using an RC smoothing lap [111].

3.4 Special Considerations for Aspheres and Freeforms

3.4.1 Aspheres

Aspheres are rotationally symmetric surfaces that deviate from a sphere in a prescribed manner. Variations of general even asphere equation, shown in Equation 3.4, have been used to describe the aspheric surfaces dating back to Schwarschild's use of the conic constant [117,118]. Changes to the conic constant, k, will result in commonly known shapes such as the hyperbola ($k < -1$), parabola ($k = -1$), ellipse ($-1 < k < 0$), sphere ($k = 0$) and an oblate spheroid ($k > 0$) [117].

$$z = \frac{ch^2}{1 + \sqrt{1 - (1+k)c^2h^2}} + h^4 \sum_{m=0}^{M} a_m h^{2m} \tag{3.4}$$

In 2007, Forbes introduced an alternative method to describe the rotationally symmetric aspheric form. His method, commonly referred to as Forbes Q-polynomials, is shown in Equation 3.5 [119,120]. Forbes Q-polynomials are normalized and orthagonalized which lead to normalized coefficients whose magnitude is meaningful to its contribution to the surface form. Large coefficients mean large departure and small coefficients mean small departure, very small values have little effect and can be set to zero. These meaningful coefficients result in easier qualitative assessment of the aspheric form and fewer digits to input into software packages for CNC generation, deterministic polishing and metrology platforms [119,121].

$$z = \frac{ch^2}{1 + \sqrt{1 - (1+k)c^2h^2}} + u^4 \sum_{m=0}^{M} a_m Q_m^{con}(u^2) \tag{3.5}$$

Advances in deterministic generation and sub-aperture polishing described earlier in this chapter have greatly increased the ability to create higher precision aspheres with increasing amounts of aspheric departure from a best fit sphere. In aspheric manufacturing, the departure from a best fit sphere often drives the process selection. For example, for an asphere with little aspheric departure (<10 µm) from a best fit sphere, the surface can be generated and fine ground as a spherical surface and then have the aspheric surface departure polished in using deterministic sub-aperture polishing. Higher aspheric departure (>50 µm) requires the aspheric surface to be produced in the generation stage. Because aspheres utilize sub-aperture polishing, they often result in mid-spatial frequency errors and may require mitigation as described earlier in this chapter.

3.4.2 Freeforms

Freeforms are optical surfaces that lack symmetry. They have been incorporated into optical systems since 1972, when Baker and Plummer incorporated a freeform eyelens to correct for tilt, astigmatism and coma in a Polaroid camera [122]. The freeform mold was fabricated using custom specialized grinders and hand polishing in the final form that was measured on a custom profilometer [123]. Significant progress in freeform definition, design, fabrication and testing has happened, the increase in freeform use in optical designs have paralleled the increase in computing power [124]. Increased computing power allows for faster computing speed for optical design software, data analysis, CNC processing and metrology analysis. Freeform surface definition and optical design continue to be active areas of research; a few of the methods include using polynomial equations [125,126], Zernike polynomials [127] and NURBS (non-uniform rotational basis-spline) [128]. These methods offer the options for freeform design with a complete lack of symmetry. Freeforms offer unique advantages in optical designs such as system miniaturization, ability to design off-axis systems and lighter-weight systems. Figure 3.22 depicts the increasing lack of symmetry from a flat surface to a freeform surface. As indicated in Fig. 3.22, the fabrication and testing process transitions from traditional grinding and polishing to a process that requires CNC deterministic fabrication methods and novel testing techniques. Freeform fabrication is very similar to complex asphere fabrication where small-tool, single-point grinding and polishing techniques are used to fabricate freeforms starting with surface generation. Freeforms have no symmetry and require raster tool paths or tool paths capable of polishing non-symmetric geometries.

The general freeform manufacturing process is shown in Figure 3.23 for a single freeform surface. Each surface requires special consideration, and it is critical to have the optical designer and optical fabricator collaborate closely in order to ensure that the optical system works as designed. Some of the

Optical Fabrication 91

Flat/Plano	Spheres	Aspheres	Freeforms
No curvature	Constant radius	Varying radius	Non-constant curvature
Radial symmetry	Radial symmetry	Radial symmetry	Little to no symmetry

Traditional processing methods → CNC processing methods

Traditional interferometric testing → Novel testing methods

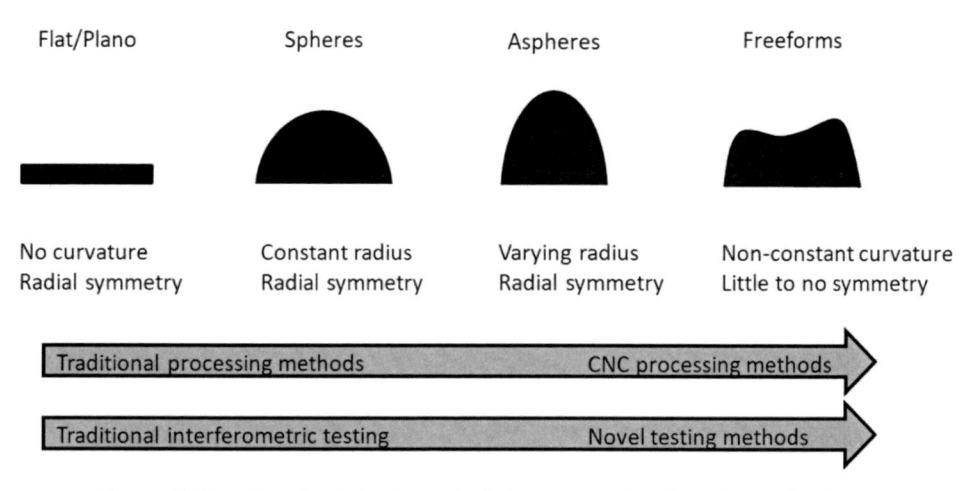

Figure 3.22 Sketch of simple optical shapes ranging from flat to freeform.

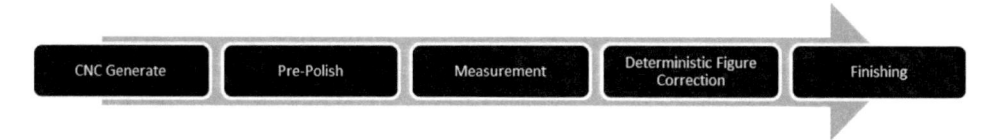

CNC Generate → Pre-Polish → Measurement → Deterministic Figure Correction → Finishing

Figure 3.23 General freeform manufacturing process flow.

consideration that should be considered in this collaboration is the surface prescription, surface extension, consideration about surface features and total error specification that includes fiducials.

When communicating with the optical fabricator, the fabricator needs to have the surface prescription (e.g., equation), and in addition, optical manufacturers require a solid model for grinding platforms and coordinate measuring machines (CMMs), cloud of points for custom polishing platforms and commercial sub-aperture polishing platforms and optical models for testing freeforms with computer generated holograms (CGH) if that option exists. Options for making a solid model to sub-micron and sub-nanometer level accuracy is an area of active research [129]. In viewing the optical model, the CNC toolpaths can be created and the surface can be evaluated for complexity and manufacturability (ex. rapid slope change within or outside the clear aperture) [130]. Total error for a freeform is a combination of surface form error and positional error. Small alignment errors (ex. 0.5° rotational offset) for freeform surfaces can manifest themselves as significant measurement errors (ex. 19 μm peak-to-valley astigmatism). In order to ensure freeforms are manufactured with low total error, fabricators measure surfaces relative to a global coordinate system, also referred to as fiducials [131,132]. Freeform fabrication and testing methods continue to be an active area of research by several companies, universities and organizations [133].

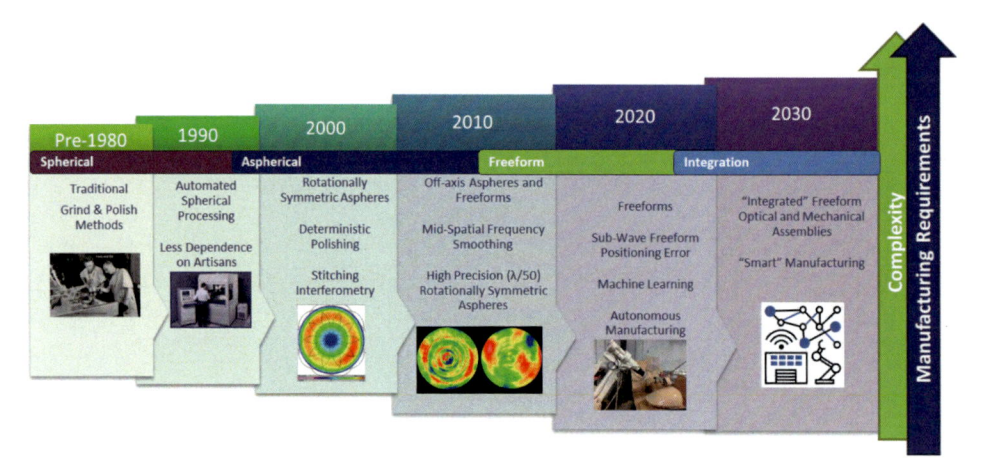

Figure 3.24 Advancement of precision optical manufacturing today (2020) and a glimpse into future possibilities.

3.5 Concluding Remarks

Optical fabrication has continued to evolve for centuries; however, in the past thirty years, the industry has transitioned from artisan, craft-based precision polishing to harnessing the advantages provided by computer-controlled grinding and polishing platforms. Figure 3.24 highlights just a few of the innovations that have spanned the last few decades in optical fabrication. As the industry continues to evolve, the complexity and manufacturing requirements will increase, but the benefit to optical systems will be realized ten-fold as we light our way into the future.

References

[1] F. Twyman, *Prism and Lens Making: A Textbook for Optical Glassworkers*, Routledge (1988).
[2] K. Airam, *Miracle of Islamic Science*, Knowledge House Publishers (1992).
[3] M. Bellis, "History of the microscope," *Retrieved May* **22**, 2004 (2004).
[4] I. Newton, *Opticks, or, a treatise of the reflections, refractions, inflections & colours of light; Based on the 4th Ed., London, 1730*, Courier Dover Corporation (1979).
[5] T. H. Maiman et al., "Stimulated optical radiation in ruby," *Nature* (1960).
[6] D. Malacara, "Optical fabrication and testing: a historical review," *Optics and Photonics News* **1**(6), 12–14 (1990).
[7] F. Preston, "The theory and design of plate glass polishing machines," *Journal of Glass Technology* **11**(44), 214–256 (1927).

[8] T. Izumitani, "American Institute of Physics Translation Series, Optical Glass," *American Institute of Physics*, New York, 9 (1986).

[9] A. Evans and D. Marshall, "Wear mechanisms in ceramics," in *Fundamentals of friction and wear of materials*, American Society for Metals (1980).

[10] B. R. Lawn, A. G. Evans, and D. Marshall, "Elastic/plastic indentation damage in ceramics: the median/radial crack system," *Journal of the American Ceramic Society* **63**(9-10), 574–581 (1980).

[11] J. C. Lambropoulos, S. Xu, and T. Fang, "Loose abrasive lapping hardness of optical glasses and its interpretation," *Applied optics* **36**(7), 1501–1516 (1997).

[12] A. Evans, "Fracture toughness: the role of indentation techniques," in *Fracture mechanics applied to brittle materials*, ASTM International (1979).

[13] M. Buijs and K. Korpel-van Houten, "Three-body abrasion of brittle materials as studied by lapping," *Wear* **166**(2), 237–245 (1993).

[14] M. Buijs and K. Korpel-van Houten, "A model for lapping of glass," *Journal of materials Science* **28**(11), 3014–3020 (1993).

[15] J. C. Lambropoulos, Y. Li, P. D. Funkenbusch, and J. L. Ruckman, "Noncontact estimate of grinding-induced subsurface damage," in *Optical manufacturing and testing III*, **3782**, 41–50, International Society for Optics and Photonics (1999).

[16] P. P. Hed, D. F. Edwards, and J. B. Davis, "Subsurface damage in optical materials: origin, measurement and removal," in *Presented at the Optical Fabrication and Testing Workshop*, (1988).

[17] V. Rupp, "The development of optical surfaces during the grinding process," *Applied Optics* **4**(6), 743–748 (1965).

[18] A. Lindquist, S. D. Jacobs, and A. Feltz, "Surface preparation technique for rapid measurement of sub-surface damage depth," *Optical Computing* **9**, SMC3-1 (1989).

[19] S. D. Jacobs, "Progress at the center for optics manufacturing," in *Intl Symp on Optical Fabrication, Testing, and Surface Evaluation*, **1720**, 169–174, International Society for Optics and Photonics (1992).

[20] J. A. Randi, J. C. Lambropoulos, and S. D. Jacobs, "Subsurface damage in some single crystalline optical materials," *Applied optics* **44**(12), 2241–2249 (2005).

[21] S. N. Shafrir, J. C. Lambropoulos, and S. D. Jacobs, "Subsurface damage and microstructure development in precision microground hard ceramics using magnetorheological finishing spots," *Applied optics* **46**(22), 5500–5515 (2007).

[22] X. Tonnellier, P. Morantz, P. Shore, A. Baldwin, R. Evans, and D. Walker, "Subsurface damage in precision ground ule® and zerodur® surfaces," *Optics Express* **15**(19), 12197–12205 (2007).

[23] J. A. Menapace, P. J. Davis, W. A. Steele, L. L. Wong, T. I. Suratwala, and P. E. Miller, "Mrf applications: measurement of process-dependent subsurface damage in optical materials using the mrf wedge technique," in *Laser-Induced Damage in Optical Materials: 2005*, **5991**, 599103, International Society for Optics and Photonics (2006).

[24] J. Lambropoulos, "Viscoelastic properties of polishers in optics manufacturing," (1993).

[25] B. E. Gillman and F. Tinker, "Fun facts about pitch and the pitfalls of ignorance," in *Optical Manufacturing and Testing III*, **3782**, 72–79, International Society for Optics and Photonics (1999).

[26] S. P. Sutton, "Development of new synthetic optical polishing pitches," in *Optical Fabrication and Testing*, OTuA2, Optical Society of America (2004).

[27] G. T. Beilby, "Surface flow in crystalline solids under mechanical disturbance," *Proceedings of the Royal Society of London* **72**(477–486), 218–225 (1904).

[28] L. Holland, "The properties of glass surfaces," (1964).

[29] W. Koehler, "Multiple-beam fringes of equal chromatic order. part vii. mechanism of polishing glass," *JOSA* **45**(12), 1015–1020 (1955).

[30] A. Kaller, "The basic mechanism of glass polishing," *Naturwissenschaften* **87**(1), 45–47 (2000).

[31] T. Izumitani, "Polishing, lapping and diamond grinding of optical glasses, in treatise on materials science and technology, edited by m. tomozawa and r.h. doremus," 115–172 (1979).

[32] T. Izumitani, "Polishing of optical glass: Mainly investigated from the view point of material science," 1–10 (1971).

[33] T. Izumitani and S. Harada, "Polishing mechanism of optical glasses," *Glass Technology* **12**(5), 131–135 (1971).

[34] L. M. Cook, "Chemical processes in glass polishing," *Journal of non-crystalline solids* **120**(1–3), 152–171 (1990).

[35] M. J. Cumbo, "Chemo-mechanical interactions in optical polishing," *Ph. D. Thesis* (1993).

[36] M. Cumbo, D. Fairhurst, S. Jacobs, and B. Puchebner, "Slurry particle size evolution during the polishing of optical glass," *Applied Optics* **34**(19), 3743–3755 (1995).

[37] O. W. Fähnle, "Process optimization in optical fabrication," *Optical Engineering* **55**(3), 035106 (2016).

[38] T. Suratwala, *Materials science and technology of optical fabrication*, Wiley Online Library (2018).

[39] D. F. Edwards and P. P. Hed, "Optical glass fabrication technology. 1: Fine grinding mechanism using bound diamond abrasives," *Applied optics* **26**(21), 4670–4676 (1987).

[40] D. Walker, A. Baldwin, R. Evans, R. Freeman, S. Hamidi, P. Shore, X. Tonnellier, S. Wei, C. Williams, and G. Yu, "A quantitative comparison of three grolishing techniques for the precessions process," in *Optical manufacturing and testing VII*, **6671**, 66711H, International Society for Optics and Photonics (2007).

[41] H. M. Pollicove and D. T. Moore, "Technology development at the center for optics manufacturing," in *Advanced Optical Manufacturing and Testing II*, **1531**, 174–178, International Society for Optics and Photonics (1992).

[42] H. Pollicove and D. Golini, "Computer numerically controlled fabrication," in *International Trends in Applied Optics*, ed. by A.H. Guenther, 5, SPIE Press (2002).

[43] OptiPro Systems, 6368 Dean Parkway, Ontario, NY 14519, USA, https://www.optipro.com.

[44] Moore Nanotechnology Systems, LLC, 230 Old Homestead Hwy, Swanzey, NY, 03446, USA, https://nanotechsys.com.

[45] Optimax Systems Inc., 6367 Dean Parkway, Ontario, NY, 14519, USA, https://www.optimaxsi.com.

[46] E. Loh, "Modern production methods in manufacturing optical components," in *Advances in Optical Production Technology I*, **109**, 40–43, International Society for Optics and Photonics (1977).

[47] Satisloh GmbH, Wilhelm-Loh-Straße 2-4, 35578 Wetzlar, Germany, https://www.satisloh.com.

[48] OptoTech Optikmaschinen GmbH, Sandusweg 2-4, 35435 Wettenberg, Germany, https://www.optotech.de.

[49] SCHNEIDER GmbH & Co. KG, Biegenstr. 8-12, 35112 Fronhausen, Germany, https://www.schneider-om.com.

[50] Y. Hayashi, Y. Fukuda, and M. Kudo, "Investigation on changes in surface composition of float glass—mechanisms and effects on the mechanical properties," *Surface science* **507**, 872–876 (2002).

[51] DMG Mori Co., Ltd., Nakamura-ku, Nagoya, Aichi 450-0002, Japan, https://www.dmgmori.co.jp/corporate/en/.

[52] Z. Yang, L. Zhu, G. Zhang, C. Ni, and B. Lin, "Review of ultrasonic vibration-assisted machining in advanced materials," *International Journal of Machine Tools and Manufacture*, 103594 (2020).

[53] E. Fess, M. Bechtold, F. Wolfs, and R. Bechtold, "Developments in precision optical grinding technology," in *Optifab 2013*, **8884**, 88840L, International Society for Optics and Photonics (2013).

[54] M. Cahill, M. Bechtold, E. Fess, T. Stephan, and R. Bechtold, "Ultrasonic grinding of optical materials," in *Optifab 2017*, **10448**, 1044807, International Society for Optics and Photonics (2017).

[55] J. E. DeGroote, A. E. Marino, J. P. Wilson, A. L. Bishop, J. C. Lambropoulos, and S. D. Jacobs, "Removal rate model for

magnetorheological finishing of glass," *Applied optics* **46**(32), 7927–7941 (2007).

[56] C. Maloney and W. Messner, "Extending magnetorheological finishing to address short radius concave surfaces and mid-spatial frequency errors," in *Optifab 2019*, **11175**, 111750N, International Society for Optics and Photonics (2019).

[57] D. D. Walker, D. Brooks, A. King, R. Freeman, R. Morton, G. McCavana, and S.-W. Kim, "The 'precessions' tooling for polishing and figuring flat, spherical and aspheric surfaces," *Optics Express* **11**(8), 958–964 (2003).

[58] C. Miao, J. C. Lambropoulos, and S. D. Jacobs, "Process parameter effects on material removal in magnetorheological finishing of borosilicate glass," *Applied optics* **49**(10), 1951–1963 (2010).

[59] M. Ando, M. Negishi, M. Takimoto, A. Deguchi, and N. Nakamura, "Super-smooth polishing on aspherical surfaces," *Nanotechnology* **6**(4), 111 (1995).

[60] M. Negishi, A. Deguchi, M. Ando, M. Takimoto, and N. Nakamura, "A high-precision coordinate measuring system for super-smooth polishing," *Nanotechnology* **6**(4), 139 (1995).

[61] J. Askinazi, A. Estrin, A. Green, and A. N. Turner, "Recent advances in the application of computer-controlled optical finishing to produce very high-quality transmissive optical elements and windows," in *Window and Dome Technologies VIII*, **5078**, 97–108, International Society for Optics and Photonics (2003).

[62] J. H. Burge, "Simulation and optimization for a computer-controlled large-tool polisher," in *Fabrication and Testing of Aspheres*, FT2, Optical Society of America (1999).

[63] S. West, R. Angel, B. Cuerden, W. Davison, J. Hagen, H. Martin, D. W. Kim, and B. Sisk, "Development and results for stressed-lap polishing of large telescope mirrors1," in *Optical Fabrication and Testing*, OTh2B–4, Optical Society of America (2014).

[64] H. Martin, R. Allen, J. H. Burge, J. Davis, W. Davison, M. Johns, D. Kim, J. S. Kingsley, K. Law, R. Lutz, et al., "Production of primary mirror segments for the giant magellan telescope," in *Advances in optical and mechanical technologies for telescopes and instrumentation*, **9151**, 91510J, International Society for Optics and Photonics (2014).

[65] H. Martin, R. Allen, J. H. Burge, B. Cuerden, W. Gressler, W. Hubler, D. Ketelsen, D. Kim, J. S. Kingsley, K. Law, et al., "Manufacture of the combined primary and tertiary mirrors of the large synoptic survey telescope," in *Advances in Optical and Mechanical Technologies for Telescopes and Instrumentation*, **9151**, 915125, International Society for Optics and Photonics (2014).

[66] FANUC, A.H.W.H.R., Rochester Hills, MI 48309, https://www.fanucamerica.com/products/robots.

[67] ABB Ltd Group Headquarters: Affolternstrasse 44, Z., Switzerland, https://new.abb.com/products/robotics.

[68] D. Walker, C. Dunn, G. Yu, M. Bibby, X. Zheng, H. Y. Wu, H. Li, and C. Lu, "The role of robotics in computer controlled polishing of large and small optics," in *Optical Manufacturing and Testing Xi*, **9575**, 95750B, International Society for Optics and Photonics (2015).

[69] M. Rinkus, "Robotic polishing in asphere manufacturing," in *Optifab 2019*, **11175**, 111750B, International Society for Optics and Photonics (2019).

[70] D. R. Brooks, M. Brunelle, T. Lynch, and K. Medicus, "Manufacturing of a large, extreme freeform, conformal window with robotic polishing," in *Optical Manufacturing and Testing XII*, **10742**, 107420L, International Society for Optics and Photonics (2018).

[71] J. Márquez, J. Pérez, J. Rıos, and A. Vizán, "Process modeling for robotic polishing," *Journal of Materials Processing Technology* **159**(1), 69–82 (2005).

[72] S. Killinger, J. Liebl, and R. Rascher, "First steps towards an automated polishing process chain using one robot," in *Seventh European Seminar on Precision Optics Manufacturing*, **11478**, 114780I, International Society for Optics and Photonics (2020).

[73] L. Li, J. Zhang, C. Song, X. Zhang, X. Yin, and D. Xue, "New generation magnetorheological finishing polishing machines using robot arm," in *AOPC 2019: Space Optics, Telescopes, and Instrumentation*, **11341**, 1134118, International Society for Optics and Photonics (2019).

[74] Zeeko Limited, 4 Vulcan Way, Vulcan Court, Hermitage Industrial Estate Coalville, Leicestershire, LE67 3FW, UK, https://www.zeeko.co.uk/.

[75] D. D. Walker, R. Freeman, R. Morton, G. McCavana, and A. Beaucamp, "Use of the 'precessions'™ process for prepolishing and correcting 2D & 2½D form," *Optics Express* **14**(24), 11787–11795 (2006).

[76] D. D. Walker, D. Brooks, R. Freeman, A. King, G. McCavana, R. Morton, D. Riley, and J. Simms, "First aspheric form and texture results from a production machine embodying the precession process," in *Optical manufacturing and Testing IV*, **4451**, 267–276, International Society for Optics and Photonics (2001).

[77] S. D. Jacobs, "International innovations in optical finishing," in *Current Developments in Lens Design and Optical Engineering V*, **5523**, 264–272, International Society for Optics and Photonics (2004).

[78] E. Fess, J. Ross, and G. Matthews, "Grinding and polishing of conformal windows and domes," in *Window and Dome Technologies*

and Materials XV, **10179**, 101790Q, International Society for Optics and Photonics (2017).

[79] S. DeFisher, F. Wolfs, M. Cahill, P. Bechtold, R. Colavecchia, and D. Mohring, "Recent advances in machines, software and processes for manufacturing complex optical components," in *Window and Dome Technologies and Materials XVI*, **10985**, 109850J, International Society for Optics and Photonics (2019).

[80] C. Bouvier, S. M. Gracewski, and S. J. Burns, "Contact mechanics models and algorithms for dome polishing with ultraform finishing (uff)," in *Window and Dome Technologies and Materials X*, **6545**, 65450R, International Society for Optics and Photonics (2007).

[81] S. Killinger, J. Liebl, and R. Rascher, "Mid-spatial frequency errors in feed direction occurring in adapt polishing," in *Sixth European Seminar on Precision Optics Manufacturing*, **11171**, 111710G, International Society for Optics and Photonics (2019).

[82] E. M. Leitz, C. Stroh, and F. Schwalb, "Reproducible and deterministic production of aspheres," in *Optifab 2015*, **9633**, 96330I, International Society for Optics and Photonics (2015).

[83] S. D. Jacobs, D. Golini, Y. Hsu, B. E. Puchebner, D. Strafford, I. V. Prokhorov, E. M. Fess, D. Pietrowski, and W. I. Kordonski, "Magnetorheological finishing: a deterministic process for optics manufacturing," in *International Conference on Optical Fabrication and Testing*, **2576**, 372–382, International Society for Optics and Photonics (1995).

[84] D. Golini, S. D. Jacobs, W. I. Kordonski, and P. Dumas, "Precision optics fabrication using magnetorheological finishing," in *Advanced Materials for Optics and Precision Structures: A Critical Review*, **10289**, 102890H, International Society for Optics and Photonics (1997).

[85] QED Technologies, 1040 University Avenue, Rochester, NY 14607 USA, https://www.qedmrf.com.

[86] S. Jacobs and S. Arrasmith, "Overview of magnetorheological finishing(mrf) for precision optics manufacturing," *Ceram. Trans* **102**, 185–199 (1999).

[87] W. I. Kordonski and S. Jacobs, "Magnetorheological finishing," *International Journal of modern physics B* **10**(23n24), 2837–2848 (1996).

[88] J. E. DeGroote, *Surface interactions between nanodiamonds and glass in magnetorheological finishing (MRF)*. PhD thesis, University of Rochester. Institute of Optics (2007).

[89] A. Shorey, *Mechanism of material removal in magnetorheological finishing of glass*. PhD thesis, University of Rochester (2000).

[90] V. Bulsara, Y. Ahn, S. Chandrasekar, and T. Farris, "Mechanics of polishing," (1998).

[91] M. Tricard, P. Dumas, G. Forbes, and M. DeMarco, "Recent advanced in sub-aperture approaches to finishing and metrology," in *2nd International Symposium on Advanced Optical Manufacturing and Testing Technologies: Advanced Optical Manufacturing Technologies*, **6149**, 614903, International Society for Optics and Photonics (2006).

[92] J. A. Menapace, B. Penetrante, P. E. Miller, T. Parham, M. Nichols, J. Peterson, D. Golini, and A. Slomba, "Combined advanced finishing and uv-laser conditioning for producing uv-damage-resistant fused silica optics," in *Optical Fabrication and Testing*, OMB4, Optical Society of America (2002).

[93] C. Maloney and P. Dumas, "New meter-class mrf platforms offer multiple size and capability options," in *Optifab 2019*, **11175**, 111750Q, International Society for Optics and Photonics (2019).

[94] A. Kulawiec, W. Kordonski, and S. Gorodkin, "New approaches to mrf in optical fabrication and testing," *Imaging Applied Optics Technical Digest (online), Paper No. OM3D* **3** (2012).

[95] C. Maloney, J. P. Lormeau, and P. Dumas, "Improving low, mid and high-spatial frequency errors on advanced aspherical and freeform optics with mrf," in *Third European Seminar on Precision Optics Manufacturing*, **10009**, 100090R, International Society for Optics and Photonics (2016).

[96] T. Arnold, G. Boehm, H. Paetzelt, and F. Pietag, "Ion beam and plasma jet based methods in ultra-precision optics manufacturing," in *Optics and Measurement Conference 2014*, **9442**, 944204, International Society for Optics and Photonics (2015).

[97] C. Weingarten, A. Schmickler, E. Willenborg, K. Wissenbach, and R. Poprawe, "Laser polishing and laser shape correction of optical glass," *Journal of laser applications* **29**(1), 011702 (2017).

[98] L. L. Taylor, J. Xu, M. Pomerantz, T. R. Smith, J. C. Lambropoulos, and J. Qiao, "Femtosecond laser polishing of germanium," *Optical Materials Express* **9**(11), 4165–4177 (2019).

[99] M. Brunelle, I. Ferralli, R. Whitsitt, and K. Medicus, "Current use and potential of additive manufacturing for optical applications," in *Optifab 2017*, **10448**, 104480P, International Society for Optics and Photonics (2017).

[100] B. G. Assefa, M. Pekkarinen, H. Partanen, J. Biskop, J. Turunen, and J. Saarinen, "Imaging-quality 3d-printed centimeter-scale lens," *Optics express* **27**(9), 12630–12637 (2019).

[101] S. Steenhusen, F. Burmeister, H.-C. Eckstein, and R. Houbertz, "Two-photon polymerization of hybrid polymers for applications in micro-optics," in *Laser 3D Manufacturing II*, **9353**, 93530K, International Society for Optics and Photonics (2015).

[102] M. Thiel, Y. Tanguy, N. Lindenmann, F. Niesler, M. Schmitten, and A. Quick, "3d printing of polymer optics," in *The European Conference on Lasers and Electro-Optics*, JS_LIM_CLEO_Europe_2_3, Optical Society of America (2017).

[103] T. Yang, N. Takaki, J. Bentley, G. Schmidt, and D. T. Moore, "Efficient representation of freeform gradient-index profiles for non-rotationally symmetric optical design," *Optics express* **28**(10), 14788–14806 (2020).

[104] R. Dylla-Spears, T. D. Yee, K. Sasan, D. T. Nguyen, N. A. Dudukovic, J. M. Ortega, M. A. Johnson, O. D. Herrera, F. J. Ryerson, and L. L. Wong, "3d printed gradient index glass optics," *Science advances* **6**(47), eabc7429 (2020).

[105] J. E. Harvey and A. K. Thompson, "Scattering effects from residual optical fabrication errors," in *International conference on optical fabrication and testing*, **2576**, 155–174, International Society for Optics and Photonics (1995).

[106] D. M. Aikens, J. E. DeGroote, and R. N. Youngworth, "Specification and control of mid-spatial frequency wavefront errors in optical systems," in *Optical Fabrication and Testing*, OTuA1, Optical Society of America (2008).

[107] C. R. Dunn and D. D. Walker, "Pseudo-random tool paths for cnc sub-aperture polishing and other applications," *Optics express* **16**(23), 18942–18949 (2008).

[108] F. Tinker and K. Xin, "Aspheric finishing of glass and sic optics," in *Optical Fabrication and Testing*, OM4D–6, Optical Society of America (2012).

[109] D. W. Kim, H. M. Martin, and J. H. Burge, "Optical surfacing process optimization using parametric smoothing model for mid-to-high spatial frequency error control," in *Optifab 2013*, **8884**, 88840B, International Society for Optics and Photonics (2013).

[110] D. W. Kim and J. H. Burge, "Rigid conformal polishing tool using non-linear visco-elastic effect," *Optics express* **18**(3), 2242–2257 (2010).

[111] J. H. Burge, D. W. Kim, and H. M. Martin, "Process optimization for polishing large aspheric mirrors," in *Advances in Optical and Mechanical Technologies for Telescopes and Instrumentation*, **9151**, 91512R, International Society for Optics and Photonics (2014).

[112] C. Klinger, "Vibe: a new process for high-speed polishing of optical elements," in *Optifab 2007: Technical Digest*, **10316**, 103161G, International Society for Optics and Photonics (2007).

[113] J. D. Nelson, B. Light, D. Savage, B. Wiederhold, and M. Mandina, "Vibe finishing to remove mid-spatial frequency ripple," in *Optical Fabrication and Testing*, OWE2, Optical Society of America (2010).

[114] T. S. Mast, J. E. Nelson, and G. E. Sommargren, "Primary mirror segment fabrication for celt," in *Optical Design, Materials, Fabrication,*

and Maintenance, **4003**, 43–58, International Society for Optics and Photonics (2000).

[115] J. Lubliner and J. E. Nelson, "Stressed mirror polishing. 1: A technique for producing nonaxisymmetric mirrors," *Applied optics* **19**(14), 2332–2340 (1980).

[116] J. E. Nelson, G. Gabor, L. K. Hunt, J. Lubliner, and T. S. Mast, "Stressed mirror polishing. 2: Fabrication of an off-axis section of a paraboloid," *Applied Optics* **19**(14), 2341–2352 (1980).

[117] G. Spencer and M. Murty, "General ray-tracing procedure," *JOSA* **52**(6), 672–678 (1962).

[118] R. Wilson, "Karl schwarzschild and telscope optics.(karl schwarzschild lecture 1993)," in *Reviews in Modern Astronomy*, **7**, 1–30 (1994).

[119] G. Forbes, "Shape specification for axially symmetric optical surfaces," *Optics express* **15**(8), 5218–5226 (2007).

[120] G. Forbes, "Robust, efficient computational methods for axially symmetric optical aspheres," *Optics express* **18**(19), 19700–19712 (2010).

[121] D. Aikens, "A better aspheric equation," *Rochester Regional Photonics Cluster Newsletter* **6**(2) (2010).

[122] W. T. Plummer, J. G. Baker, and J. Van Tassell, "Photographic optical systems with nonrotational aspheric surfaces," *Applied optics* **38**(16), 3572–3592 (1999).

[123] W. T. Plummer, "The origins of commercial free-form optics," *SPIE Newsroom* 38 (2016).

[124] J. Nelson, "The evolution of freeform fabrication and testing: lessons learned and the roadmap to higher precision," *EPIC Meeting on Precision and Freeform Optics at WZW-OPTIC* (2019).

[125] G. Forbes, "Characterizing the shape of freeform optics," *Optics express* **20**(3), 2483–2499 (2012).

[126] A. Bauer and J. P. Rolland, "Design of a freeform electronic viewfinder coupled to aberration fields of freeform optics," *Optics express* **23**(22), 28141–28153 (2015).

[127] K. Fuerschbach, G. E. Davis, K. P. Thompson, and J. P. Rolland, "Assembly of a freeform off-axis optical system employing three φ-polynomial zernike mirrors," *Optics Letters* **39**(10), 2896–2899 (2014).

[128] M. P. Chrisp, B. Primeau, and M. A. Echter, "Imaging freeform optical systems designed with nurbs surfaces," *Optical Engineering* **55**(7), 071208 (2016).

[129] F. Wolfs, E. Fess, S. DeFisher, J. Torres, and J. Ross, "Freeform grinding and polishing with prosurf," in *Optifab 2015*, **9633**, 96331G, International Society for Optics and Photonics (2015).

[130] S. Powers, M. Brunelle, and M. Novak, "Design and manufacturing considerations for freeform optical surfaces," in *Freeform Optics*, JW1A–2, Optical Society of America (2019).

[131] M. Brunelle, J. Yuan, K. Medicus, and J. D. Nelson, "Importance of fiducials on freeform optics," in *Optifab 2015*, **9633**, 963318, International Society for Optics and Photonics (2015).

[132] J. D. Nelson, M. J. Brunelle, T. F. Blalock, and D. Brooks, "Using total error measurements of freeforms during manufacturing to aid in alignment," in *Advances in Optical and Mechanical Technologies for Telescopes and Instrumentation IV*, **11451**, 114510U, International Society for Optics and Photonics (2020).

[133] "Freeform optics topical meeting," *Proceedings Optical Design and Fabrication 2021 (Freeform, OFT, IODC, Flat Optics)* (2021).

Jessica DeGroote Nelson is the Director of Technology and Strategy at Optimax Systems, Inc. She graduated from The Institute of Optics at the University of Rochester with her B.S., M.S, and Ph.D. in 2002, 2004 and 2007, respectively. Dr. Nelson returned to the University of Rochester to earn her M.B.A. from the Simon Business School in 2013. In addition to working full time at Optimax, she is currently an adjunct faculty member at The Institute of Optics teaching undergraduate and graduate courses in optical materials, fabrication and testing. She is also an active member and presenter with The Optical Society (OSA) and SPIE.

Chapter 4
Metrology

Daewook Kim
James C. Wyant College of Optical Sciences, 1630 E. University Blvd., Tucson, AZ 85721, USA

Isaac Trumper and Logan R. Graves
ELE Optics, 405 E. Wetmore #117 #260, Tucson, AZ 85705, USA

4.1 Introduction

Humanity has been interested in optics for quite some time, the desire to expand our powers of observation being a powerful motivator for discovery and engineering. One example of this is the Nimrud lens, an oval-shaped rock crystal plate discovered by English archaeologist Henry Layard in 1853. The crystal, dated between 900-700 BCE, is bi-convex and has been theorized to have uses ranging from a simple magnifying glass to being a lens in a telescope [1]. Figure 4.1 demonstrates the Nimrud lens as it appears today, along with the surface profile, which is not the perfectly symmetrical shape we are so familiar with today in modern optics.

As civilizations grew and changed, so too did optics. By the fifteenth and sixteenth centuries, the optical sciences had advanced enough that convex and concave lenses for eyeglasses were at least available, if not common place [2]. Of course, these optics were not the precision lenses we are familiar with today. By studying works of art from the same period, it has been observed that some painters used optical elements as an aid in their endeavors. This observation was enabled by the fact that said optical elements had significant aberrations and artifacts that carried over into the final works [3].

In just a few hundred years, a short time span when compared to the scope of humanity, scientists rapidly evolved their capabilities when it came to making optics. While the astral bodies and mysteries of space have been inspiring the search for knowledge since the beginning of humanity, in the sixteenth century, optical devices began to make great strides to expand our ability to peer into this vast frontier of the until then unknown. In 1609,

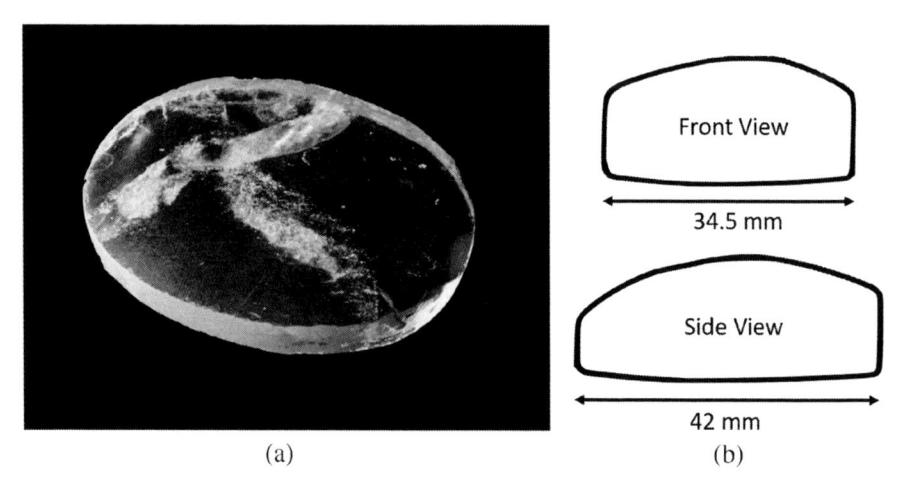

Figure 4.1 The Nimrud lens (a), a crystal optical surface dated to 900-700 BCE, is one of the earliest examples of an optical lens. It is a biconvex optic with loose symmetry (b) [1]. Image Credit: © The Trustees of the British Museum.

Galileo broke ground by delivering the 1.5-cm telescope. By 1908 the 60-inch (~1.5 m) telescope was rapidly eclipsed by the Hooker telescope, delivering a whopping 2.5 m aperture in 1917. Today, these numbers pale in comparison to large observatory telescopes, including the two 8.4-m primary mirrors in the Large Binocular Telescope and the 10.4-m aperture Gran Telescopio Canarias. Soon even these will be considered small with the new generation of extremely large telescopes offering apertures 20 to 40 meters in size [4]. Figure 4.2 demonstrates the evolution of the growth of optical telescope apertures from the 1700s to today.

However, simply delivering larger optics, or smaller as in the case for various optical fields, does not capture the true evolution of optical

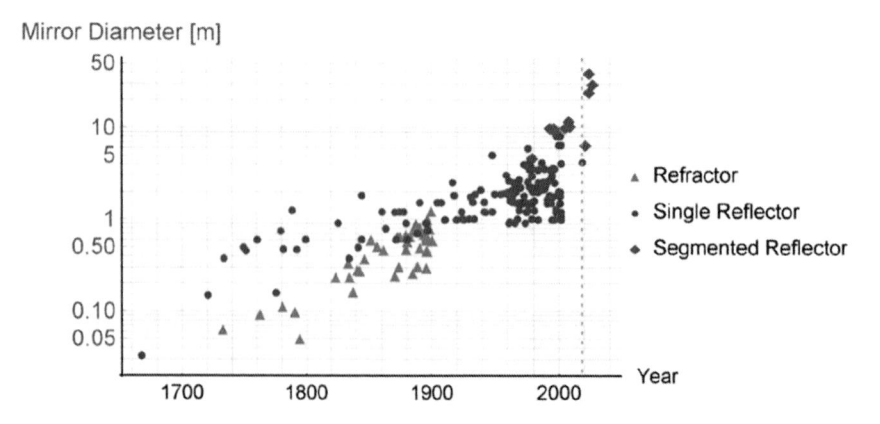

Figure 4.2 From the 1700s to today, there has been rapid growth in the size of the optics (i.e., refractive lens and/or reflective mirror) in telescopes.

fabrication. When one considers the Nimrud lens, it takes a moment to recognize it as a lens (no offense meant to the original creator). No one would trade their eyeglasses made today for ones fabricated in the 1500s, or at least no one who wanted to see things clearly. And when we consider telescope optics from only a hundred years ago, we recognize their place is in a museum and not in the observatory. The true change has been the quality of the optics made and the repeatability in the quality. It is one thing to deliver a precision optical surface by chance; it is quite another to reliably and consistently know how to fabricate the accurate shape and surface quality requested. This is where the craft transforms to the science of fabrication and metrology. While there have been incredible improvements in fabrication methods, evolving from the art of hand polishing to precision robotic computer-controlled lapping techniques or laser-driven surface figuring, the assessment and knowledge of the optics have undergone equally impressive and revolutionary advancements. It is this area, metrology, that is addressed in this chapter.

We have broken this chapter into three primary sections, with the first covering standard metrology techniques which measure with an accuracy from the micron to coarse millimeter range. The second section covers precision metrology, which includes techniques that range from nanometer to micrometer height resolution. These are often the metrology solutions to guide the optics manufacturing processes. The final section covers high-precision quality check or final verification metrology methods, which often report sub-nanometer features of surfaces. The techniques listed cover the most common metrology techniques utilized for optics; it is not, however, meant to cover every metrology method. Further, this chapter is meant as a source of information on the existing metrology techniques, but it is unfortunately not a substitute for deep experience with these metrology methods, which requires dedicated use to truly become familiar with.

4.2 Standard Opto-Mechanical Metrology

The Daniel K. Inouye Solar Telescope (DKIST) primary mirror is a 4.2-meter monolithic piece of Zerodur glass. This glass was, over years, transformed from a giant rough chunk into a surface which deviated, across the full aperture, from an ideal off-axis parabola by only 23 nanometers root-mean-square (RMS) figure error [5]. Unfortunately, this relationship is challenging to mentally picture. Even more unfortunate, the ratio is small enough that it is not easy to find a suitable metaphor; for instance, if we considered the 4.2-meter mirror was actually the size of a full football pitch (soccer field) that was perfectly smooth, those 23 nanometers of RMS height error would be equivalent to half a micron, or roughly 1/10th the size of a red blood cell, or 1/150th the width of an average human hair. The error is vanishingly minuscule, so tiny the surface earned the prestigious title of 'super smooth'.

The DKIST primary mirror did not start out life as a super smooth surface. Indeed, after the monolithic glass piece had been cast, it was delivered to an eagerly awaiting fabrication shop with a surface roughness in excess of 50 micrometers. It was a rough, non-specular reflective surface that was anything but the perfect mirror it was meant to be. However, as is common in the earliest days of life, the DKIST primary was about to undergo the most rapid shape changes of its life. During a roughly 3-month period known as the grinding phase, the DKIST primary mirror swiftly transformed from a rough, non-reflective (in the visible spectrum) surface to a far smoother, 1-micron RMS error from the ideal off-axis parabola it was meant to be. By the end of the three months, a careful observer could see a reflection from the surface.

During those three months, this incredible transformation of potential into reality was guided by a class of metrology we refer to as Standard Metrology. This area of metrology is concerned with measurements of optical surfaces whose roughness ranges from millimeters down to microns. As the description above indicates, these surfaces are typically not reflective in the visible spectrum, making many of the more well-known metrology methods (interferometry, for example) challenging if not impossible. Instead, optics that require coarse metrology can handle a more direct approach, using metrology tools that physically touch the surface to get a lay of the land, such as the coordinate measurement machine or the robotic arm techniques. Alternatively, projected light that scatters from the rough surface can be used to infer the surface shape down to microns of accuracy across large surface height ranges with structured light projection. Pictures with cameras in different angles can be used to determine a 3D measurement of the test surface using machine vision. For particularly large optics, such as the DKIST primary mirror mentioned above, tracking a laser spot at points across the surface can provide a sampling of surface points that can be used to generate a surface point cloud measurement. Knowing the correct tool for the job is important in maintaining an efficient metrology process for the optic under test when it comes to coarse metrology. Section 4.2 discuss the various standard metrology tools and their common use cases.

4.2.1 Coordinate measuring machines

A coordinate measuring machine (CMM) is defined as a point-by-point sampling method that records the 3D coordinates of a point defined in a coordinate frame. We discuss the use of CMMs in the field of optical metrology when making measurements with an accuracy of ~1 to 100 μm, which is typical for this class of instrument. We further break the usage of CMMs into direct surface measurements and optical component alignment. The role of the metrology device is significantly different in these two scenarios, and as such they have different considerations. Metrology of surfaces using a CMM typically involves fiducials in or around the optical surface that help with mechanical alignment, which can then be used to register the optical surface to the mounting surfaces during alignment and assembly. In this way, the two usages of this metrology tool can be connected but are typically employed in different stages during production.

4.2.1.1 Direct surface measurements

Coordinate measuring machines that are commercially available and in use at optical production companies can achieve around 1 μm accuracy [6] over parts of approximately 0.5 m in diameter. A primary benefit of utilizing a CMM measurement is that the output data is a surface map of deviations from a defined origin, which can be fiducials and datums on the surface used later for alignment, or the specified surface itself. This flexibility in choosing the reference makes the CMM a very versatile and flexible tool that is well suited to complex geometries with multiple optical components and parts. The CMM also excels at measuring low-order surface shapes, which can be especially problematic with other metrology techniques, and therefore can complement other tools nicely.

The sampling density of the CMM is usually limited by the time taken to measure the surface. Therefore, it is typically not able to measure high-spatial frequency surface errors, which a full aperture metrology method is more likely to record [7]. Furthermore, there is a risk of damaging the optical surface due to the mechanical contact of the probe tip and the force applied, which can range from millinewtons to tenths of newtons (unless a non-contact optical probe is used). The probe tip of the CMM is a critical component, as its geometric and thermal characterization contributes to systematic errors [8]. Measurements of the surface from the different styles of probe also impact the data processing, which needs to consider the local surface normals.

4.2.1.2 Optical alignment and assembly

A CMM may be used to help align and assemble optical components with their mechanical mounting structure. Since the CMM can also reference the optical surface, it has a great benefit of measuring the interface between these two systems in the same device. Depending on the application, the CMM may need to be portable, which typically increases the measurement accuracy to tens of μm. Such a portable CMM is invaluable for the alignment of large optical assemblies [9].

Mechanical fiducials can be created and placed on the optic using the CMM [10]. The fiducials are then useful for aligning the optic to the mechanical housing. The mechanical mating surfaces can also be measured using the CMM to verify that there is no wedge or large surface irregularity in the mechanics. This step of verifying the mechanical structures can reduce the complexity of aligning a fully assembled optical system by isolating the components and identifying the error stack up. Figure 4.3 demonstrates the

Figure 4.3 With custom optical mounts, it can be essential to measure the position of all components relative to one another to assure proper alignment. A CMM arm uses an optical probe tip to measure the coordinates of mechanical fiducials on a custom optic mount. Reprinted with permission from [11] © The Optical Society.

use of a non-contact optical probe tip, known as a point source microscope (PSM), being used on a CMM to measure spherical mounted retroreflectors (SMR) which are acting as opto-mechanical fiducials on a custom optical mount.

4.2.1.3 Robotic arm metrology

Robotic arm instruments exist which allow for full and accurate range of motion of a robotic controlled arm. These devices can move the probe tip in multiple degrees of freedom, allowing for a fixed arm to access various areas of the surface to be tested, which can also be useful for alignment and mounting purposes. Figure 4.4 demonstrates a robotic arm in use for freeform surface measurements. Particularly useful is the ability to program an automated measurement (available for some advanced robotic arm instruments) of a surface, providing repeatable and accurate coordinate measurements.

4.2.1.4 Practical tips

It almost goes without saying that any metrology tool that will be in physical contact with an optical surface requires special care. However, mistakes have been made by the best of us; therefore, special care is required with a CMM, as it will be in physical contact with your optical surface unless you use a

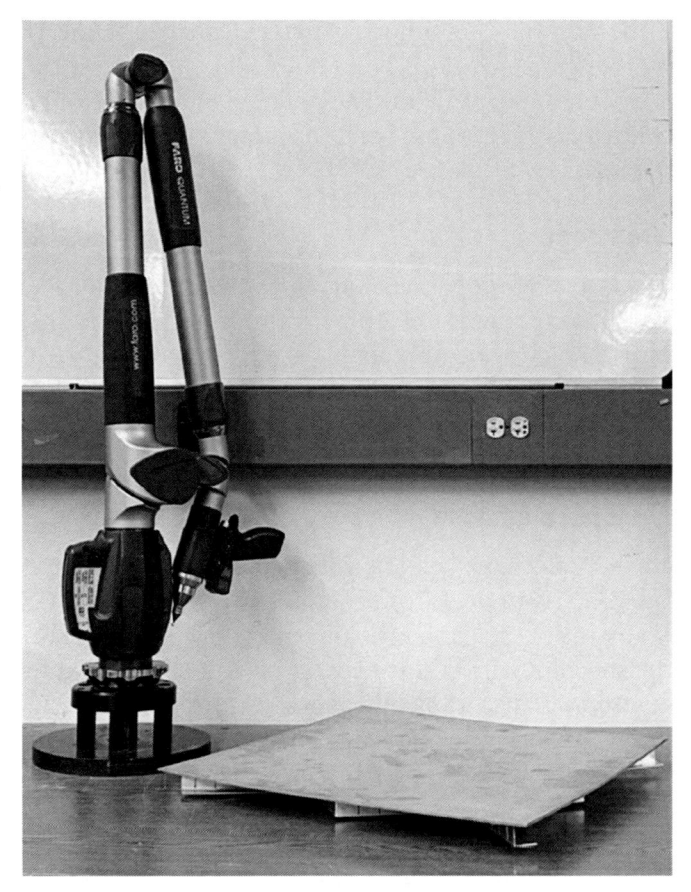

Figure 4.4 Robotic arm CMM solution called FaroArm. A robotic arm allows for a flexible positioning and movement for measurements with high precision.

non-contact optical probe. The probe tip can be easily damaged on a CMM, and certain optical materials can be even more easily damaged (and far harder to replace); therefore, check multiple times without your optic in place to be familiar and comfortable with the amount of force you are applying with the CMM probe arm.

With the probe in mind, it is in many devices calibrated at a fixed angle. While the probe tip may be able to be rotated and tipped, once it is calibrated these motions will undo your hard work and require starting over. Therefore, prior to starting any measurement, briefly check to ensure your probe arm can reach all the surfaces and datums required to accomplish the measurement you need.

Finally, the CMM measurement results, like other metrology methods listed, can only be reasonably relied upon to be accurate if the CMM is properly calibrated prior to testing. A common calibration method involves measuring a precision sphere (e.g., steel ball) with the probe tip several times to determine the accuracy of the measurement device. If the probe tip is a dragged probe, which is meant to be in physical contact with the surface and pulled across to obtain a profile measurement, a standard can be obtained to verify the linearity of response and accuracy of the probe across the measurement range.

4.2.2 Machine vision

Machine vision is a very wide area, encompassing practically any system that uses a camera and processing unit to perform an inspection or measurement task. The hardware is typically off-the-shelf components (e.g., telecentric camera) with potentially custom software written to control the devices. The systems help automate processes, speed up manufacturing, and increase yield by replacing a task done by a human that can fatigue and make mistakes.

4.2.2.1 Surface defect inspection

Machine vision systems have become widely used in production environments for their automation of an operator's manual inspection process. Such systems can be used to quantify the geometric dimensions of mechanical features [12]. More recently, machine vision has been applied to the assessment of optical surface quality (scratch dig), typically a task performed by an expert operator. The automation of surface quality inspection affords a repeatable and standardized assessment at the cost of single-purpose systems that may not catch all defects [13]. A further method to increase the dynamic range of such an inspection technique makes use of sub-aperture stitching, resolving finer details but requiring multiple images [14].

4.2.2.2 Alignment using photogrammetry

Photogrammetry is a technique that makes use of calibrated camera and test targets in order to perform dimensional metrology and has been applied to the alignment of the James Webb Space Telescope (JWST) hardware in a vacuum environment [15]. Careful assessment and simulations of the photogrammetry system showed that uncertainties on the order of 0.1 mm were possible, and were achieved during testing, demonstrating this technology for the use of coarse alignment of large-scale optical components and opto-mechanical structures.

4.2.2.3 Practical tips

Machine vision cameras require careful calibration prior to operation for accurate results, which can be cumbersome due to the software and device firmware communication. Machine vision is also a robust and large field, and there are a great variety of calibration methods that have been discovered, allowing for calibration with standards, without standards, and simply via motion of the camera itself in some known and repeatable way. For custom-built

Metrology 113

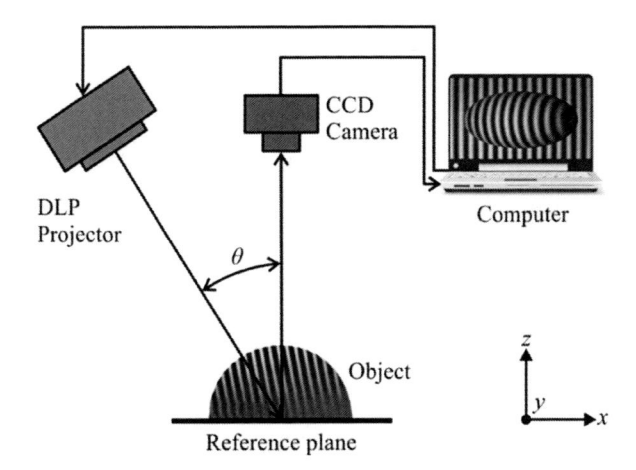

Figure 4.5 Schematic setup for a structured light projection metrology system for digital fringe projection profilometry. Reprinted with permission from [18] © The Optical Society.

machine vision systems, the reader is strongly encouraged to do thorough reading on calibration methods [16,17].

4.2.3 Structured light projection

Structured light projection (SLP) is a non-contact optical metrology method which can measure to the micron scale. The technique involves projecting a known light irradiance pattern onto the surface under test, which then scatters the light creating a 3D image on the surface. The scattered light is recorded by one or more cameras and the scattered pattern is compared to the modeled projected pattern to determine the surface profile.

For a test surface to be measurable with the SLP technique, it must be a scattering surface over the spectrum of light that is projected on to it. Typically, visible light is used for a SLP device, as the light sources and projection mechanisms are more straightforward to achieve, although another wavelength can be used when necessary. Figure 4.5 demonstrates a projected fringe pattern that was captured by a camera, and the corresponding surface that was calculated.

4.2.3.1 Surface height accuracy

The surface height features that are measurable by the SLP technique are correlated to the resolution of the pattern projected onto the test surface. The standard methods have produced surface height resolution of 2 to 10 microns RMS over several square millimeters. Also, the measurable area can often be increased at the expense of surface measurement accuracy. However, submicron accuracy has been achieved by utilizing a cube beam splitter to create a coherent fringe projection, which resulted in a 150-nm RMS surface accuracy [19].

4.2.3.2 High-speed measurements

Resolution is not where SLP offers its largest benefit; instead, the technique allows for measuring larger snapshots of test optical surfaces and its surrounding mechanical structures. Particularly in scenarios where the test surface may be changing rapidly, SLP can offer metrology snapshots of the surface on the millisecond scale, depending on the capture rate of the camera(s) used. In order to capture such fast data sets, a stroboscopic fringe-pattern method has been developed that 'freezes' the test surface in place. This stroboscopic pattern was able to measure a precision flat to 6.2-μm RMS in surface error [20].

4.2.3.3 Freeform surface metrology

An additional benefit of the SLP method is the ability to measure highly freeform surfaces. Multiple projectors may even be used to fully illuminate a test surface that otherwise may be impossible to fully project onto due to the test surface itself blocking light ray paths [21]. Using this multi-projector technique, a precision ball was able to be fully measured with only several microns of RMS error. In some scenarios the freeform surface may instead return interreflections, where the projected light onto one surface reflects from another surface and returns to the camera. Epipolar imaging and regional fringe projection can help to accommodate these issues, although accuracy may suffer [21].

4.2.3.4 Practical tips

A phrase often muttered in optics is 'you can never have too much light', and this is particularly true when it comes to structured light applications. Take care to reduce stray or ambient light in any test setup, and don't be afraid to use a powerful light source if you find it difficult to get a good measurement. Further, if you have time, repetition can be your friend. Multiple measurements can help to average out temporal random noise.

Also, if you can, measure a standard that you know well to calibrate your system. The projected light will be imperfect, and your camera may also have aberrations such as distortions. Properly accounting for them is essential in getting an accurate geometry of the measurement configuration. If it is possible, you can try a comparative measurement (e.g., quality check application), although this is not always feasible. Additionally, if you have a rotationally symmetric optic under test, you can perform a rotation calibration to reduce systematic errors [22].

Lastly, lock down your parts for this heavily geometry-dependent optical metrology solution. It can take quite a lot of time to properly configure and position your system and all the components; however, it only takes a moment to bump an optical table and move something. Take the time to properly secure all components in place, not only so that they remain secure during testing but also so that you can repeatably position your optic under test if necessary.

4.2.4 Laser tracker

For larger optics or optical system metrology, a laser tracker can provide sparse sampling points across the test surface or structures, which can allow for low-spatial-frequency surface metrology [23]. The technique uses a laser tracker, a device that has two angular encoders and a distance-measuring interferometer which measure the position of a spherically mounted retroreflector (SMR) in 3D space by aiming a laser beam at the reflector (i.e., corner cube centered in a precision spherical ball) and measuring the return signal. The distance-measuring interferometer provide accurate distance metrology, and new improvements, such as using a mode-locked laser, have further improved the measurement accuracy [24].

4.2.4.1 Calibration

A laser tracker alone is sometimes not enough to measure the surface shape of a test optic under certain vibrations or fluctuations since the point-by-point surface scanning process takes time to finish the measurement. Four (or more) independent distance-measuring interferometers (DMIs) are positioned to measure retroreflectors at the edge of the test surface continuously. In doing so, the rigid body motion (i.e., 6 degrees of freedom) is well defined for the object under test for each sampling point measurement, and air refractive index variations can also be accounted for. The independent DMI measurements additionally serve to calibrate the laser tracker measurement.

4.2.4.2 Measurement

In the actual data collection of the test surface, a SMR can be positioned at known points on the test surface. Ensuring dense enough sampling is taken across the full aperture is essential to guarantee an accurate full-aperture measurement is achieved without missing any key mid- to high-spatial-frequency features. Once all data points are measured, they can be combined and fitted modally to form a surface map of the surface under test. This technique has been used on the first 8.4-m off-axis segment of the Giant Magellan Telescope and was able to provide independent corroboration of low- to mid-spatial-frequency metrology results for the surface with an accuracy exceeding 1-μm RMS.

4.2.4.3 Practical tips

Laser tracker measurements, depending on the situation, can often require two people to achieve. One to control the laser tracker, which can be many meters away from the unit under test, and the other to carefully place the SMR and provide instructions for where to point the laser. Prior to starting any measurement, try to plan how many sample points are required for the surface shape frequencies you are trying to measure, and once you know the required sampling, take care where you choose to measure across the optical surface.

Various laser trackers can be guided by automated programming, which can make acquisition of the data easier and more repeatable. If the surface is to be measured several times, or over the course of its fabrication life, it may well be worth spending the time creating an automated metrology process sequence.

4.2.5 Infrared scanning system

A method to obtain more detailed high-spatial-sampling data has been implemented for surfaces further along in the generation stage when their surface roughness is <10 μm RMS. This method is an infrared (IR) deflectometry system and can produce height maps with much higher spatial resolution than other grinding stage metrology methods while achieving around 1-μm-level accuracy [25].

The principle of deflectometry, a reverse Hartmann test, uses an optical source and an imaging camera to measure local surface slopes across the optical surface. Deflectometry fundamentally relies on a 'ray' of light which is deflected by an optic; however, practically in most test systems the ray is in fact reflected by the optical surface, as shown in Figure 4.6. Specifically, the surface must specularly reflect the ray, as the directionality of the reflected light must be maintained, requiring a minimum of scattering, and also that the source light itself acts as a source radiating over a large solid angle, such that it does not have directionality (i.e., a laser beam). The camera in a deflectometry test system is set to focus on the surface under test, requiring a depth of field sufficiently large to encompass the surface height range. The camera detector pixels are mapped to the test surface and represent local discrete areas over which the surface slope will be determined. By modulating the source emission over a known location, the precise ray origin location can be well known. Correlating the recorded intensity of every pixel with the source emission as a function of location allows for determining the start and end ray locations, and the mapped camera pixel areas on the test surface fill in the final puzzle piece by providing a 3D location where the ray interacted with the test surface.

Deflectometry is a non-null test method that achieves a high dynamic range. In the case of IR deflectometry, the source emits in the long-wave infrared (e.g., 10-μm wavelength). An example of such a source is a long metal ribbon that is heated to a steady thermal state. By knowing the wire location in 2D space and measuring with the IR camera how the reflected image of the wire changes as the wire is translated in two perpendicular directions, the surface slope is calculated by triangulation and ray path

Metrology

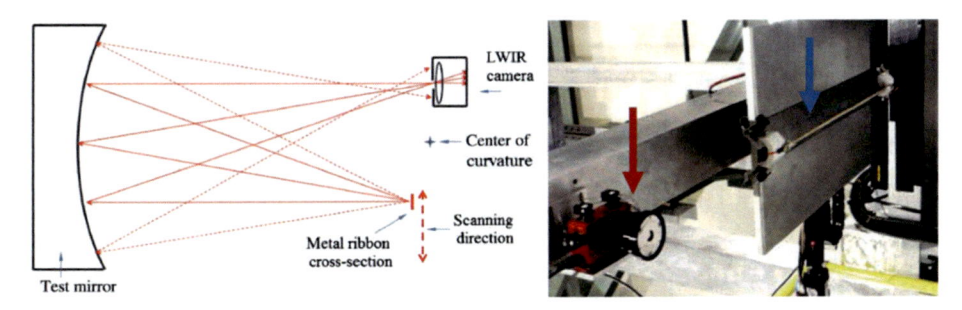

Figure 4.6 Overview of one implementation of an infrared deflectometry system used for grinding-stage metrology (left) and the hardware implementation of the system (right). The long-wave infrared (LWIR) wavelength permits testing of rougher surfaces (earlier in the fabrication process), which can greatly reduce the total time of manufacturing by providing metrology data to guide the high removal processes of grinding. Reprinted with permission from [26] © The Optical Society.

tracing. An integration step is then required to reconstruct the surface height information from the slope data. A schematic of an IR deflectometry setup used to measure large optical surfaces is shown in Figure 4.6.

Infrared deflectometry was used with much success on the 4.2-m Daniel K. Inouye Solar Telescope (DKIST) during the grinding stage where loose abrasives as large as 25 μm were used. This metrology tool guided the fabrication from 25- to 12-μm loose abrasive grinding, which resulted in a 1-μm RMS surface [5]. After the surface reaches this level of roughness, a clear reflection is obtained by a minimal polishing-out run, and other more traditional test methods may be used in conjunction with finer fabrication methods.

In the LWIR, a hot scanning source, such as a heated tungsten ribbon or heated ceramic rod, provides radiation in the thermal-infrared spectrum that can specularly reflect from surfaces with surface roughness as high as 50-μm RMS [25]. A separate test of a 6.5-meter diameter mirror captured during 40-μm grit loose abrasive grinding phase, shown in Figure 4.7, demonstrates the rough surface measurement capabilities during an optics manufacturing process.

An infrared scanning Shack-Hartmann system (SSHS) was employed as a full-aperture test on the JWST by Tinsley. It operates at a wavelength of 9.3-μm [28]. A Shack-Hartmann system has a high dynamic range and can characterize the mid-to-high spatial frequencies on the mirror substrate. The system is scanned over the surface making sub-aperture measurements, which are then stitched together to form a high-resolution surface map of the full aperture. This type of infrared system guides the grinding stage fabrication in an efficient manner, allowing for higher quality control over the final product. A sample measurement of an aluminum surface used to test the capabilities of the SSHS is shown in Figure 4.8, where residual lathe turning marks are

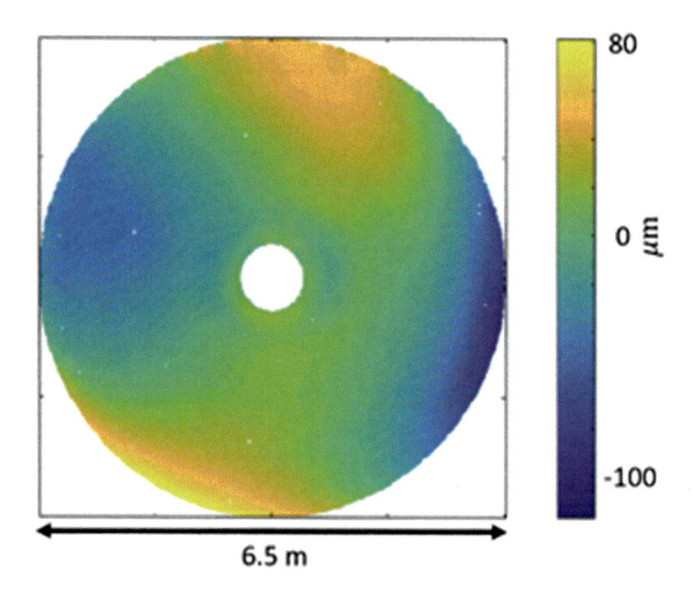

Figure 4.7 The grinding phase of fabrication of an optic allows for extremely rapid convergence to the desired surface shape. Above, after grinding a 6.5-meter optic with a 40-μm grit, the surface is tested with an improved infrared deflectometry system which uses a heated ceramic rod as the source and the surface map is reconstructed [27]. Image Credit: Hyemin Yoo, et al., "Improvements in the scanning long-wave optical test system," in Optical Manufacturing and Testing XII, 10742, 1074216, International Society for Optics and Photonics (2018).

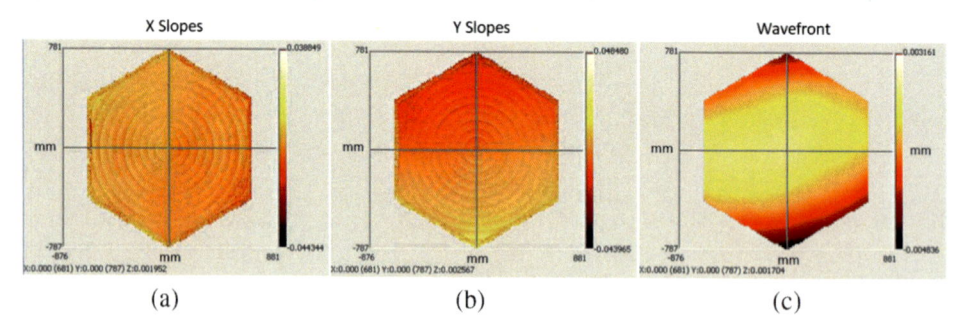

Figure 4.8 Sample measurement results of an aluminum mirror measured with the infrared scanning Shack-Hartmann system (SSHS) developed for grinding phase metrology of the JWST mirrors [28]. Shown in (a) and (b) are the x and y slopes (unitless) of the surface, respectively. These slopes are integrated to obtain the wavefront shown in (c) in units of millimeters. Image Credit: Craig Kiikka, et al., "The JWST infrared scanning Shack Hartman system: a new in-process way to measure large mirrors during optical fabrication at Tinsley," in Space Telescopes and Instrumentation I: Optical, Infrared, and Millimeter, 6265, 62653D, International Society for Optics and Photonics (2006).

clearly visible in the x and y slope maps (a) and (b), respectively, but becomes dominated by the low-order shapes in the wavefront map (c). The machined aluminum surface from the lathe was too rough to obtain quality data, so the surface was ground with automotive body shop methods [28].

4.3 Precision Process Guiding Metrology

Optical components range from things we take for granted every day, such as camera lenses in our mobile phone, to precision optics like biomedical microscope objectives and extremely large astronomical telescopes. One commonality to these optical elements is their quality of fabrication. When we say quality, we typically are referring to their surface shape as it compares to the ideal designed form. Planar optics are referenced to ideal flats and spherical optics to ideal spheres. The industry standard today is to achieve a peak-to-valley (PV) departure from the ideal shape of at least $\lambda/4$, where for most optical elements λ is referring to a wavelength of 632.8 nm, corresponding to a PV of ~150 nm.

For precision elements, the surface figure accuracy requires less than $\lambda/20$, or 30-nm RMS, which is a more statistical criterion. At this level, the metrology options become more limited, as the methods that can achieve nanometers of precision are harder to come by. We must consider that vibrations can cause mechanical shifts in excess of 30 nm. Thermal gradients can shift and deform mechanical structures, or even the optical surface under test, far more than 30 nm. Combining these effects makes for an even more challenging test experience.

To give one scenario, while testing the DKIST primary mirror, which was located at the time 3 stories underground in the basement of the Wyant College of Optical Sciences in Tucson Arizona, a test tower held a variety of test instruments, including interferometers and deflectometry devices. To obtain the final surface finish of the mirror, which ended up pushing from the range of precision to ultra-precision, an extremely large and sensitive test pathway was set up. The metrology tools were positioned near the center of curvature of the optic, about 15 meters above the surface in the test tower. Recall that the metrology of the precision scale is sensitive to the slightest of variations. Tucson is famous for a variety of things, but one well-known characteristic is extremely hot summers, with average temperatures reaching above 110° F (43° C) and large temperature drops at night due to the dry climate. This leads to impressive thermal loads and shifts, which on the scale of a test device housed in a tower 15 meters above a mirror can cause millimeters of shift just from temperature gradients. This is coupled with another contributor to Tucson's fame, the presence of the well-respected Arizona Wildcats collegiate basketball team, who play in a stadium located roughly 130 meters from the College of Optics. Eager fans line up to watch these games early in the day, and much like the terrifying videos of pedestrians hitting the resonant frequency of a bridge and causing ripples across an architectural structure assumed to be static, their mild footsteps were sufficient to cause vibrations as far away as the test tower, making testing impossible during games (of course, this also affords the metrology team a good excuse to take a break to watch the sporting event). Despite all of these factors, due to a variety of metrology tools, ranging from contact to non-contact solutions, and robust calibration methods, metrology proceeded largely unabated and was able to guide the fabrication process and verify the final surface shape, producing the world's largest off-axis aspheric primary mirror for solar observations.

With how sensitive a dedicated test lab specifically setup for creating precision optics is, it is a wonder that any metrology on the precision scale is possible, let alone common. This is where the knowledge and improvements of precision metrology have made their impact felt. There are a variety of improved metrology tools which are robust against environmental factors, although it requires proper knowledge to ensure correct calibration and setup to avoid error. Both contact and non-contact methods exist in the precision metrology range which allow for a dynamic test setup and cross-verification, further making metrology more verifiable at these fine levels. Section 4.3 discusses various precision metrology methods.

4.3.1 Contact stylus profilometry

A stylus profilometer utilizes a small stylus probe that is moved across the optical or opto-mechanical surface under test. The height variations of the stylus tip are measured to determine the test surface height profile. Stylus profilers can measure features as small as the nanometer scale up to approximately 1 mm. The stylus tips are typically made of hard materials, such as diamond or ruby. Modern stylus profilers feature a tip radius of curvature between 0.05 mm to 50 mm [29], which influences the lateral spatial resolution. The stylus tip load can be varied from 0.1 mg up to greater than 50 mg of force, allowing for the selection of an appropriate tip load to accurately measure surface features without applying too great a force that could deform or scratch the surface under contact. Figure 4.9 illustrates a schematic diagram of a transducer-based stylus profilometer.

4.3.1.1 Stylus tip geometry

The stylus tip can come in a variety of profile shapes, but a cone and pyramid are two most common shapes. The stylus tip shape and angle determine the as-measured depth of measurement of the instrument, with the output of a profiler being the convolution of the stylus tip with the test surface profile. For a cone-shaped stylus tip, a conventional shape is a conical 2 μm radius with a cone angle of 60°. The pyramid-shaped stylus is asymmetrical, which allows

Metrology   121

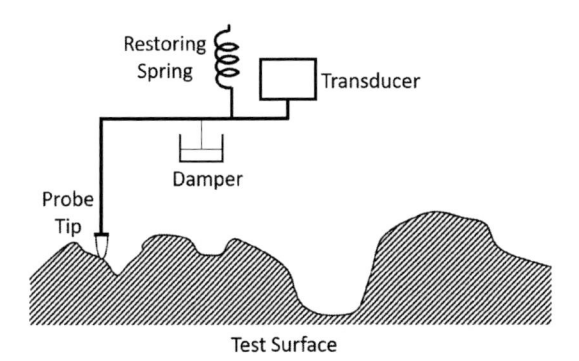

Figure 4.9 A schematic diagram showing a stylus profilometer utilizes a stylus tip which is dragged across a test surface. The height motion of the stylus tip can be measured using a variety of methods, but a transducer is a common approach [30]. Reprinted with permission from Springer Nature: Springer, Surface Characterization and Roughness Measurement in Engineering by J. Whitehouse, COPYRIGHT (2000).

for added strength in the tip against shock and reduces pressure. It is important to remember that the smaller dimension, referred to as 'a', should move perpendicular to the surface features of the surface under test. Figure 4.10 demonstrates the common cone and pyramid stylus tip shapes.

If too large a tip is used, high-frequency surface features will be filtered out of the measurement due to the convolution effect. However, too fine or sharp of a tip can also negatively impact measurements, as the force of the tip load can become too large over the local stylus tip area, which can cause surface deformations or scratches. If the surface deformation is elastic, it results in an inaccurate surface measurement, while a plastic deformation can lead to permanent changes in the surface profile, which are undesirable. Figure 4.11 demonstrates plastic deformations in a crystal surface by conical, pyramidal, and triangular tipped stylus tips on a crystal surface.

Typical of conic stylus tips is a 60° cone angle, which provides a 1:1 aspect ratio. However, as it has become more common to have surfaces with deep

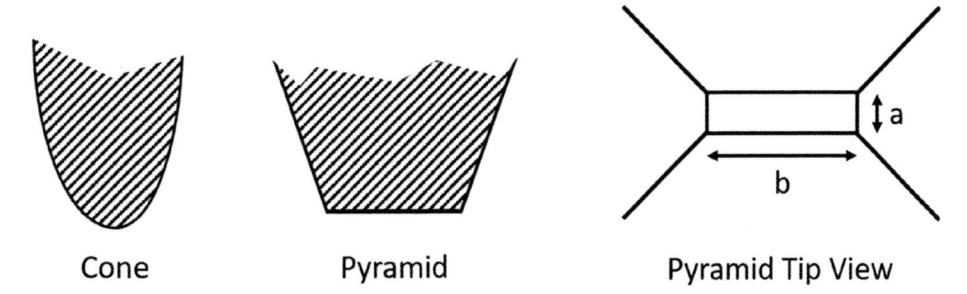

Figure 4.10 The stylus tip geometry of a surface profiler can come in many different shapes. A cone and asymmetrical pyramid are two of the common shapes used in many commercially available profilometers [30]. Reprinted with permission from Springer Nature: Springer, Surface Characterization and Roughness Measurement in Engineering by J. Whitehouse, COPYRIGHT (2000).

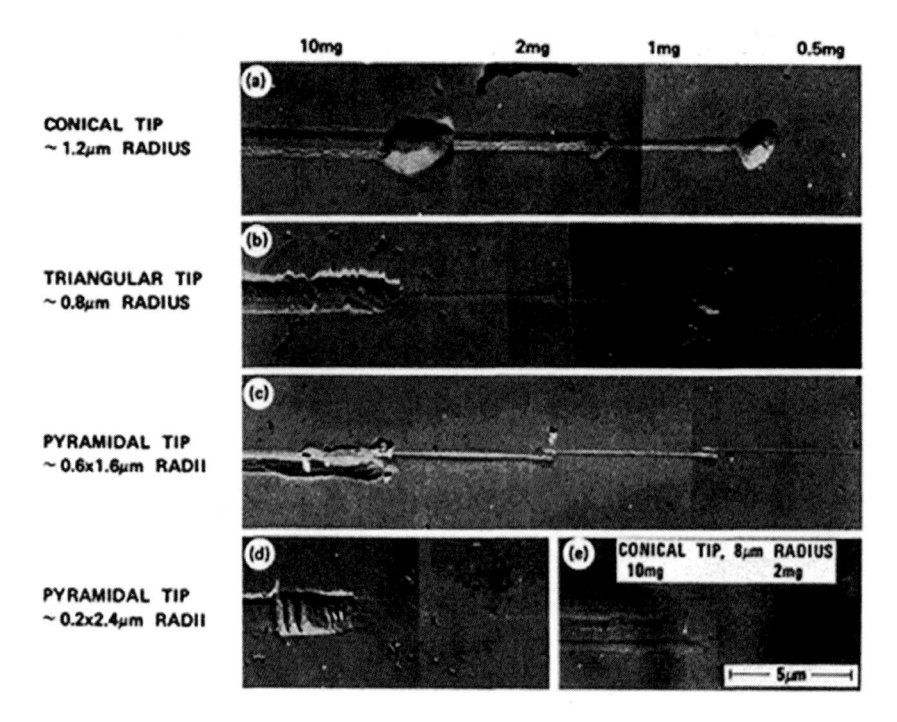

Figure 4.11 Too great a stylus load can cause permanent deformations in the test surface. Above, tracks were made in a single-crystal KCl material with conical (a), triangular (b), and pyramidal (c, d) tipped stylus with loadings from 10, 2, 1, and 0.5 mg. Reprinted with permission from [31] © The Optical Society.

trenches such as in newer lithographically printed chips or freeform surface features, sharp styluses have been created with aspect ratios as high as 10:1. Custom tip shapes can also be generated for measuring unique features with high precision on test optical surfaces and structures.

4.3.1.2 Load force

Modern instruments offer load forces as small as 0.1 mg, allowing for low force in testing to avoid deformation of the test surface. With such small forces, the instruments must be well isolated from outside forces, such as vibrations and air currents. The time required for an accurate measurement also increases as small force and small tipped probes are used, which can result in a measurement of a few thousand data points taking several minutes.

The stylus tip geometry, as well as the sampling intervals between data points, determines the lateral resolution of the instrument. For a spherical stylus tip measuring a sinusoidal surface, the shortest sinusoid period that can be measured is dependent upon the stylus radius, r, and the amplitude of the sinusoid, a [31]. The shortest measurable period d is

$$d = 2 * \pi * \sqrt{ar} \tag{4.1}$$

This informs the required stylus tip radius when considering features to be measured.

4.3.1.3 Reference datum

A reference datum is often measured to assure an accurate measurement of the test surface. This can come from utilizing a skid, which is moved across the surface with the stylus. However, when utilizing a skid, false surface features can be measured due to an apparent depression being seen when the skid moves over a bump feature on the real surface.

Utilizing a separate reference can provide greater accuracy but also can impose limits on the scan length and the maximum height variation. Reference optical flats with flatness of $\lambda/20$ are readily available and can serve as a stable reference. While this provides a reference during measurement, calibration is often done prior to measurement with various standards. Traceable height and roughness standards are readily available. They are often made from chrome on fused glass or etched in silicon. Measuring of the step features periodically ensures proper calibration and allows for determining the linearity of the height responsivity across the full measurement range.

4.3.1.4 Practical tips

Today's stylus profilers can measure surface roughness with a root-mean-square (RMS) on the sub-angstrom scale and with lateral resolutions as small as 100-200 nanometers. The instruments can measure 100-nm step heights with a repeatability of 6 Å and a 60-μm step height with 7-nm repeatability [29].

Surface profilers can serve as a robust metrology device for a wide range of applications for optics manufacturing and assembling processes. The devices often are utilized when the profile of a long measurement (up to 200 mm) are needed. Longer measurements can be achieved by stitching individual scans together. The instruments are often utilized for metrology of etched wafers, measuring wafer planarity, etch depth, feature profile, and surface roughness. Custom systems have also been utilized to measure surface sphericity and lower spatial frequency features [32].

If possible, much like a new household cleaner, try to measure an inconsequential region on your test surface with the tip and load force you plan to use prior to measuring the actual test surface area to ensure no damage will occur. Also consider what level of surface shapes you intend to measure and what tip type will be required. Stylus profilometer instruments can be extremely sensitive to outside vibrations; therefore, always try to limit sources of noise, control the environment, and take multiple measurements, if possible, to help average out the random effects.

4.3.2 Non-contact slope sensor scanning

Further methods of measuring mid-to-high spatial frequencies also involve using high resolution optical profilometry using a precision mechanical scanning stage and an optical slope sensing probe. Optical profilometry tools also measure slope but cover the spatial dimension by scanning the device on mechanical stages. The device measures the local slope of the surface, and records over time as it is moved across the entire aperture. Many samples are taken as the device is scanned, leading to very high spatial resolution. To measure the local slope, a light source illuminates a small area on the surface and the reflection is monitored. The change in the reflected angle is used to calculate the local slope value. With these instruments, sub-microradian slope accuracy is achievable, which results in height errors on the order of nanometers RMS [33,34], as shown in Figure 4.12.

4.3.2.1 Practical tips

In order to obtain high-accuracy mid-spatial-frequency data, calibration of the instrument is paramount. Due to the sensitivity of these devices, multiple repeated measurements may be required to average out vibrational or random noise. Finally, it must be noted that while this technique uses non-contact sensors, it nonetheless relies on mechanical stages and motion to capture a full measurement. Thus, the technique is only as accurate as its weakest link, such as systematic bias or errors. Fortunately, high-precision mechanical devices are available. Also, multiple sensors in a specific array can be utilized simultaneously to provide a self-calibration capability during a scanning path.

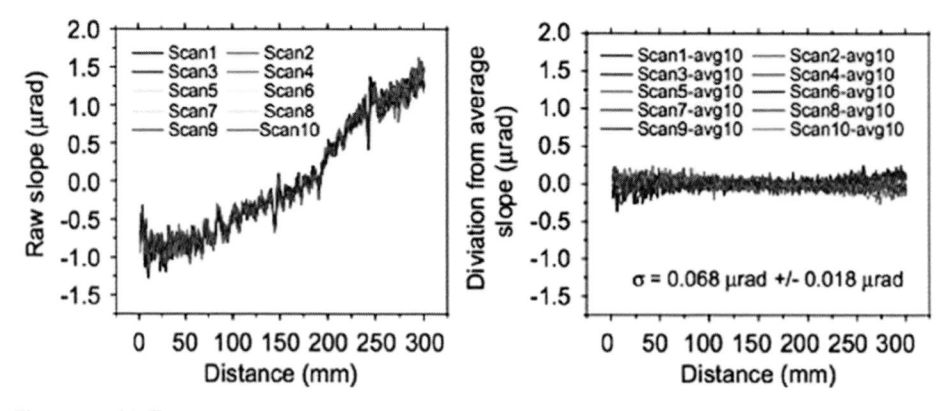

Figure 4.12 Ten measurements of the same surface with the raw slopes on the left, and the deviation from the averaged profile on the right. We see a sub-microradian slope accuracy [33]. Reprinted from *Nuclear Instruments and Methods in Physics Research Section A: Accelerators, Spectrometers, Detectors and Associated Equipment*, J. Qian et al., Performance of the APS optical slope measuring system, Copyright (2013), with permission from Elsevier.

4.3.3 On-machine metrology

On-machine metrology is the next step in the evolution of manufacturing. It reduces manufacturing cycle time, mitigates mounting errors, and guarantees better quality parts produced. However, implementing a precision measurement within the constraints of the manufacturing environment is greater when the two systems must be integrated. Further, since a host of tests to verify requirements must be done per workpiece, sometimes requiring complex and custom test configurations, developing a generalized system is a large challenge.

4.3.3.1 Co-mounted instruments

A co-mounted measuring instrument leverages the machine as a platform to perform metrology after a process has been completed, as shown in Figure 4.13. This is most like the conventional methods of fabrication and testing as two separate processes but combined into a single platform. This saves time by not removing the part from the machine and reduces errors due to differences in test and fabrication geometry.

Machines with high-precision motion about multiple axes are strong existing candidates for this type of upgrade, and diamond turning machines

Figure 4.13 By integrating an on-machine metrology tool onto a fabrication platform, it saves time and reduces remounting-induced errors. Above, an aluminum mirror is being fabricated, with an on-machine deflectometry metrology tool, consisting of a screen (solid arrow) and camera (dashed arrow) waiting covered to measure the surface after the fabrication run is completed.

are a prime example [35]. The metrology tools used to enable this type of measurement must be engineered carefully to account for the extra environmental uncertainties [36]. Many interferometric methods for on-machine metrology of surface finish and form error exist [37,38,39]. Profilometry and other contact based metrology methods can be used during the grinding stages, also [40,41].

4.3.3.2 In-process measurements

In-process measurements are made at the same time as the fabrication process, continuously monitoring the fabrication to provide real time feedback. Most of the work on in-process metrology has been monitoring the surface roughness. Confocal microscopy [42] or direct laser scattering [43] have been shown to provide well-correlated surface roughness measurements to the standard white-light scanning interferometry.

4.3.3.3 Practical tips

When developing a precision on-machine metrology tool, the environment must be considered carefully. Larger forces, vibrations, water, and removed materials are hazards that can degrade the quality of measurement and the integrity of the metrology system. The motion of the fabrication platform must also be well-understood because it can lead to mechanical interferences or blockages and, at worst, damage to the tooling or part being fabricated.

4.4 High-Precision Quality Check or Verification Metrology

At the cusp of transitioning from our 'macro' world, where we deal with large, solid optical components and surfaces, and the atomic world, where light begins to have wave-like behavior and features only a few nanometers in height are considered standard requirements, sits high precision metrology. Applications still cover a wide range: ultra-large optics and semiconductor wafers both can require sub-nanometer precision that only high-precision metrology offers.

We have followed the fabrication journey of the DKIST 4.2-meter primary mirror throughout this chapter, from its humble beginning as a rough, monolithic piece to a smooth, reflective surface. However, due to the amounts of light, resolution, and other pressing requirements of the system, the DKIST surface not only had to be measured and verified to have the correct surface shape (and off-axis parabola), but it must also be nearly perfectly smooth across that surface. Specifically, the requirements for the optic called for a surface roughness, a common metrology requirement in the high-precision region, of just 2-nm RMS. Across the full 4.2 meters of the optic, the glass surface must be smooth to within several atoms in height. By leveraging a combination of various metrology methods, and cross-comparing with a white-light-interferometer surface measurement (a technique discussed below), the super-smooth surface finish was achieved and verified.

Now, consider any electronic device you have used recently: your mobile phone, computer, television, smartwatch, etc. Most of these modern devices utilize advanced silicon wafers to achieve impressive computation power. These silicon wafers are constructed from features that are several nanometers in height, and their location relative to one another is extremely important. The same metrology techniques applicable for the DKIST primary mirror can also be applied to these advanced silicon wafers (although of course not in the exact same way). Two totally different systems, with different surface features and requirements, both require high-precision metrology.

This region of metrology tools, which include birefringence measurements, contact profilometry, the industry-standard interferometry, and other methods, are what allow for guidance and verification of the very smallest and most precise features required in modern optics. These metrology techniques come in a variety of modalities, and each has unique applications and limitations. This section highlights some of the most common high-precision metrology methods and their applications.

4.4.1 Birefringence measurements

Injection molded high-optical-quality components are now commonplace in high-volume production environments. Along with the surface-quality metrology solutions, injection molding can induce stress birefringence in the part. The flow rates at injection and thermal gradients during cooling phase can cause stress in the plastic, creating a difference in refractive index along different axes, which is birefringence [44]. At high volume, when parts are being quickly produced, these effects can cause serious production quality issues [45]. Assessing the quality of the parts as they are produced requires either a visual inspection by a trained operator or to use another machine to automatically make quantitative measurements.

4.4.1.1 Qualitative inspection

A traditional qualitative measurement is done using a polarization microscope [46], which requires a trained operator to assess the resulting visual image. The basic principles for analyzing birefringence within a material are to use a set of polarizers, as shown in Figure 4.14, where unpolarized light is made polarized, sent through the sample and then analyzed with a second polarizer that is set to block the light from the first polarizer. Anisotropy and birefringence in the refractive index of the sample will cause a change of the polarization state and therefore allow some of the light to pass through the analyzing polarizer.

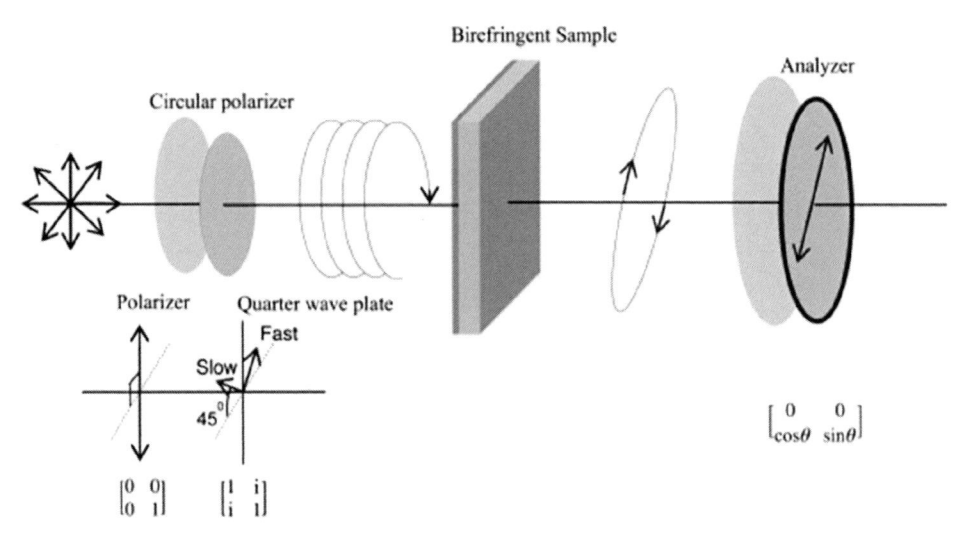

Figure 4.14 A polarimeter consists of a light source, whose polarization state is set using a polarizer. The polarized light passes through a birefringent sample, where due to anisotropy in the refractive index some of the polarization state will be rotated. A final polarizer, with a polarization state that would block light from the first polarizer, is then placed in front of an analyzer. Analyzing what amount of light, and where, on the sample makes it through can reveal much of the sample [47]. Reprinted from *Theoretical and Applied Mechanics Letters*, Vol. 6, A. Adhikari, T. Bourgade, and A. Asundi, Residual stress measurement for injection molded components, Copyright (2016), with permission from Elsevier.

4.4.1.2 Quantitative measurements

If quantitative measurements of the birefringence are required, or more detailed data about the distribution, another instrument is required. There are many point scanning systems that have been used to assess the quality of optical disks and LCDs and, more recently, 2D images. Using a photo elastic modulator and a polarized HeNe laser, a scanned 2D map of a sample can be created [48]. 2D imaging polarimeters can be used across different wavelengths by selecting the appropriate hardware, which offer faster data acquisition than the scanning methods for the polarization-based refractive of index or birefringence metrology [49].

4.4.1.3 Practical tips

Assessing the polarization behavior of an optical component or system in a quantitative manner requires much more consideration than a pass/fail test, because many factors (e.g., coatings, temperature, surface roughness) influence the behavior. Further, achieving a full aperture measurement of the birefringence of an element can be time consuming if quantitative comprehensive data is required.

Metrology 129

4.4.2 Swing arm profilometry

Another technique for measuring surface figure profile uses a swing arm profilometer (SAP), as shown in Figure 4.15. The technique uses a highly accurate probe which is mounted on a characteristic rotating arm whose axis of rotation passes through the center of curvature of the optic under test [50]. The probe trajectory defines an arc which lies on a spherical surface defined by said center of curvature. The SAP device then measures the optical surface departure from the spherical surface, and can measure convex, concave, and flat optical surfaces. This allows for repeatable and extremely accurate measurements of the deviation from an ideal sphere of a surface under test. This technique is especially useful for a metrology of large convex optics, which cannot be easily measured using an interferometric or deflectometric systems.

4.4.2.1 Optical probe

By utilizing robust calibration and high-resolution mechanical positioning and scanning stages, a SAP can achieve accuracy as good as 40-nm RMS [51].

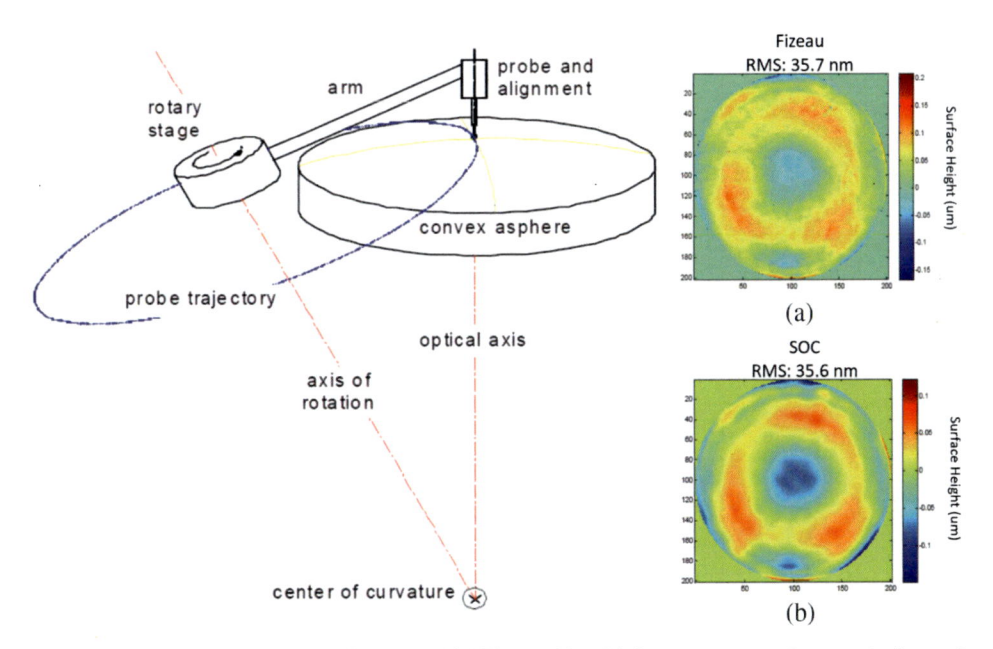

Figure 4.15 A swing-arm profilometer (SAP) provides high-accuracy surface metrology of test surfaces by moving a probe across a test surface. When compared to a Fizeau interferometry test of a test surface (a), a custom SAP method called swing-arm optical coordinate-measuring-machine (SOC) was able to achieve highly similar metrology results (b). The direct subtraction shows a difference of only 9-nm RMS [52]. Image Credit: Peng Su, et al., "Swing arm optical CMM: self calibration with dual probe shear test," in Optical Manufacturing and Testing IX, 8126, 81260W, International Society for Optics and Photonics (2011).

The swing arm profilometer does not necessarily require a contacting touch probe; a non-contact optical probe can be used in what is known as a Swing-arm Optical Coordinate-measuring-machine (SOC) test. Using an optical probe in the swing-arm device can produce surface figure results comparable to a Fizeau interferometry test of the same surface [52]. In order to achieve this level of accuracy, the SOC test makes use of a dual-probe shearing method which calibrates the system. This dual-probe shear calibration allows for calibrating out systematic errors in the measurement arm, which will always have errors due to the imperfection of arm bearing. Both probes will observe the same bearing errors while measuring separate and unique areas on the test surface, allowing for the isolation and removal of the systematic error. Figure 4.15 demonstrates the SOC process as well as the accuracy of the SOC system as compared to a Fizeau test of a 1.4-meter diameter aspheric surface with 300 μm of aspheric departure.

An additional variation on the SAP technique is to place the surface under test on a rotation table and rotate the part under a fixed probe. This test, termed Swinging part profilometer has the benefit of being able to be deployed for *in situ* metrology of a part during the fabrication process [53]. The swinging part profilometer approach has its own unique configuration challenge; however, it has been successfully used to test optical flats and exceeded the expected probe accuracy of 300 nm.

4.4.2.2 Practical tips

Substituting an optical probe can lead to highly impressive, and non-contact, profiling of a test surface for a swing arm profilometer. However, not all test surfaces are specularly reflective, which is a requirement for the most optical probes. Even worse, some surfaces may have areas that are specularly reflective and others that are less than ideally reflective, leading to non-continuous data limiting the ability to measure low-order shapes. For the highest accuracy, an optical probe tip may be required, but it also requires a specularly reflective surface fully across the test area.

4.4.3 Deflectometry

Deflectometry is a surface slope measurement technique, which relies on measuring the deflection of a ray of light that originates from a known location. The deflected ray must be deflected into a camera which records the final ray position. Knowledge of the ray start location, the approximate deflection position, and the final ray location allows for a trigonometric equation to be solved to determine the surface slope where the ray was deflected [54]. The technique distinguishes itself as a non-null metrology method which can measure a large range of surface slopes, and has been successfully used to measure optics ranging from freeform optics [55] through

Metrology 131

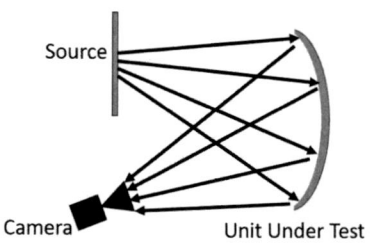

Figure 4.16 Deflectometry has been used recently for metrology on several of the NGT optics, including the 4.2-m DKIST primary mirror and 8.4-m GMT primary mirror segments. The method uses a source which emits light at a known location. The light deflects from the optic under test and is recorded by a camera. By knowing the precise 3D location of all components, the local slope of the optic under test can be determined, and the surface sag is determined by integrating the local slopes.

highly convex surfaces [56]. A traditional deflectometry setup demonstrating the concept is shown in Figure 4.16.

There are two key metrics that must be met to measure a full-aperture test surface. First, as mentioned the test surface must be able to either specularly reflect the light, if the deflection occurring is a reflection, or, it must be able to transmit the ray, if the deflection is in a transmissive setup. Second, to measure any given location on a test surface requires that light emitted from some point on the source area must be deflected into the camera. This second metric refers to the testable slope dynamic range of a deflectometry system, which depends on the hardware setup. If a clear line of sight can be established from the camera to a test location on the surface, and similarly from the source to the surface, and if the ray, obeying either the law of reflection of refraction, can travel from the source to the camera after deflection, then it is within the testable dynamic range. If the above two metrics are met, a surface can be tested with a deflectometry system.

4.4.3.1 Phase shifting deflectometry

Smoother surfaces allow for a wider range of more traditional, and more accurate sources to be used in a deflectometry system. One example is that for surface that specularly reflect in the visible region, a traditional digitally modulated electronic display can be used, such as a liquid crystal display. With a source that can be precisely modulated over the entire surface, more advanced source methods can be used as well. One such method is phase shifting deflectometry (PSD), which is based on a sinusoidal pattern being displayed across the source area and being phase shifted, allowing for phase unwrapping to be done in post-processing to determine the relative phase across the test surface. The phase-shifted sinusoidal image must be presented in orthogonal directions to measure orthogonal slopes. This technique has been used in testing multiple large precision optics [54,57,58]. With proper calibration, the PSD technique can be extremely accurate. The technique has been able to measure x-ray mirrors down to 1-nm RMS in surface height accuracy [57] and has achieved 300-nanoradian RMS slope precision [59]. Figure 4.17 demonstrates the acquisition and calibration process for a PSD system employed on a recently produced 6.5-meter mirror, as well as the surface test results compared to an interferometric test.

PSD methods do not need to be temporally phase shifted if a multiplexing approach is applied. Leveraging the three color channels in a RGB display and camera allows for encoding a phase shift of the sinusoidal pattern in each

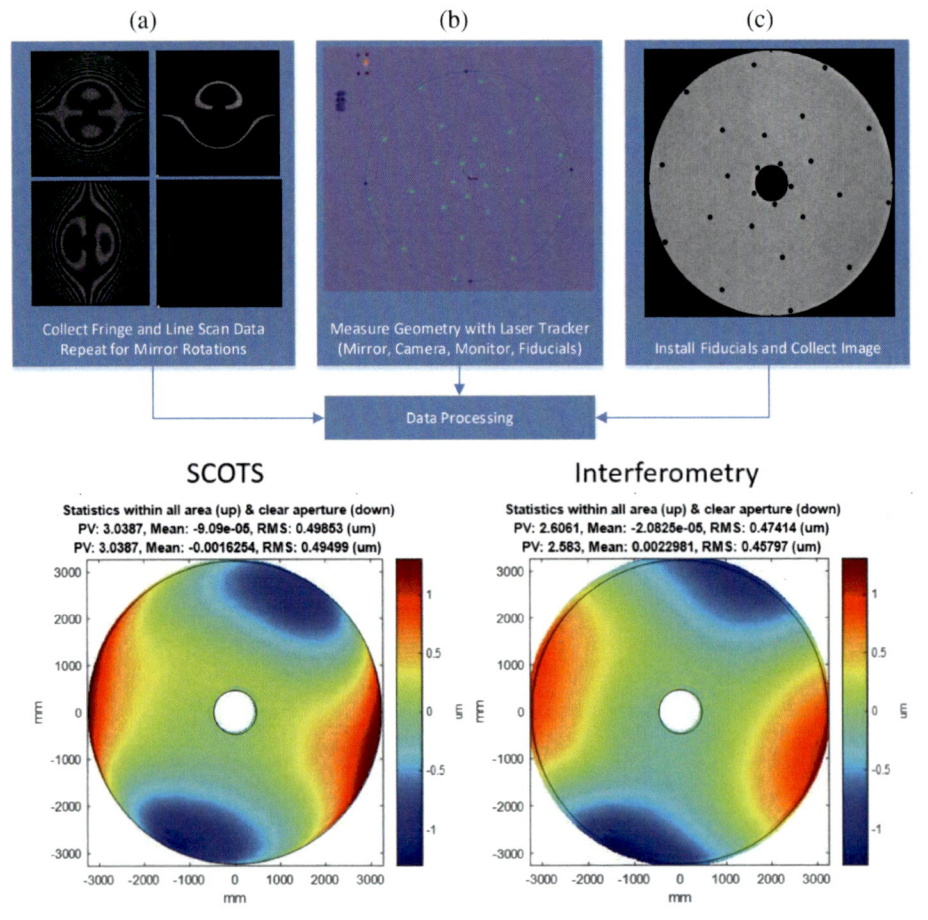

Figure 4.17 Phase Shifting Deflectometry (PSD) and line scanning (a) are two common deflectometry source approaches. For the highest precision, the geometry of the deflectometry test system is measured with a laser tracker (b), and fiducials are applied to the optic to map the camera distortion (c). Using this calibration approach, a PSD measurement (bottom left) achieved similar results to an interferometric measurement (bottom right) of a 6.5-m optic [61]. Image Credit: Andrew E. Lowman, et al., "Measurement of large on-axis and off-axis mirrors using software configurable optical test system (SCOTS)," in Advances in Optical and Mechanical Technologies for Telescopes and Instrumentation III, 10706, 107061E, International Society for Optics and Photonics (2018).

color channel, and further using a Fourier encoding method can produce a multiplexed phase image on the source which allows for a single shot capture of a surface that results in an instantaneous PSD measurement of a test optic [60].

4.4.3.2 Practical tips

Deflectometry is extremely versatile due to its non-null test geometry proving a large measurable dynamic range (especially compared to nulling interferometry) but can lead to uncertainties in the low-order surface shape unless a careful calibration is performed because the measurement process is very sensitive to the positioning and alignment uncertainty of the metrology configuration. It excels at differenced measurements, where a golden reference surface can be used to compare against further test parts in the final quality check process of mass production. Calibration methods to reduce the low-order uncertainties are critical and typically require physical contact with the surface to achieve maximum accuracy. The software to process a deflectometric measurement typically impact the final measurement data quality more than the system hardware, which can be off-the-shelf components (e.g., LCD display).

4.4.4 Null interferometry

Interferometry is a mature, high-precision metrology method which leverages the wave nature of light to produce an interference pattern between two optical paths. One path is an established reference path, while the other path is the test path configured to incorporate a test surface, either via reflection or transmission. When the two paths combine, the path length information from both paths is combined into the irradiance fringe pattern, which can be analyzed to interpret the optical path difference (OPD). The irradiance variation, or interreference fringe pattern, is recorded by a digital detector in most commercial devices, and the analysis of the variation in the irradiance is computed via software to interpret the OPD. Alternatively, visualization of the fringe pattern can be done by a skilled observer, and the fringes can be interpreted to obtain order of magnitude information (e.g., quarter-wave optics) of the path length differences with no computers required. There exist a wide variety of instrument designs for an interferometric test, and each setup can greatly impact what types of measurements can be made, the operating conditions required, and the dynamic measurable height range, to name a few.

4.4.4.1 Radius of curvature measurement

Interferometers can be robustly deployed for alignment purposes, measuring the correct positioning of optical elements as well as low-spatial-frequency features of optics under test, such as the radius of curvature (RoC). Fizeau interferometers are one of the most common configurations for full-aperture interferometry devices, which feature a common path allowing for wavefront error cancellation and simple robust optical design. Measurement of the radius of curvature of an optic can be achieved in a Fizeau setup by using a converging beam and placing the optic under test at the 'cat's Eye' (a.k.a. cateye) position, which is the location where the beam comes to focus at the surface of the optic under test. The rays with approximately focus to a point (ignoring diffraction) and return along the same path, returning what is known as the cat's eye reflection. The optic under test is then mechanically shifted along the optical axis to the point where it is equal to the radius of curvature of the test surface away from the cat's eye location. At this position, the test wavefront is reflected directly back onto itself, which results in a null fringe. By measuring the distance between the two, the radius of curvature can be well known. Figure 4.18 demonstrates the basic Fizeau interferometry setup for RoC metrology.

Mechanical motion of the optic under test is limited in accuracy to the mechanical positioning systems, and because the optic is exposed to the environment, even subtle thermal effects can degrade accuracy. One method has been created which utilizes a short coherence dynamic Fizeau interferometer design with internal path matching to measure the radius of curvature. In this technique, an optical delay mechanism composed of a polarization Twyman-Green interferometer produces two separate over-lapping, orthogonally polarized beams, s-polarized and p-polarized. A moving mirror M1 is on a motion stage which allows the optical path length of the exiting p-polarized beam to be adjusted relative to the s-polarized beam, which originates at mirror M2, and is fixed. Because a short-coherence-length source is used, only beams whose optical paths match from the source to the camera will interfere. This means that the optical path delays internal to the interferometer between M1 and M2 is directly related to the optical paths outside of the interferometer. This leads to significantly improved accuracy and stability in the measurement of the distance between the cat's eye position and the confocal position. Early results demonstrated that for a 10-mm radius

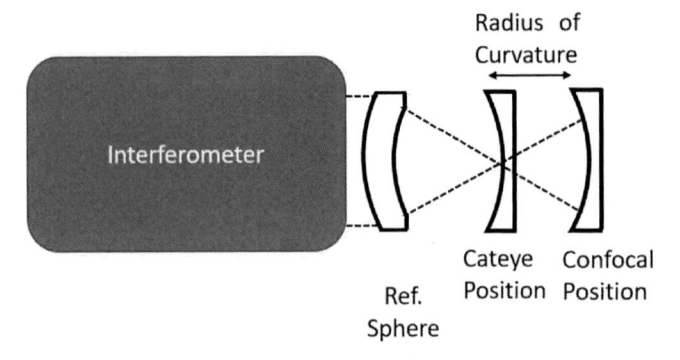

Figure 4.18 Fizeau setup for measuring the radius of curvature (RoC) of an optic under test by measuring the distance between the cat's eye and confocal positions [62].

of curvature optic, the technique results in a standard deviation of less than a micron in error, and the thermal stability of the measurement process can also be well calibrated [62].

4.4.4.2 Nulling interferometry for aspheric and freeform metrology

One of the highest-precision optical metrology results that can be achieved is via full-aperture nulling interferometry. As interferometry is a null-test method, it requires a null configuration to obtain the best test results. This typically can be achieved by using a customized optic to null out the test optic wavefront; however, the process becomes more complicated for a freeform optic. To achieve the highest level of testing accuracy for freeform surfaces, the use of a custom computer-generated hologram (CGH) is a commonly preferred method. Figure 4.19 below demonstrates a custom CGH and the resulting fringe pattern.

One additional benefit of CGHs is that they can provide advanced alignment of the null configuration by utilizing additional alignment holograms outside of the main testing aperture of the CGH. Because the external references are generated at the same time as the main null pattern is lithographically patterned, they are aligned to the accuracy of the lithographic process that created the CGH [63]. The result is a better-aligned, higher-performance custom null configuration which can achieve extraordinary

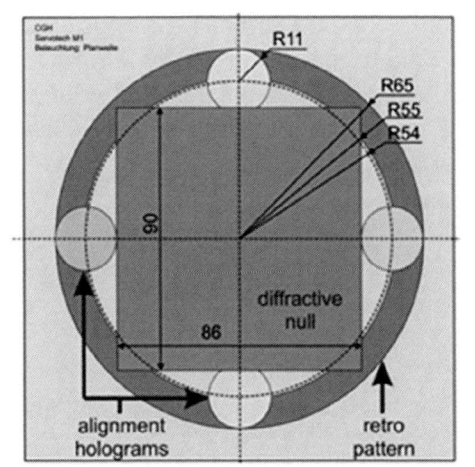

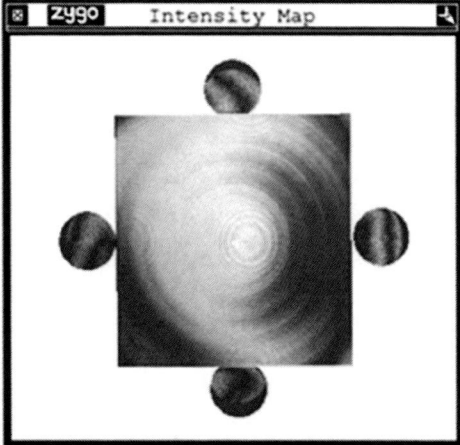

Figure 4.19 For freeform optics, a custom null optic is required to achieve adequate fringe density over the entire optic aperture. A computer-generated hologram (CGH) is an accurate way to generate extremely-high-precision custom null optics and can feature helpful alignment features (left). When used, they can provide a proper fringe density for testing even over a highly freeform surface (right) [63]. Image Credit: Sebastian Scheiding, et al., "Freeform mirror fabrication and metrology using a high performance test CGH and advanced alignment features," in Advanced Fabrication Technologies for Micro/Nano Optics and Photonics VI, 8613, 86130J, International Society for Optics and Photonics (2013).

precision metrology of the aspheric and freeform optics. Even with the improved alignment and nulling features a CGH presents, the exact implementation for testing some of the described optics may introduce distortion in the interferometric setup which must be modeled, mapped, and calibrated to achieve a high-accuracy metrology result [64].

CGHs and custom null optics are limited in the sense that they are static null optics. While it is possible to achieve an extremely precise null to an optic with the correct custom null, if the optic under test deviates significantly from the custom null for any reason, such as a manufacturing error, the null is no longer useable. This has led to explorations of dynamic null testing methods. The options for dynamic null methods are diverse. The use of spatial light modulators (SLMs) as a reconfigurable phase CGH has been explored. This technique in simulation was able to measure the form and mid spatial frequencies of a freeform mirror with a sag departure of 150 μm, with a theoretical form uncertainty of only 50.62-nm RMS [65]. Alternatively, a Hybrid Refractive and Diffractive Variable Null (HRDVN) has been developed which also can dynamically adjust the null profile to measure extreme freeform surfaces. The device is composed of a refractive aspheric null lens and a liquid crystal SLM. It was able to measure a freeform surface with rotationally symmetric departure of nearly 110 μm and non-rotationally symmetric departure of 14.5 μm, with moderate accuracy [66].

SLM based devices are not the only option for dynamic null devices. Deformable mirrors (DM) have been incorporated as a dynamic nulling component in interferometric configurations. The DM can be rapidly adjusted with excellent surface shaping abilities with fundamentally zero chromatic aberrations. The DM surface is optimized using a stochastic parallel gradient descent algorithm, which determines the optimal surface shape to achieve the best null for the optic under test. To verify the shape of the DM, a deflectometry system is incorporated which watches the surface changes of DM. Because the relative surface to the start surface shape (a flat) is the target in the deflectometry test, calibration needs are limited and an extremely accurate relative surface change can be obtained, informing the users of the DM null shape. The method was able to measure a freeform optic with ~15 μm peak to valley of departure from a flat and resulted in only 18.07-nm RMS difference from an commercial interferometric measurement of the same surface [67].

The traditional high-precision metrology technique of interferometry is undergoing improvements. One important area that has made great strides is sub-aperture stitching interferometry. This method is essential when testing large convex optics, which are featured in several next generation telescope designs. Recently, a method which utilizes a reconfigurable null test was devised which utilizes a multi-axis platform and two rotating CGHs to minimize systematic errors in sub-aperture measurements of a large convex asphere [68]. The acquisition and stitching process is demonstrated in Figure 4.20.

4.4.4.3 Ghost reflection management

Interferometers are often prone to parasitic reflection issues when testing transparent optics. One solution to this ghost reflection issue is to use a broadband source which allows for localized high-contrast interference fringes. This idea forms the basis of a white-light interferometer, which are typically limited to small-area measurements. Broad band Fizeau interferometer designs allow for localized fringes which can remove the issue of parasitic reflections degrading the desired target surface interferometry measurement.

Methods for achieving a broad band Fizeau design have a wide range. Some designs combine a broadband source with a Fabry-Perot cavity to achieve control of the spectrum, thereby allowing for localized fringes in the Fizeau configuration [69,70]. Other approaches allow for higher-order

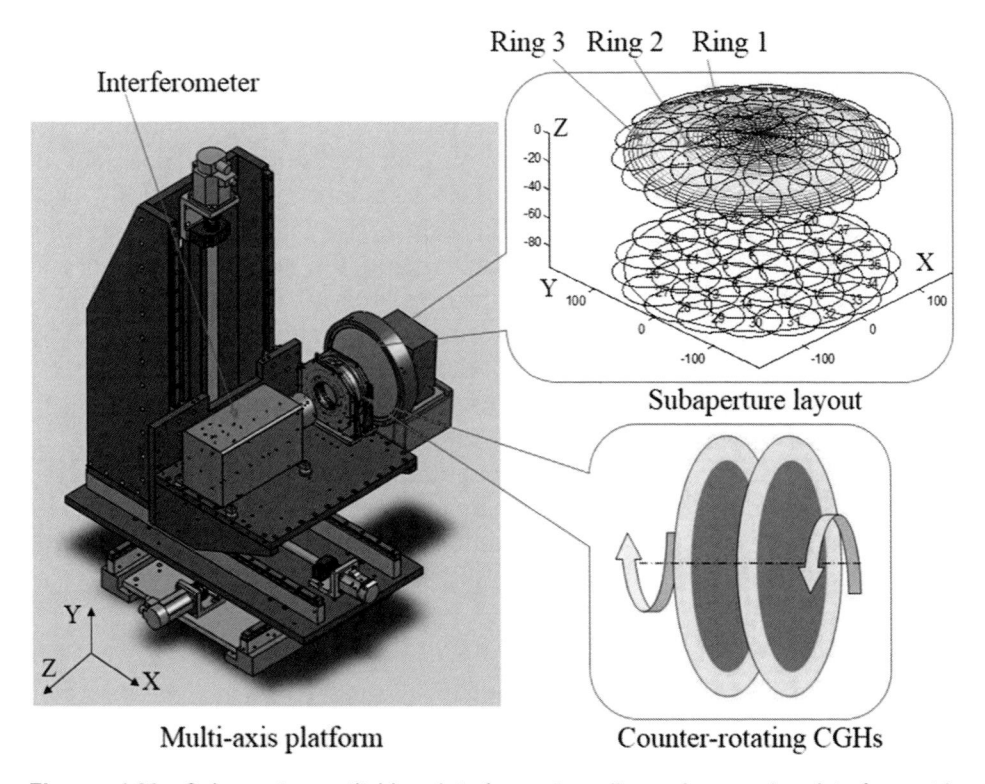

Figure 4.20 Sub-aperture stitching interferometry allows for precise interferometric measurements of large surface that cannot be measured using a standard interferometric test. By utilizing a multi-axis platform to carefully adjust the measured sub-aperture (left) and counter rotating CGHs (bottom right), a full map of the optic surface is obtained (top right). The sub-apertures are stitched, and system errors are calibrated out to produce a highquality map [68]. Reprinted from *Optics & Laser Technology*, Vol. 91, S. Chen et al., Subaperture stitching test of convex aspheres by using the reconfigurable optical null, Copyright (2017), with permission from Elsevier.

wideband interferometry by leveraging a moving cavity to control the source spectrum [71,72]. In either modality, a comb-like coherence function is produced by the mechanical motion of the cavity or the reference surface.

Another method for creating localized fringes in a Fizeau device is the method called spectrally controlled interferometry. The key of this method is that an extended bandwidth source modulated by a sinusoidal function creates a convolution of the nominal coherence function with delta functions at non-zero locations [73,74]. This allows for the creation of localized fringes without mechanical motion in the interferometry device, creating a more stable and mechanically simple method to achieve fringe localization. The method was successfully employed to measure an uncoated right-angle prism, as well as a 600-μm thin window, demonstrating its capabilities at achieving challenging measurements typically plagued by parasitic reflections.

4.4.4.4 Practical tips

As mentioned before, parasitic ghost reflection issues can be a practical challenge when testing transparent optics. While it can be possible to mitigate parasitic reflections by coating or treating the surfaces, including the use of special index matching solutions or even simple Vaseline coating on the back surface, this can be time consuming and possibly dangerous to the optic.

Interferometry has also been deployed for measuring the positions of components in an optical assembly. This technique has particularly been leveraged in larger optical assemblies, such as for telescopes. By utilizing custom CGH components, which can code the tip, tilt, and even distance, and incorporating them into the assembly process, a non-contact extremely high precision alignment process can be performed. This technique was successfully utilized to align the 4-mirror prime focus corrector for the Hobby-Eberly Telescope, achieving <±1.5 μm of transverse alignment precision [11,75]. In 2017, a group described the use of a CGH alignment process with an integrated Fizeau reference flat surface and Fizeau interferometer to align the camera lens assembly for the European Space Agency's Euclid Telescope. The overall alignment accuracy achieved an impressive sub-micron accuracy, with an alignment sensitivity in the range of 0.1 μm [76].

4.4.5 Instantaneous dynamic interferometry

One common challenge in an interferometry test is the presence of vibrations or turbulence negatively impacting a measurement. While averaging and isolation systems can help to mitigate these issues, a multiplexed interferometry method, which is called 'instantaneous interferometry', can remove this issue by capturing a full-aperture measurement in a single shot. This is achieved by using a pixelated detector with a phase mask over the pixels. The phase-mask encodes a high-frequency spatial interference pattern on two collinear, orthogonally polarized reference and test beams. The phase difference between the two beams can then readily be calculated by using an N-bucket or spatial convolution method, allowing for a single shot to measure all the data needed for a phase-shifted interferometry measure of the optic under test [77]. This technique has been robustly developed into a variety of commercial instruments which offer the standard precision afforded by commercial interferometry devices (nanometer-level accuracy) in a single-shot, instantaneous measurement tool.

Another metrology system that must be noted is the highly unique cryogenic center of curvature test system used for the JWST [78]. The test system consists of a multi-wavelength instantaneous interferometer, a calibration system, and a reflective null. Due to the long path length required for the center of curvature test, the instantaneous interferometry becomes essential to eliminate random variations arising from vibrations, which with multiple tests can be averaged out. The system achieved a wavefront error (WFE) repeatability of 10.8-nm RMS. Figure 4.21 demonstrates the test results for a segment of the JWST using the cryogenic center of curvature test.

4.4.6 Microscopic white-light interferometry

Microscopic white-light interferometry (WLI) broadly classifies white-light interference optical profilers which utilize axial scanning. The interference irradiance distribution is attenuated by the coherence envelope in the vertical scanning direction, with the peak being located near the best-focus position [79]. White-light interferometry instruments map the surface profile of a test surface by locating the vertical peak position of the coherence envelope across the test surface. WLI instruments are well established as the leading tool for

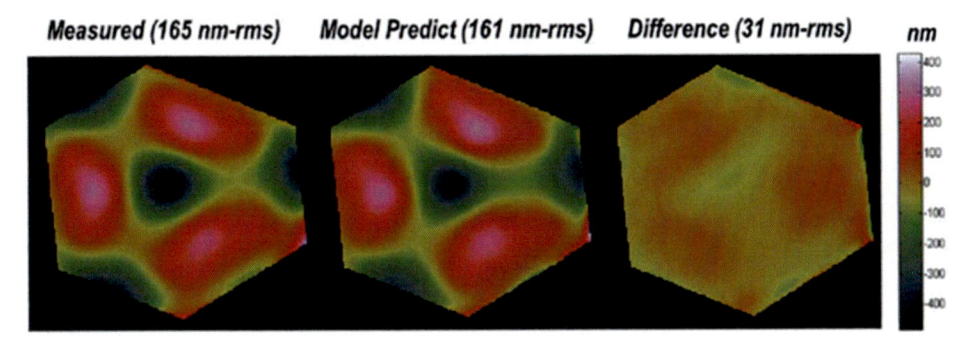

Figure 4.21 The JWST mirror segments have extremely tight surface figure tolerances. Thus, a multi-wavelength instantaneous interferometer was an essential tool for measuring surface figure. The measured figure of one segment (left) is comparted to a model-precited surface (middle) and the difference of 31-nm RMS across the surface (right) was determined [78]. Image Credit: James B. Hadaway, et al., "Performance of the primary mirror center-of-curvature optical metrology system during cryogenic testing of the JWST pathfinder telescope," in Space Telescopes and Instrumentation 2016: Optical, Infrared, and Millimeter Wave, 9904, 99044E, International Society for Optics and Photonics (2016).

measuring various engineered surfaces with small structures with high depth or very rough surfaces, such as a small Fresnel lens, MEMS devices, CGHs, or binary optics. The vertical resolution of WLI devices can be as good as single wavelength PSI (Phase Shifting Interferometry) methods (0.3 nm) but is more often closer to 3 nm [80]. Both the PSI and WLI are widely used as a micro roughness (a.k.a. surface finish) measurement solution for the surface scattering control purpose.

The source used in a WLI device is a broadband source whose spectral profile spans the visible spectrum. The broadband spectral profile creates a low temporal coherence, and the light source is not considered a point source, leading to low spatial coherence. The temporal and spatial coherence characteristics creates fringes localized in space. The white light source provides several benefits over a single wavelength setup, including no spurious fringes, the ability to measure large step sizes (multiple-wavelength operation), and ease of determining focus.

The difference wavelengths from the source spectrum are mutually incoherent. White light fringes are created by the superimposition of the individual wavelength's fringes. These white light fringes are observed with a monochromatic detector, which records the sum of all the fringes simultaneously. The maximum irradiance from the sum of the fringes will align at only one point in the axial scan, where the OPD is zero for all the wavelengths. As the axial distance moves away from this point, summed irradiance from the fringes from all wavelengths falls off rapidly; hence why the fringes are said to be localized. Figure 4.22 demonstrates the individual and summed irradiance profile seen by a detector in a WLI. The maximum contrast fringe, that is the fringe which corresponds to the zero OPD, is called the zero-order fringe. This fringe can easily be located across an object under

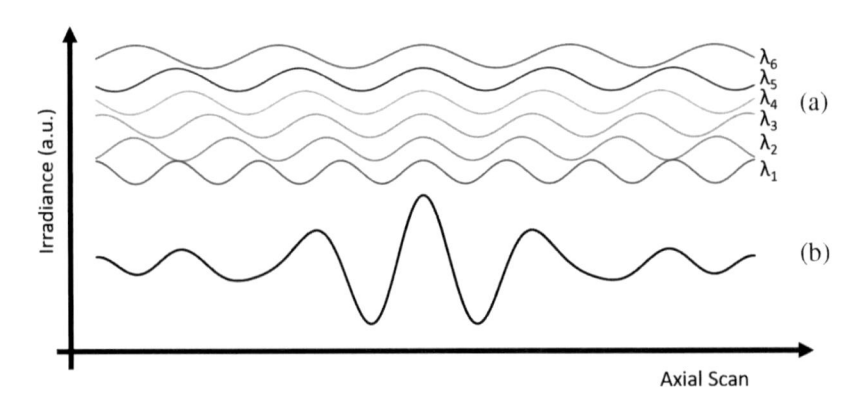

Figure 4.22 In a white light interferometer, the irradiance from each wavelength is captured by the sensor (a). The sensor takes the sum irradiance, which results in a single irradiance signal that has an envelope with a peak at the center, corresponding to the zero OPD fringe (b) [80].

Table 4.1 Comparison of common white-light interferometer objectives.

Objective	Magnification	Central Obscuration	Limited Working Distance	Numerical Aperture
Michelson	Low	No	Yes	Limited
Mirau	Medium	Yes	No	Limited
Linnik	Large	No	No	Large

test, which allows for determining the surface shape without ambiguity. The ability to eliminate ambiguity in numbering fringes allows WLI devices to measure discontinuous surfaces or surfaces that are rough.

The three most common interference objective configurations used in a WLI device are Michelson, Mirau, and Linnik [79]. Each objective has their own advantages and disadvantages, which are summarized in Table 4.1 below.

4.4.6.1 Practical tips

For homogenous materials, the spectral phase does not cause a problem. However, when a test surface features different materials side by side, each material will introduce a different phase shift upon reflection for different wavelengths, which will create an incorrect measured height different at the boundary where the materials meet. It is possible to correct for this difference by knowing the optical constants of the different materials for the wavelengths used. The optical constants for various materials can be found in various sources; one excellent repository is The Handbook of Optical Constants of Solids by Palik [81]. However, what can be a better solution for non-homogenous surfaces, if the scenario permits, is to coat the material with a thin layer of an opaque material, such as a metal coating.

4.5 Concluding Remarks

Optical components today can range from extremely large, such as the 8.4-m primary mirrors segments of the Giant Magellan Telescope, to the micron level with micro-optics and lenslet arrays. If we are even more liberal with our description, modern silicon wafers and other ultra-precision devices can contain features on the nanometer scale. In order to assure that these features are accurate, metrology methods are required. These range from the industry standard interferometry techniques, which leverage the coherent wave nature of light to inspect surfaces to sub nanometer accuracy without ever physically touching the surface, to stylus profilometry instruments which can detect miniscule height changes and steps by placing a stylus in contact with the optic. Of course, not all optics require precision metrology. Indeed, during fabrication, many optics undergo an earlier stage, named the generating and grinding phase, where they transform from raw material into rough surfaces approximately the shape of the final goal surface. Other optics, such as diffractive, Fresnel lenses, or freeform optics, may have surface features that depart from a best-fit-sphere from nanometers to millimeters across the optical aperture. In these cases, a coarse metrology device may be more applicable, capturing the larger surface features that may be outside of the measurement dynamic range of more precision instruments. Thus, it becomes apparent that to create the optics of today, a variety of metrology tools are required, and that these tools have their own strengths and weaknesses.

If there is one statement to remember with respect to metrology, it is that there is no single metrology solution which will meet all measurement needs. However, there is an optimal metrology solution for any given optical manufacturing, alignment, or process control application. Each metrology technique has certain unique advantageous and limitations, and it is the responsibility of the inspector to determine the correct technique for the job. Our goal with this chapter is to introduce the reader to the broad range of metrology techniques for optical components and systems and provide practical guidance and up to date performance metrics, and to offer a list of references for further, in-depth study for any given method. We hope this information will allow the reader to make a more informed and optimal selection of metrology methods for any given application.

References

[1] R. Willach, "The long route to the invention of the telescope," *Transactions of the American philosophical society* **98**(5) (2008).

[2] A. M. Smith, *From sight to light: The passage from ancient to modern optics*, University of Chicago Press (2014).

[3] D. Hockney and C. M. Falco, "Optical insights into renaissance art," *Optics and Photonics News* **11**(7), 52–59 (2000).

[4] L. R. Graves, G. A. Smith, D. Apai, and D. W. Kim, "Precision optics manufacturing and control for next-generation large telescopes," *Nanomanufacturing and Metrology* **2**(2), 65–90 (2019).

[5] C. J. Oh, A. E. Lowman, G. A. Smith, P. Su, R. Huang, T. Su, D. Kim, C. Zhao, P. Zhou, and J. H. Burge, "Fabrication and testing of 4.2 m off-axis aspheric primary mirror of Daniel K. Inouye Solar Telescope," in *Advances in Optical and Mechanical Technologies for Telescopes and Instrumentation II*, **9912**, 99120O, International Society for Optics and Photonics (2016).

[6] T. Blalock, K. Medicus, and J. D. Nelson, "Fabrication of freeform optics," in *Optical Manufacturing and Testing XI*, **9575**, 95750H, International Society for Optics and Photonics (2015).

[7] K. Medicus, S. DeFisher, M. Bauza, and P. Dumas, "Round-robin measurements of toroidal window," in *Optifab 2013*, **8884**, 88840Y, International Society for Optics and Photonics (2013).

[8] F. Fang, X. Zhang, A. Weckenmann, G. Zhang, and C. Evans, "Manufacturing and measurement of freeform optics," *CIRP Annals* **62**(2), 823–846 (2013).

[9] L. A. Crause, D. E. O'Donoghue, J. E. O'Connor, and F. Strümpfer, "Use of a faro arm for optical alignment," in *Modern Technologies in Space-and Ground-based Telescopes and Instrumentation*, **7739**, 77392S, International Society for Optics and Photonics (2010).

[10] E. F. Howick, D. Cochrane, and D. Meier, "Using a co-ordinate measuring machine to align multiple element large optical systems," in *Optical System Alignment and Tolerancing*, **6676**, 66760L, International Society for Optics and Photonics (2007).

[11] C. J. Oh, E. H. Frater, L. Coyle, M. Dubin, A. Lowman, C. Zhao, and J. H. Burge, "Alignment of four-mirror wide field corrector for the hobby-eberly telescope," in *Optical System Alignment, Tolerancing, and Verification VII*, **8844**, 884403, International Society for Optics and Photonics (2013).

[12] K. Harding, *Handbook of optical dimensional metrology*, CRC Press (2013).

[13] A. Schöch, P. Perez, S. Linz-Dittrich, C. Bach, and C. Ziolek, "Automating the surface inspection on small customer-specific optical elements," in *Optics, Photonics, and Digital Technologies for Imaging Applications V*, **10679**, 1067915, International Society for Optics and Photonics (2018).

[14] Z. Cao, F. Cui, and C. Zhai, "Vision system with high dynamic range for optical surface defect inspection," *Applied Optics* **57**(34), 9981–9987 (2018).

[15] S. R. Lunt, C. Wells, D. Rhodes, and A. DiAntonio, "Use of close range photogrammetry in James Webb Space Telescope alignment testing under cryogenic conditions," *Journal of Astronomical Telescopes, Instruments, and Systems* **6**(1), 018005 (2020).

[16] A. Hornberg, *Handbook of machine vision*, John Wiley & Sons (2006).

[17] L. S. Feng, Y. P. Huang, Z. K. Xu, and Y. Zhang, "Image calibration for machine vision inspection system," in *Applied Mechanics and Materials*, **556**, 2841-2845, Trans Tech Publ (2014).

[18] A. Silva, A. Muñoz, J. L. Flores, and J. Villa, "Exhaustive dithering algorithm for 3D shape reconstruction by fringe projection profilometry," *Applied Optics* **59**(13), D31–D38 (2020).

[19] J. A. Rayas, M. León-Rodríguez, A. Martínez-García, K. Genovese, O. M. Medina, and R. R. Cordero, "Using a single-cube beam-splitter as a fringe pattern generator within a structured-light projection system for surface metrology," *Optical Engineering* **56**(4), 044103 (2017).

[20] G. Yang, C. Sun, P. Wang, and Y. Xu, "High-speed scanning stroboscopic fringe-pattern projection technology for three-dimensional shape precision measurement," *Applied optics* **53**(2), 174–183 (2014).

[21] H. Zhao, Y. Xu, H. Jiang, and X. Li, "3d shape measurement in the presence of strong interreflections by epipolar imaging and regional fringe projection," *Optics express* **26**(6), 7117–7131 (2018).

[22] C. J. Evans, R. J. Hocken, and W. T. Estler, "Self-calibration: reversal, redundancy, error separation, and 'absolute testing'," *CIRP annals* **45**(2), 617–634 (1996).

[23] T. L. Zobrist, J. H. Burge, and H. M. Martin, "Laser tracker surface measurements of the 8.4 m GMT primary mirror segment," in *Optical Manufacturing and Testing VIII*, **7426**, 742613, International Society for Optics and Photonics (2009).

[24] Y.-S. Jang and S.-W. Kim, "Distance measurements using mode-locked lasers: a review," *Nanomanufacturing and Metrology* **1**(3), 131–147 (2018).

[25] T. Su, W. H. Park, R. E. Parks, P. Su, and J. H. Burge, "Scanning long-wave optical test system: a new ground optical surface slope test system," in *Optical Manufacturing and Testing IX*, **8126**, 81260E, International Society for Optics and Photonics (2011).

[26] D. W. Kim, C. J. Oh, P. Su, and J. H. Burge, "Advanced technology solar telescope 4.2 m off-axis primary mirror fabrication," in *Optical Fabrication and Testing*, OTh2B–3, Optical Society of America (2014).

[27] H. Yoo, G. A. Smith, C. J. Oh, A. E. Lowman, and M. Dubin, "Improvements in the scanning long-wave optical test system," in *Optical Manufacturing and Testing XII*, **10742**, 1074216, International Society for Optics and Photonics (2018).

[28] C. Kiikka, D. R. Neal, J. Kincade, R. Bernier, T. Hull, D. Chaney, S. Farrer, J. Dixson, A. Causey, and S. Strohl, "The JWST infrared Scanning Shack Hartman System: a new in-process way to measure large mirrors during optical fabrication at tinsley," in *Space Telescopes and Instrumentation I: Optical, Infrared, and Millimeter*, **6265**, 62653D, International Society for Optics and Photonics (2006).

[29] J. Schmit, K. Creath, and J. Wyant, "Surface profilers, multiple wavelength, and white light interferometry," *Optical Shop Testing* **667755** (2007).

[30] D. J. Whitehouse, *Surface Characterization and Roughness Measurement in Engineering*, 413–461. Springer Berlin Heidelberg, Berlin, Heidelberg (2000).

[31] J. M. Bennett and J. Dancy, "Stylus profiling instrument for measuring statistical properties of smooth optical surfaces," *Applied Optics* **20**(10), 1785–1802 (1981).

[32] P. Scott, "Recent developments in the measurement of aspheric surfaces by contact stylus instrumentation," in *Optical Design and Testing*, **4927**, 199–207, International Society for Optics and Photonics (2002).

[33] J. Qian, J. Sullivan, M. Erdmann, A. Khounsary, and L. Assoufid, "Performance of the APS optical slope measuring system," *Nuclear Instruments and Methods in Physics Research Section A: Accelerators, Spectrometers, Detectors and Associated Equipment* **710**, 48–51 (2013).

[34] X. Xu, S. Ma, Z. Shen, Q. Huang, and Z. Wang, "High accuracy measurement of power spectral density in middle spatial frequency range of optical surfaces using optical profiler," in *8th International Symposium on Advanced Optical Manufacturing and Testing Technologies: Subnanometer Accuracy Measurement for Synchrotron Optics and X-Ray Optics*, **9687**, 968707, International Society for Optics and Photonics (2016).

[35] D. Li, B. Wang, Z. Qiao, and X. Jiang, "Ultraprecision machining of microlens arrays with integrated on-machine surface metrology," *Optics express* **27**(1), 212–224 (2019).

[36] M. Yamauchi and K. Hibino, "Measurement of air turbulence for on-machine interferometry," *Applied optics* **42**(34), 6869–6876 (2003).

[37] D. Wang, X. Fu, P. Xu, X. Tian, O. Spires, J. Liang, H. Wu, and R. Liang, "Compact snapshot dual-mode interferometric system for on-machine measurement," *Optics and Lasers in Engineering* **132**, 106129 (2020).

[38] D. Li, X. Jiang, Z. Tong, and L. Blunt, "Development and application of interferometric on-machine surface measurement for ultraprecision turning process," *Journal of Manufacturing Science and Engineering* **141**(1) (2019).

[39] D. Kapusi, T. Machleidt, K.-H. Franke, and R. Jahn, "White light interferometry in combination with a nanopositioning and nanomeasuring machine (npmm)," in *Optical Measurement Systems for Industrial Inspection V*, **6616**, 661607, International Society for Optics and Photonics (2007).

[40] F. Chen, S. Yin, H. Huang, H. Ohmori, Y. Wang, Y. Fan, and Y. Zhu, "Profile error compensation in ultra-precision grinding of aspheric surfaces with on-machine measurement," *International Journal of Machine Tools and Manufacture* **50**(5), 480–486 (2010).

[41] M. S. Rahman, T. Saleh, H. Lim, S. Son, and M. Rahman, "Development of an on-machine profile measurement system in elid grinding for machining aspheric surface with software compensation," *International Journal of Machine Tools and Manufacture* **48**(7-8), 887–895 (2008).

[42] S. Fu, F. Cheng, T. Tjahjowidodo, Y. Zhou, and D. Butler, "A non-contact measuring system for in-situ surface characterization based on laser confocal microscopy," *Sensors* **18**(8), 2657 (2018).

[43] O. Faehnle, R. Rascher, C. Vogt, and D. W. Kim, "Closed-loop laser polishing using in-process surface finish metrology," *Applied optics* **57**(4), 834–838 (2018).

[44] S. Bäumer, *Handbook of plastic optics*, John Wiley & Sons (2011).

[45] M. D. Chidley, T. Tkaczyk, R. Kester, and M. R. Descour, "Flow-induced birefringence: the hidden psf killer in high performance injection-molded plastic optics," in *Endoscopic Microscopy*, **6082**, 60820E, International Society for Optics and Photonics (2006).

[46] "Molecular expressions microscopy primer: Specialized microscopy techniques - polarized light microscope configuration." https://micro.magnet.fsu.edu/primer/techniques/polarized/configuration.html. Accessed: 2020-08-13.

[47] A. Adhikari, T. Bourgade, and A. Asundi, "Residual stress measurement for injection molded components," *Theoretical and Applied Mechanics Letters* **6**(4), 152–156 (2016).

[48] A. Adhikari and A. Asundi, "Birefringence characterization of injection molded microplates," in *International Conference on Experimental Mechanics 2014*, **9302**, 93022Q, International Society for Optics and Photonics (2015).

[49] R. A. Chipman, "Advances in polarization metrology," in *Novel Optical Systems Design and Optimization VI*, **5174**, 43–50, International Society for Optics and Photonics (2003).

[50] D. S. Anderson and J. H. Burge, "Swing-arm profilometry of aspherics," in *Optical manufacturing and testing*, **2536**, 169–179, International Society for Optics and Photonics (1995).

[51] M. Callender, A. Efstathiou, C. King, D. Walker, A. Gee, A. Lewis, S. Oldfield, and R. Steel, "A swing arm profilometer for large telescope mirror element metrology," in *Optomechanical Technologies for Astronomy*, **6273**, 62732R, International Society for Optics and Photonics (2006).

[52] P. Su, Y. Wang, C. J. Oh, R. E. Parks, and J. H. Burge, "Swing arm optical cmm: self calibration with dual probe shear test," in *Optical Manufacturing and Testing IX*, **8126**, 81260W, International Society for Optics and Photonics (2011).

[53] P. Zhang, J. Li, G. Yu, and D. D. Walker, "Development of swinging part profilometer for optics," in *Optics and Measurement International Conference 2016*, **10151**, 101510B, International Society for Optics and Photonics (2016).

[54] M. C. Knauer, J. Kaminski, and G. Hausler, "Phase measuring deflectometry: a new approach to measure specular free-form surfaces,"

in *Optical Metrology in Production Engineering*, **5457**, 366–376, International Society for Optics and Photonics (2004).

[55] R. Huang, P. Su, T. Horne, G. B. Zappellini, and J. H. Burge, "Measurement of a large deformable aspherical mirror using scots (software configurable optical test system)," in *Optical Manufacturing and Testing X*, **8838**, 883807, International Society for Optics and Photonics (2013).

[56] L. Graves, H. Quach, H. Choi, and D. Kim, "Infinite deflectometry enabling 2π-steradian measurement range," *Optics express* **27**(5), 7602–7615 (2019).

[57] R. Huang, "High precision optical surface metrology using deflectometry," (2015).

[58] D. W. Kim, J. H. Burge, J. M. Davis, H. M. Martin, M. T. Tuell, L. R. Graves, and S. C. West, "New and improved technology for manufacture of gmt primary mirror segments," in *Advances in Optical and Mechanical Technologies for Telescopes and Instrumentation II*, **9912**, 99120P, International Society for Optics and Photonics (2016).

[59] A. V. Maldonado, P. Su, and J. H. Burge, "Development of a portable deflectometry system for high spatial resolution surface measurements," *Applied optics* **53**(18), 4023–4032 (2014).

[60] I. Trumper, H. Choi, and D. W. Kim, "Instantaneous phase shifting deflectometry," *Optics Express* **24**(24), 27993–28007 (2016).

[61] A. E. Lowman, G. A. Smith, L. Harrison, S. C. West, and C. J. Oh, "Measurement of large on-axis and off-axis mirrors using software configurable optical test system (scots)," in *Advances in Optical and Mechanical Technologies for Telescopes and Instrumentation III*, **10706**, 107061E, International Society for Optics and Photonics (2018).

[62] M. N. Morris and E. Frey, "Short coherence dynamic fizeau interferometer with internal path matching for radius of curvature measurement," in *Interferometry XIX*, **10749**, 1074910, International Society for Optics and Photonics (2018).

[63] S. Scheiding, M. Beier, U.-D. Zeitner, S. Risse, and A. Gebhardt, "Freeform mirror fabrication and metrology using a high performance test cgh and advanced alignment features," in *Advanced Fabrication Technologies for Micro/Nano Optics and Photonics VI*, **8613**, 86130J, International Society for Optics and Photonics (2013).

[64] P. Zhou, H. Martin, C. Zhao, and J. H. Burge, "Mapping distortion correction for gmt interferometric test," in *Optical Fabrication and Testing*, OW3D–2, Optical Society of America (2012).

[65] R. Chaudhuri, J. Papa, and J. P. Rolland, "System design of a single-shot reconfigurable null test using a spatial light modulator for freeform metrology," *Optics letters* **44**(8), 2000–2003 (2019).

[66] S. Xue, S. Chen, G. Tie, Y. Tian, H. Hu, F. Shi, X. Peng, and X. Xiao, "Flexible interferometric null testing for concave free-form surfaces using a hybrid refractive and diffractive variable null," *Optics letters* **44**(9), 2294–2297 (2019).

[67] L. Huang, H. Choi, W. Zhao, L. R. Graves, and D. W. Kim, "Adaptive interferometric null testing for unknown freeform optics metrology," *Optics Letters* **41**(23), 5539–5542 (2016).

[68] S. Chen, S. Xue, Y. Dai, and S. Li, "Subaperture stitching test of convex aspheres by using the reconfigurable optical null," *Optics & Laser Technology* **91**, 175–184 (2017).

[69] J. Schwider, "White-light fizeau interferometer," *Applied optics* **36**(7), 1433–1437 (1997).

[70] J. Schwider, "Coarse frequency comb interferometry," in *Interferometry XIV: Techniques and Analysis*, **7063**, 706304, International Society for Optics and Photonics (2008).

[71] I. Kozhevatov and E. K. Kulikova, "Interferometric methods for surface testing. high-order white-light interferometer," *Instruments and Experimental Techniques* **44**(1), 84–87 (2001).

[72] I. E. Kozhevatov, E. A. Rudenchik, N. P. Cheragin, and E. H. Kulikova, "A new in situ method for testing the optical thickness of removed transparent elements," in *Current Research on Holography and Interferometric Methods for Measurement of Object Properties: 2000-2002*, **5134**, 50–53, International Society for Optics and Photonics (2003).

[73] C. Salsbury and A. G. Olszak, "Spectrally controlled interferometry," *Applied Optics* **56**(28), 7781–7788 (2017).

[74] C. Salsbury, *Spectrally Controlled Interferometry: Methods and Applications*. PhD thesis, The University of Arizona (2018).

[75] E. Frater, "Optical alignment with cgh phase references," (2016).

[76] J.-M. Asfour, F. Weidner, C. Bodendorf, A. Bode, A. G. Poleshchuk, R. K. Nasyrov, F. Grupp, and R. Bender, "Diffractive optics for precision alignment of euclid space telescope optics," in *Astronomical Optics: Design, Manufacture, and Test of Space and Ground Systems*, **10401**, 104010V, International Society for Optics and Photonics (2017).

[77] J. Millerd, N. Brock, J. Hayes, M. North-Morris, B. Kimbrough, and J. Wyant, "Pixelated phase-mask dynamic interferometers," in *Fringe 2005*, 640–647, Springer (2006).

[78] J. B. Hadaway, C. Wells, G. Olczak, M. Waldman, T. Whitman, J. Cosentino, M. Connolly, D. Chaney, and R. Telfer, "Performance of the primary mirror center-of-curvature optical metrology system during cryogenic testing of the jwst pathfinder telescope," in *Space Telescopes and Instrumentation 2016: Optical, Infrared, and Millimeter Wave*, **9904**, 99044E, International Society for Optics and Photonics (2016).

[79] A. Harasaki, J. Schmit, and J. C. Wyant, "Improved vertical-scanning interferometry," *Applied optics* **39**(13), 2107–2115 (2000).
[80] J. Schmit, K. Creath, and J. C. Wyant, *Surface Profilers, Multiple Wavelength, and White Light Intereferometry*, ch. 15, 667–755. John Wiley & Sons, Ltd (2006).
[81] E. D. Palik, *Handbook of Optical Constants of Solids*, Academic Press (1997).

Daewook Kim is an associate professor of optical sciences and astronomy at the University of Arizona. He has been working in the field of optical engineering, mainly focusing on very large precision optics such as the 25 m diameter Giant Magellan Telescope primary mirrors and space telescope such as the Aspera UV space telescope NASA mission. His research area spans precision freeform optics fabrication and various metrology options, such as interferometric test systems using computer-generated holograms, direct curvature measurements, and dynamic deflectometry systems. He is the chair of SPIE's Optical Manufacturing and Testing conference, SPIE's Astronomical Optics: Design, Manufacture, and Test of Space and Ground Systems conference, and OSA's Optical Fabrication and Testing conference. He has published over 150 journals/conference papers. He is a SPIE fellow and a senior member of OSA, and served as an associate editor of OSA's *Optics Express* journal.

Logan Rodriguez Graves is the Chief Operating Officer at ELE Optics. He has worked in optical engineering for over 10 years, with his current work focused on optical science software for team collaboration. He received his Ph.D. from the Wyant College of Optical Sciences at the University of Arizona in 2019, where his main research area covered precision optical metrology, focusing on deflectometry hardware and software development for improved freeform measurement capabilities. His prior work focused on biomedical optics, with research including autofluorescent imaging of colorectal lesions and mapping nonlinear rod-cone interactions at the retinal level.

Isaac Trumper is a cofounder and the Chief Technology Officer at ELE Optics, an optical software startup formed in 2019. He previously graduated with a Ph.D. from the Wyant College of Optical Sciences in 2019. His main research topics focused on metrology development for large freeform and aspheric optics, and freeform optical design using custom software algorithms. He received his B.S. in Optics from the Institute of Optics at the University of Rochester in 2015.

Chapter 5
Optical Coatings

Ronald R. Willey
Willey Optical Consultants, 13039 Cedar St, Charlevoix, MI 49720, USA

5.1 Introduction

The primary application of optical coatings on a surface is to control reflection from that surface. It might be to increase the reflection to make a mirror, or it might be to reduce the reflection like an anti-reflection (AR) coating on a camera lens, etc. The coating might also be to control the color or amount of the light reflected or transmitted at that surface like a beamsplitter. This might be a wavelength-selective coating or color-filter. An extreme case of a color-filter coating is one which only passes a narrow band. Such a narrow bandpass (NBP) filter might be produced to pass only a certain laser light while reflecting all the "noise" of other wavelengths as in a fiber-optics communication filter.

The physical principles needed to understand and produce optical thin film coatings were becoming known before the twentieth century and were particularly aided by the work of Huygens, Newton, Fresnel, Maxwell, and others. One generally wants a coating which is solid on the coated surface and which will endure whatever environmental conditions that it is expected to experience in its application. In the early twentieth century, most physical materials for coatings were heated in a vacuum until they evaporated; the atoms or molecules travelled without collisions with other atoms or molecules through a vacuum; and they condensed on a cooler surface to form the coating. This is generally referred to as physical vapor deposition (PVD). A vapor can also be formed by an arc or by sputtering. Sputtering of target materials occur when gaseous ions are formed and accelerated into the target to dislodge atoms or molecules by their momentum transfer. Arcs do similar things by putting a lot of energy like a lightning bolt on a small spot of the target and causing it to boil and throw off particles and vapors. These were already being demonstrated in laboratories in the 19th century.

Ancient mirrors, before PVD optical coatings were available, were made by polishing metal surfaces. Special alloys called speculum metal were developed for making mirrors. Chemical processes were also developed which could, for example, deposit a silver coating on glass. Van Liebig did this in 1835. This was the source of common mirrors after speculum metal mirrors, and silver had the highest visible reflectance of any metal. Silver, however, has an environmental durability problem when exposed to the atmosphere. It tarnishes in the presence of sulfur and oxygen vapors, and it loses much of its reflectance. This problem was solved by coating the silver on the back side of a clear protective plate glass and then protecting the back of that coating from air by a layer of paint. This silvering was a chemical process with no vacuum or vapor involved.

In 1930 and 1934, the sputter deposition of silver and gold were reported for roll coating on fabrics (perhaps for the fashion industry), but they do not seem to have influenced the optical coating field at that time.

In 1932, with the development of the 200-inch mirror for the Mt. Palomar Telescope, John Strong reduced to practical production the process of evaporating aluminum onto mirrors. Aluminum has only 90–95% the amount of reflectance of silver, but has significantly better environmental durability. Subsequent to that, processes were developed to apply protective overcoats, such as silicon monoxide, etc., over the aluminum for even better durability. John Strong published [1] his work on this in 1933, and Evaporated Metal Films (EMF) of Ithaca, NY was founded around 1935 to produce such coatings.

In that same era, single-layer AR (SLAR) coatings started to be deposited by PVD films of various low index of refraction materials which were one quarter wave of optical thicknesses (QWOT) at a wavelength in the visible spectrum. Magnesium fluoride (MgF_2) has evolved as the most durable of such coatings, if deposited at 250–300°C. A major milestone in this new application of AR coatings was for the debut of the film *Gone with the Wind* in 1939, where the major theaters used coated projection lenses by Bausch and Lomb to far improve the contrast with the color movie. Before AR coatings, each glass surface of a camera or projection lens reflected stray light of 4–6% back and forth in the lens to the film image, causing a very "washed out" appearance in the images on the film and on the movie screen.

Geffken [2] patented the NBP filter in 1939 which was a dielectric Fabry-Perot spacer coated with semi-transparent silver films on each side.

According to Angus Macleod [3], Smakula had started the production of AR coatings at Carl Zeiss in 1935 [4], but that work was kept secret up to and including WWII, until the Cartwright and Turner patent was released in the USA.

As of World War II, computers did not exist for the complex calculations required for the coatings with more than one layer. Generally, single-layer coatings were all that was used, such as fluoride ARs, high-index material beamsplitters, or metal mirrors such as aluminum, silver, etc.

More than one layer coatings were evolving in the late 1940s, and Mary Banning [5] published a definitive paper on multilayer coatings in 1947. Her colleague, Harry Polster [6], expanded on her work and on layer thickness control in 1950s. Layer thickness was controlled by the optical properties of the films by color and later by photoelectric signals. The quartz crystal monitor (QCM) or quartz micro-balance (QMB) was introduced in the 1960s and gained great popularity as a film-thickness monitor and controller.

The concept of a three-layer AR design using three materials had been invented by Geffken in 1942. The three materials used were of medium (M), high (H), and low (L) index of refraction and each had a thickness measured in integral QWOTs (1M, 2H, & 1L). This was because QWOTs were all that could be easily calculated and controlled in production with the technology of that day. When limited to QWOTs, the choice of indices became critical to the end results. Coating labs had cabinets full of every available material with which to experiment. In the 1950s, these 3-layer High Efficiency Coatings (HEA) became "all the rage". This technology was augmented by the invention by Rock in 1969 of the 4-layer, 2-material version which is still in common use. The 4-layer requires two initial layers which are not QWOTs, and therefore more complex calculations are needed. Fortunately, computer and software capabilities advanced rapidly from the 1960s to make this easier to handle. Both of these designs are now usually referred to as Broadband ARs (BBARs). Figure 5.1 shows the reflectance of an uncoated substrate, a

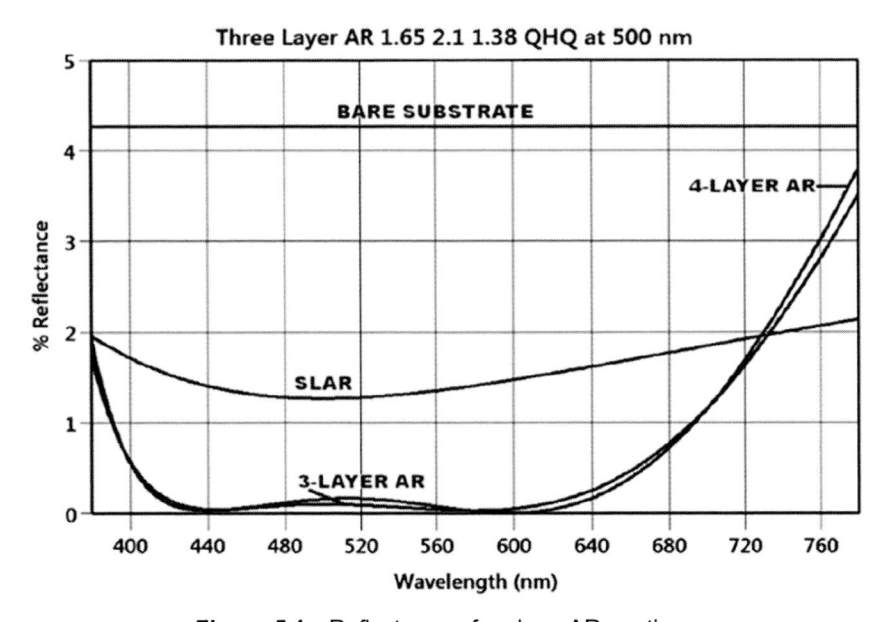

Figure 5.1 Reflectance of various AR coatings.

substrate with a single-layer AR (SLAR) of MgF$_2$, and both the 3-layer and 4-layer ARs (which are the same to within manufacturing reproducibility).

One could say that 1930–1940 was the era of the single-layer optical coatings, 1950 was the 3-layer era, the 1960s started to see the use of a dozen or so layers, and by the end of the twentieth century, over 4000 layers have been reported. This latter was perhaps an extreme case, but in 2020, 100–200 layer coatings were not uncommon. Clearly, such numbers require automation to be practical, but that is readily available in the 21st century.

Transparent conductive oxides (TCO) also became important to our lives from the 1950s until the present where they are employed in almost every display system we use. Clark Bright points out, in an extensive history of the TCO field [7], that the first TCO was reported in 1907, was advanced in 1937, and that indium oxide (IO) was reported as made by post-oxidation of the indium metal in 1954. From that time to the present, the technology has been highly refined, and other materials than the classic indium tin oxide (ITO) have been refined and applied because of the current cost of indium.

Process and equipment changes in the last half of the 20th century have been significant. In the 1950s, most processes heated materials in tantalum, molybdenum, or tungsten resistance "boats" or on tungsten wires or coils until they evaporated into a "vacuum". Here, vacuum might mean in the low 10^{-5} or high 10^{-6} Torr range, where the mean free path (MFP) of an evaporated atom/molecule was about ½ to 2 meters, or about the size of the chambers. In such cases, only about 1/3 of the atoms reached the substrate without colliding with the residual vapors (usually water) in the chamber. The rest of the molecules might have had one collision. The energy of the evaporated atoms was on the order of 0.1–0.2 eV (very lazy). The equipment was mostly an 18" diameter glass bell jar; large ones would typically be such a shape but made of metal. One premiere optical instrument company in that era had a large converted steam boiler tank which would be able to coat optics of perhaps 2 meters in diameter.

Glow discharges were generated in these chambers during the pump-down cycle by applying a high voltage when the pressure was on the order of 2 milli-Torr. These glows bombarded the surfaces in the chambers with energetic ions of nitrogen, oxygen, hydrogen, and hydroxide for cleaning and activating those surfaces.

In the 1960s, electron beam guns (EBG) started to be used which could concentrate a beam of kilowatts of electrons at ~1 amp and at several thousand volts onto an area as small as one square millimeter. The EBG had somewhat the effect of a well-controlled arc. This was enough to evaporate any material. This was of particular interest to allow the evaporation of high melting point refractory materials such as SiO_2, TiO_2, Ta_2O_5, etc., which had better hardness and durability than previously used materials. Radio frequency (RF) sputtering of dielectrics such as SiO_2 was also used in this era.

Chemical vapor deposition (CVD) started to influence the growing semiconductor industry. In that case, gaseous materials were passed over heated wafers to be coated, and reactions took place which deposited the desired films. This saw little application to optical coatings in those days, because the process temperatures were too high to use on glass optics and particularly on plastic optics.

The issues of the energy involved in the coating processes came under increased scrutiny. Heating the substrates and the process temperature had been the primary approach to controlling surface energy up until the present era, but new ways to introduce energy instead of substrate heat were starting to be investigated. In the 1950s, optical coatings were thought to be vitreous/glassy and dense like the materials which were put in the evaporation source. However, researchers like Macleod and Lissbarger were finding that many (if not most) films were porous. By the 1980s, transmission electron micrographs (TEM) showed clearly that a columnar or dendritic structure was typical and there were voids between the columns. Accelerated ions were beginning to be applied in the 1960s. These ions would add a local spike of energy to the depositing surface which could disrupt the columnar growth and provide films of greater density. Maddox introduced ion plating in 1963, and the optical industry later adopted that idea in reactive ion plating (RIP) at Balzers and a somewhat related advanced plasma source (APS) by Leybold Optics.

Sputtering was an energetic process, and ion-assisted deposition (IAD) was another key approach to provide energy to the condensing films. Plasma-enhanced CVD (PECVD) began to be more applicable for optical coatings, since some of the energy could be provided by the plasma instead of the high temperatures of the CVD processes used in the semiconductor industry.

Energy (local temperature) as a function of time seems to be a key factor in the resulting properties of the coating. In the early days this was a function of the substrate temperature when the film was depositing and also the effects of annealing of the coatings after deposition. There were also cases where some coatings such as Ta_2O_5 and TiO_2 had some grey absorption after deposition which could be reduced or eliminated by baking overnight at elevated temperatures. It is thought that this provided the oxygen necessary to complete the oxidation of the metal and achieve stoichiometry of the film.

Sputtering, which had preceded thermal evaporation by many decades, came into increased application for optical coatings in the 1970s, particularly for large-area coatings like architectural windows. The Leybold Company of Germany and the Southwall Corporation of the San Francisco Bay Area claim to have cooperated in building the first large-scale, multi-cathode sputter roll coater in 1980.

Atomic layer deposition (ALD) started to appear as a viable optical coating technique in the 1990s. It is basically a CVD process with two advantages. One advantage is that chemistries have been found which work at low enough temperatures that they can be used on glass and plastic optics. The second advantage is that many of the atomic layer processes can be self-limiting. This means that layers can be deposited in a sequence of chemical reactions wherein only one single atomic layer of the material (such as Al_2O_3) is deposited in each cycle. Thereby, the thickness is exactly controlled by only counting the number of cycles. ALD has a further benefit that the layer is conformal. Every exposed area is coated with an equal thickness of the material.

Requirements for more demanding NBP filters evolved in the 1990s from the fiber optics communications industry. These are called dense wavelength division multiplexing (DWDM) filters. These requirements were an order of magnitude tighter on bandwidth and stability than was ordinary in the industry prior to that time, and they put much greater demand on designs, processes, and production equipment. This provided the impetus for the optical coating industry to make a giant leap into the 21st century.

5.2 The New Century

As discussed above, the 20th century ushered in the vacuum coating of optical thin films, and almost all of the basic technologies that are used in the 21st century were introduced in the 1900s or even the 1800s. The progress of the 21st century has reduced these to practice for production and significantly improved our understanding of the details and behavior of these processes and resulting thin films.

5.2.1 Better understanding

5.2.1.1 Heat versus energetic atoms

It has become more obvious that the energy at the site of depositing atoms and molecules has a major effect on the resulting film. In the mid-1900s, room-temperature depositions of MgF_2 on glass resulted in films that could be wiped off of the substrates by the finger but would survive eraser rubbing if deposited at 250–300°C. Aluminum films deposited at room temperature were good but if deposited at over 150°C would become hazy due to crystalline growth at the surface.

The early process improvements for most dielectric films such as TiO_2, Ta_2O_5, etc., were to heat the substrates to ~300°C. This heated substrate provided energy to the incoming atoms with less than 0.2 eV of energy to allow them to move into more dense packing than the feathery configuration, as illustrated in the simulation of Figure 5.2 from Guenther [8], that results from atoms which stick at the first place where they make contact with the growing film for lack of energy to do otherwise. Figure 5.3 from Liu, et al. [9]

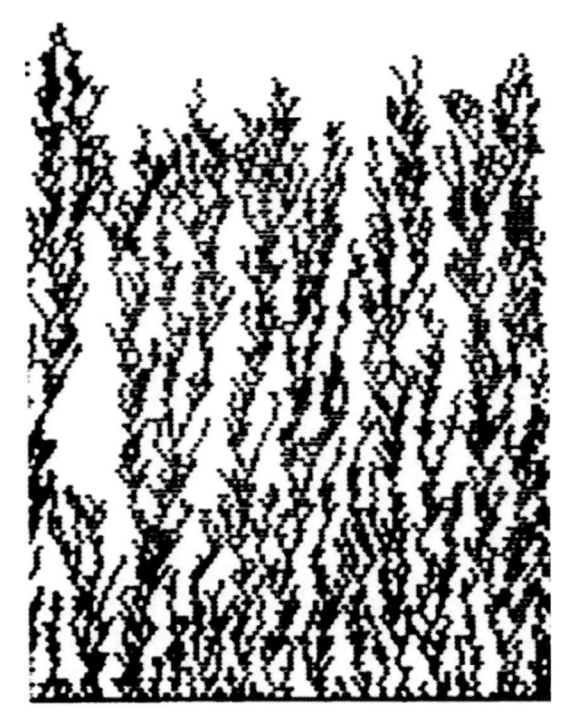

Figure 5.2 Simulated film growth at low energy. Reprinted with permission from [8] © The Optical Society.

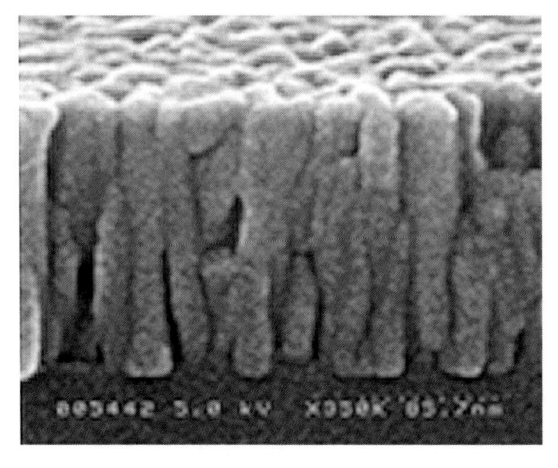

Figure 5.3 MgF$_2$ film deposited at 300°C. Reprinted with permission from [9] © The Optical Society.

shows an actual MgF$_2$ film deposited at 300°C which is relatively dense, but it does still have some voids.

The next improvement was to use energetic processes such as sputtering which gave the incoming depositing atoms energies of tens of eV. Another process of ion-assisted deposition (IAD) bombarded the depositing surface with inert (Ar) ions or reactive (O_2, N_2, etc.) ions of hundreds of eV. These both put energy into the film as it was being deposited and replaced the need for heating the substrates. This, in turn, reduced the process times required to heat and cool the substrates.

5.2.1.2 Stress

Stress in films is always a problem that must be addressed. It will lead to delamination of the films and tend to bend the substrate from its previously perfected shape. It has become clear that there are two causes of stress. One cause of stress is due to the mismatch of the coefficients of expansion between the coating material and the substrate material. A fix for this is to deposit the film at the same process temperature at which it will be used in practice (most typically at room temperature). At room temperature, most films need sputtering or IAD to induce good adhesion of the film. Such processes were perfected in the 2000s. The second cause of stress is due to the intricacies of the film growth processes. Chason et al. [10,11] show the progress made in the 21st century in the understanding of stress and what aspects of the process effect stress.

5.2.1.3 Nanomaterials

The 21st century has seen a burst of activity in nano-technology and plasmonics. The great amount of theoretical and experimental research in these fields has already led to many practical applications for the benefit of physical, chemical, and biological work. The understanding of the behavior of atoms on this nano-scale has been greatly advanced. The ability to "see" and produce structures on the nano-level has allowed the production of meta materials which can have properties (n & k, etc.) which are not available in natural bulk materials. A widespread application of nano-technology and plasmonics is the production of quantum dots which can produce almost any color desired and are used in many new television sets.

5.2.2 Design

5.2.2.1 Ideal antireflection coating

Prior to the turn of the century, Willey [12] showed empirically that the ideal AR coating had a Gaussian index of refraction profile from the substrate index to that of the medium in which the substrate is immersed. Such is illustrated in Figure 5.4, where the index ranges from index 1.52 at the substrate (crown glass) to near 1.00 at the medium of air or vacuum. The application of this principle will be discussed below with respect to "Moth Eye" AR coatings and also with respect to AR coatings for freeform optical surfaces.

Ideal Gaussian Index Profile

$$N(x) = n_i + (n_s - n_i)e^{-x^2}$$

Optical Thickness in QWOTs at Longest Wavelength

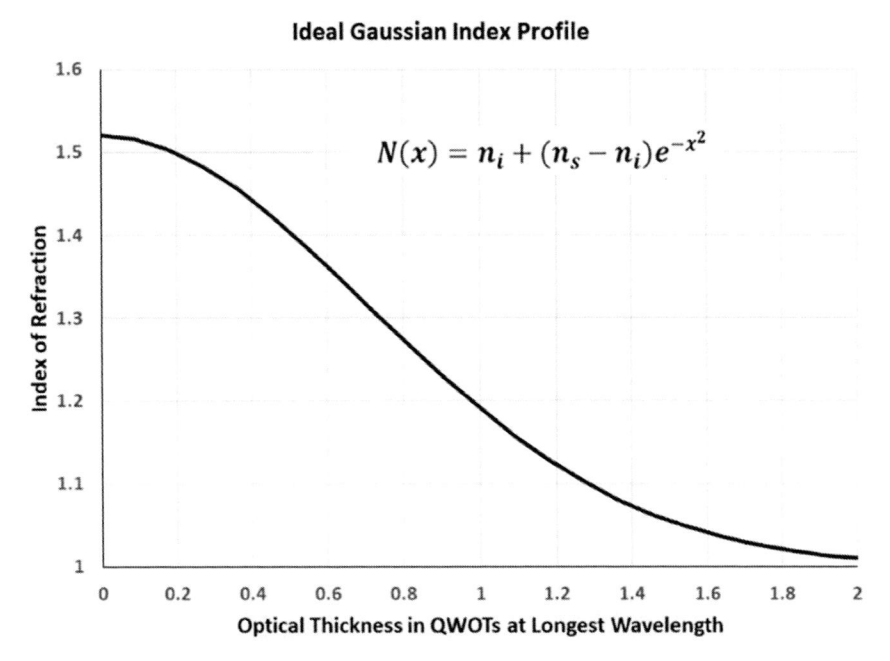

Figure 5.4 Index refraction profile versus optical thickness for an ideal antireflection coating on a substrate of index 1.52.

5.2.2.2 Thin layers

Although thinking about design in the mid-1900s was mostly confined to QWOTs because of the lack of computers to do the "heavy lifting", long before the year 2000 the effects of any thickness layer could be properly calculated and designed. The ability to control and produce layers less than 10 nm thick was, however, not common. Thin-film design contests were often limited to "no layer thinner than 10 nm". This was most likely because such layers had "startup problems" due to the variable nucleation effects which depended on the surface preparation before coating, etc. It was known in the 1900s that such things existed, but they were not well modeled until more recent times.

5.2.2.3 Index versus thickness

In the 2000s, extensive work has been done to model and design layers from a few nanometers on up. Much of the change in effective index of refraction (n & k) versus thickness and wavelength is due to the nucleation of the coating material into islands and their growth with newly arriving material. With increasing effective thickness, such film islands will grow until they touch each other, or coalesce (percolate) and eventually become continuous films. The growing islands have some height above the substrate "sea', so that the effective height or thickness as measured by a QCM is the average of the deposited mass divided by the area of the "sea". The behavior of films before the percolation point is plasmonic and somewhat like quantum dots. The detail behavior is a function of the deposition process parameters and the materials of the substrate, the coating material of the layer, and the coating material of the following layer. Software now exists to model this well enough to obtain designs which provide good production results, if the process is reproducible enough.

5.2.2.4 Design of experiments methodology

A major benefit to process development and optimization has come from the application of design of experiments (DOE) methodology which has blossomed in the 21st century, although the methodology evolved in the early 1900s for the agricultural industry. DOE makes it possible to glean the maximum amount of information about a process with a minimum number of experiments.

5.2.3 Materials

5.2.3.1 Mixed materials

Merck in Germany and Umicore in Liechtenstein are two of the key suppliers of materials for the optical thin films industry. Friz and Waibel [13] provided a helpful review of many common coating materials. They give information on special mixtures, solid solutions, or compounds of materials which have been developed by Merck and Umicore and which have improved properties over the single-material compounds. For example, Umicore's Lima and Merck's L5 are mixtures of SiO_2 and Al_2O_3 which lower stress in the deposited SiO_2 and have enhanced chemical stability. Umicore's Paso and Merck's M1 are mixtures of Al_2O_3 and Pr_6O_{11} which melt completely before evaporation to provide an even surface during evaporation.

5.2.3.2 Improved deposition techniques

Many processes have been improved over the years. Aluminum is an example where many decorative coating producers still use tungsten filaments to evaporate aluminum, as was done by Strong [1] in 1933, but aluminum sputtering equipment has been built which greatly improves the process speed and reduces the labor content of equivalent processes (see Willey, Sec. 2.2.3 [14]).

Zirconium oxide has been known for layers as a material whose index varied significantly with thickness (inhomogeneity). In recent years, ZrO_2 has become a much more uniform and thereby useful material by the addition of IAD in the process (see Willey, Sec. 2.2.16 [14]).

5.2.3.3 Graded index materials

The principle shown in Figure 5.4 could be approximated by a stack of thin homogeneous layers whose indices decreased from the substrate to the medium (air) in a step-down fashion to present an index profile as close as practical to that of Figure 5.4. It is further possible to create an approximation of each of these thin layers of various indices by three even thinner layers per each previously individual single layer by using only a single high and single low index in all of the three-layer combinations. This would be called a Herpin [15] approximation which would create the whole step-down coating by using only the two high- and low-index materials in a combination of relatively thin layers with a variety of thicknesses. (The details of this Herpin approximation are beyond the scope of this discussion.)

Yet another approach to realizing the principles of Figure 5.4 is what the industry calls a "Moth Eye" surface treatment. This can be accomplished by etching away a layer of the substrate surface in such a way as to create cones, pyramids, or more random shapes which taper from the substrate to the medium in the required manner. The ability to lithographically etch many materials on a "nano-scale" has advanced rapidly in the last 20 years. The effective index in any thin section of such coating parallel to the surface of the substrate would be the average of the area of the voids (of index 1.0) between the cones, pyramids, etc., and the area (of index n) of that section of the substrate within those shapes. If that effective index has the same profile as Figure 5.4, as long as its details are sufficiently smaller than the wavelengths of interest to the AR coating, it will create the same AR effects. Figure 5.5 illustrates on a microscopic scale the ideal profile of such an etched surface. In this figure, pyramids are illustrated whose side view, as shown in this figure, is the square root of the curve in Figure 5.4 so that the area of the pyramid at any section plus its surrounding air/vacuum would have the needed effective index of refraction for that section.

The recent two decades have seen widespread use of consumer electronics with cameras having polymer or plastic optics. AR coatings for plastic optics are limited to processes that will not exceed temperatures which will harm the polymer optics, such as 100°C. If the classical SLAR of one QWOT of MgF_2

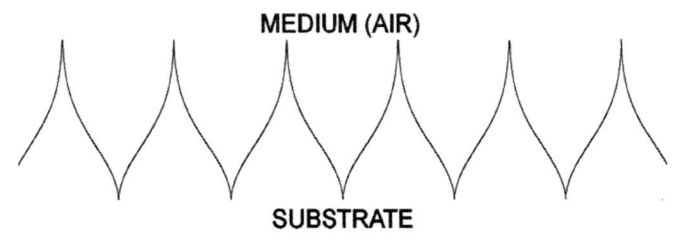

Figure 5.5 Microscopic section of a "Moth Eye" etched structure to produce a perfect AR surface/coating.

is applied to polymer surfaces at these low temperatures, it will not be dense and will not be likely to have good adhesion to the polymer. Such coatings could possibly still be used on interior lens surfaces if they are not exposed or handled by the end-user. MgF_2 coatings have historically obtained higher density and good adhesion by being deposited at 250–300°C, where the depositing atoms do not quickly lose their energy before being able to move into more consolidated positions among the already deposited atoms. Willey and Shakoury [16] overcame this temperature limitation by supplying the needed energy during deposition at room temperature by IAD with nitrogen atoms. These coatings were deposited without added process heat and had full density and hardness for potential use on exterior optical surfaces.

The micro-etching approach for plastic or polymer optics has been extensively investigated with success by Schulz, et al. [17,18,19] and others. Figure 5.6 is from reference 17. Reference 17 also has many more references to that work for the history and status of that field. Figure 5.7 is from the latest work of Schulz et al. [19].

5.2.3.4 Antireflection coatings for freeform optical surfaces

The technical problem for a freeform optical surface can be simply viewed as the same problem as that of an AR coating for a hemisphere. The hemisphere has areas that are at every possible angle of incidence (AOI) from 0° to 90° and in all azimuths. The further question is then: "at what range of AOI for incoming rays is the AR needed to perform in that particular area of the freeform surface?" In the simple case of a collimated beam of light incident on the optical axis of the hemisphere, in any given segment, it will have a single unique AOI between 0° and 90°. If the hemisphere were illuminated at some arbitrary AOI with a beam of significant divergence, each area will have a variety of AOIs. This is the same situation as with some general freeform optical surface.

Any smooth surface will reflect increasingly more light with increasing AOI (as illustrated in Figure 5.8) for a glass or plastic surface of index 1.52. The p-polarization of the incident light will reflect less light up to the Brewster angle (~57° in this case) where it reflects zero, and it then it increases to 100% at 90°. The s-polarization increases continually with AOI until it is 100% at 90°. This illustrates the challenge of an AR for a hemisphere or general freeform surface at large angles. From a practical point of view, it might be wise to only consider attempting ARs for AOI of less than 60° or 70°.

Another well-known factor is that the effects of an interference coating such as an AR will shift to shorter wavelengths with increasing AOI. Figure 5.9 shows a common 3-layer broad band AR (BBAR) coating at 0° and 60°, which has been optimized for the visible spectrum (380-780 nm), at 0° AOI. A means to overcome these limitations due to AOI effects is to have a

Optical Coatings 163

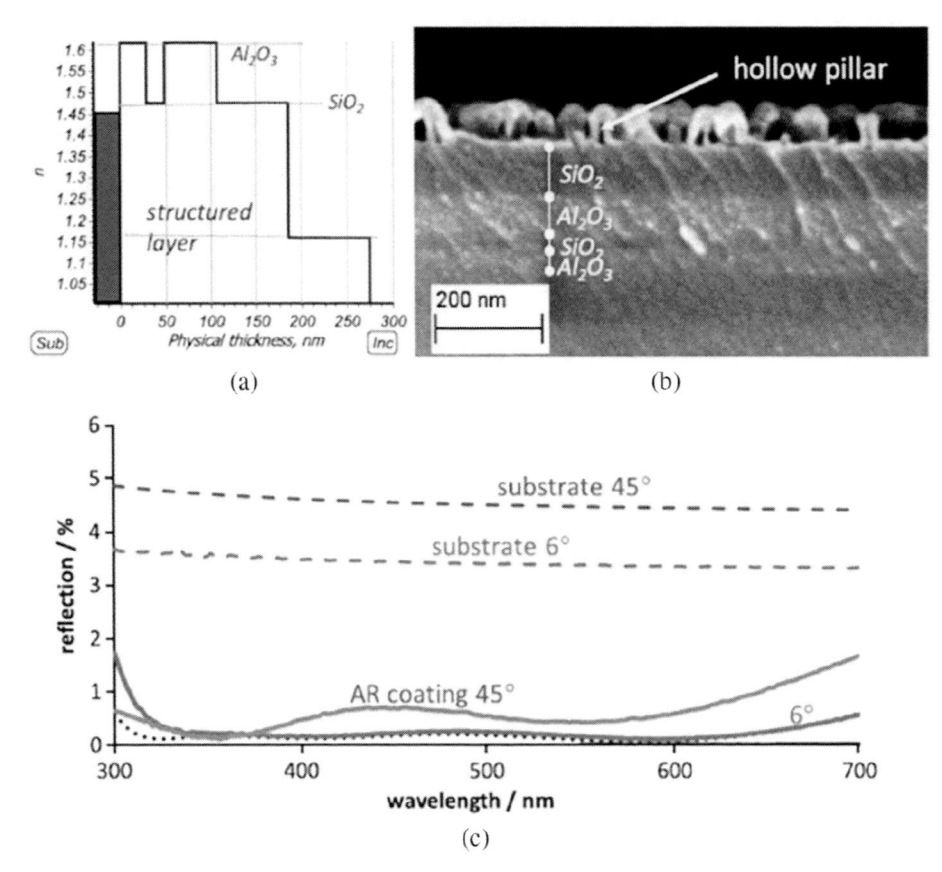

Figure 5.6 Properties of AR coating on fused silica consisting of SiO₂, Al₂O₃, and the nanostructured layer. (a) Refractive index profile of the design including the hybrid layer with n_{eff} = 1.16; (b) SEM cross section; (c) reflection for AOI 6° and 45° (average polarization) of uncoated quartz substrate (dashed lines), coated quartz (solid lines), and calculated spectrum for AOI 6° (dotted lines). All spectra without rear side.

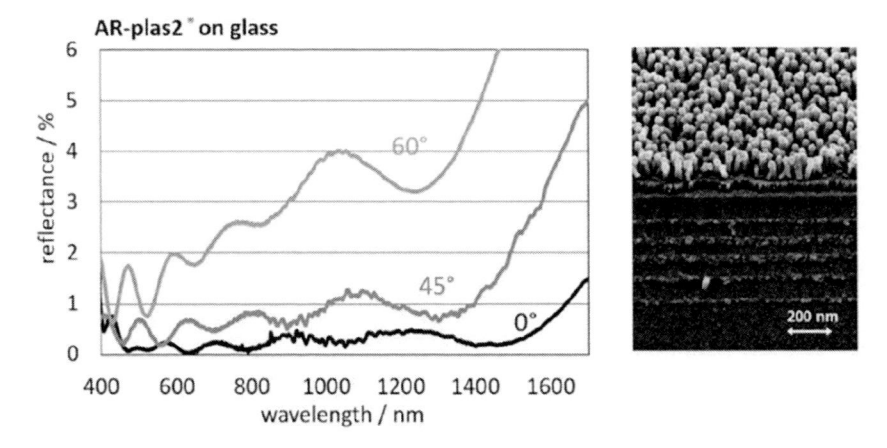

Figure 5.7 Latest version of the work of Schulz et al.

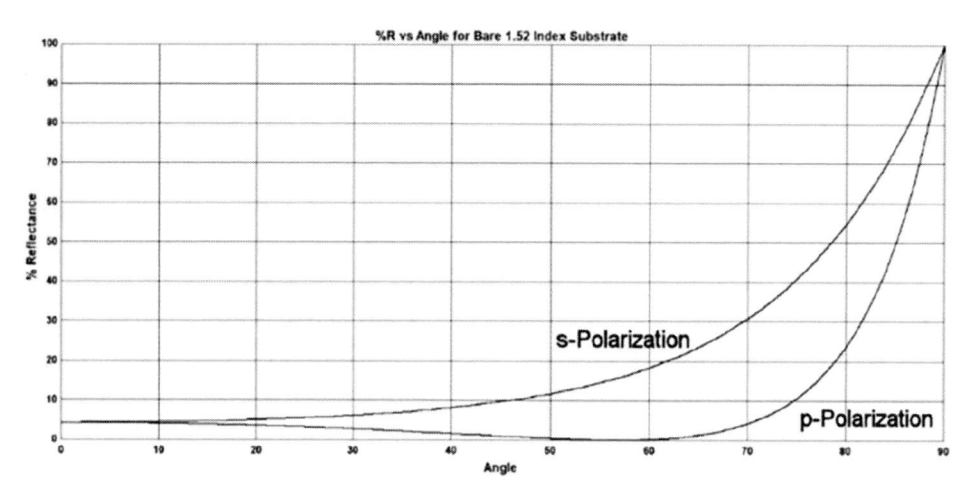

Figure 5.8 Reflectance versus AOI of a 1.52 index surface as a function of angle.

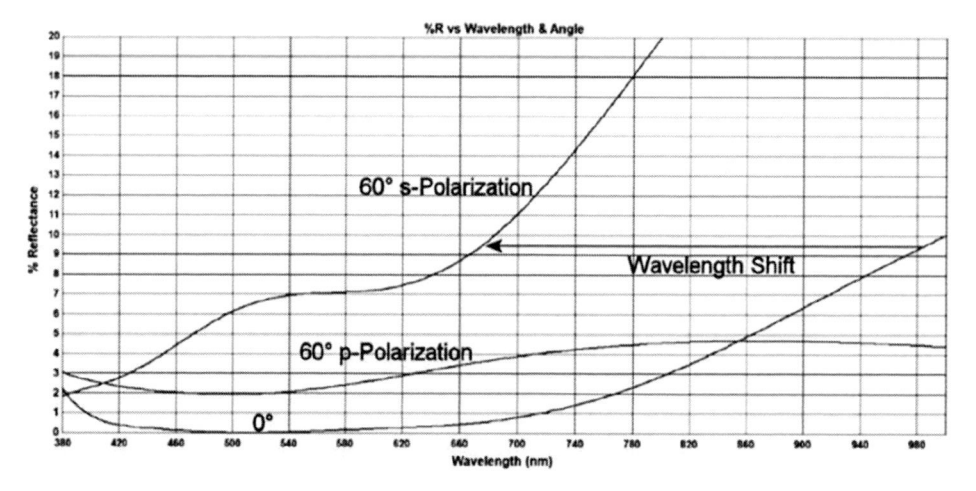

Figure 5.9 Common BBAR for visible spectrum at 0° and 60° AOI. Shows angle effects on wavelength and reflectance.

very broad band AR coating so that the long wavelength rise in %R does not move so far to the left (short waves) as to cause any problems.

As it turns out, the ideal/perfect AR discussed above will allow us to do what is needed here. This is demonstrated in Figure 5.10. This shows a case of an AR coating with less than 0.1% reflectance over the whole visible spectrum for all AOI up to 60°. Note that the %R scale of Fig. 5.10 goes from 0 to 1.0% reflectance. This is the ultimate coating concept for freeform optics.

A helpful partial solution to the freeform surface AR need can be provided by ALD. As mentioned above, ALD provides a conformal coating which is the same thickness on all exposed surfaces independent of the surface orientation. In certain classes of freeform surface AR requirements, this may

Optical Coatings 165

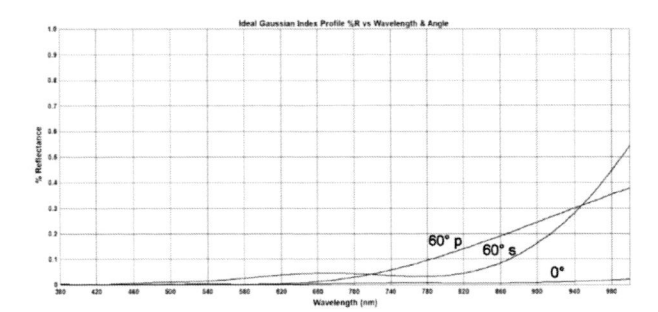

Figure 5.10 Ideal AR coating at 0° and 60°. NOTE: %R scale of this figure is maximum at 1%R.

be all that is needed. If the ALD stack could further have the ideal index profile, then this would solve almost all AR problems. However, obtaining very low indices approaching that of the air medium does not currently seem to be within the current realm of possibility.

If different areas of a freeform surface required different designs for ALD ARs, such designs could be produced by "tiling" the surface with different ALD designs by masking some areas while depositing on others.

5.2.4 Equipment

5.2.4.1 Stress measuring instruments

One of the advances in equipment is the commercial availability of *in situ* systems [20] to measure film stress during deposition. A laser is made into multiple equally spaced parallel beams by multiple reflections from two etalons with orthogonal axes. These beams reflect from the stressed surface and back to a CCD camera and processing system which measures the displacement of the beams and converts the measurements to give the radius of the stressed surface and thereby its stress.

Ennos [21] provided valuable experimental information in the early days (1966) of the maturation of the technology for optical thin films. The measurement of stress is most commonly done by deposition on thin bars or disks of known flatness and mechanical properties. The shape/bending change due to the film stress is measured during and/or after the coating deposition.

The stress can be calculated [21,22] in various ways from the measured bending of the substrate. Lechner et al. [23] give the following procedure for intrinsic stress measurements: "The residual intrinsic stress of the films is measured interferometrically (Newton fringes) by the deformation of coated circular glass discs (1.1-mm thickness, 50-mm diameter in their case, or silicon discs are often used). The radius of curvature can be determined using the fringes numbered m and j from the interference pattern with Equation (5.1), referred to as the Stoney formula [24] in Ohring [25] on page 727:

$$r = \frac{d_m^2 - d_j^2}{4\lambda(m-j)}, \qquad (5.1)$$

From the different radius of curvature before and after coating, knowing additionally the Young's modulus (E_s), Poisson's ratio (ν_s) and the corresponding thickness t_s and t_f of substrate and film, the intrinsic stress σ_f can be calculated according to Equation (5.2):

$$\sigma_f = \frac{E_s}{6(1-\nu_s)} \frac{t_s^2}{t_f} \left(\frac{1}{r_{f+s}} - \frac{1}{r_s} \right), \qquad (5.2)$$

For such measurements, nearly flat substrates showing a concentric circular Newton fringe pattern are desired. Since, however, only a few substrates have an ideal finish, substrates with a slightly elliptical deformed fringe pattern can also be accepted. To obtain a mean film stress value, the mean bending radius of the sample is determined by 45° stepwise rotated fringe diameter measurements or by an average of the large and small axes of the fringe.

5.2.4.2 Low-budget stress measurements

The measurement of stress has historically involved the use of expensive equipment and substrates. Here is shown a very-low-cost method, materials, and equipment which will allow adequate measurement of coating stress for many applications in optical thin film coating.

For the levels of stress which we normally experience in visible spectrum coatings, it is difficult and expensive to measure stress. This is because of the usual need to prepare samples that are relatively flat and thin enough with respect to their diameter to show a workable number of fringes or curvature at low stress. The substrate thickness ranges from 0.3 to 0.5 mm thick, or as stated above by Lechner et al. [23], "1.1-mm thickness, 50-mm diameter".

Inexpensive microscope cover glasses can be used to quantize this stress, albeit at a less precise level. Figure 5.11 shows an optically polished sample of 0.30-mm-thick H-K9L borosilicate crown optical glass (similar to BK7 glass) before (left) and after coating (right) with about four quarter-waves of optical thickness at 550 nm of MgF_2 deposited by physical vapor deposition onto the substrate at a process temperature of 300°C. This is known to produce hard coatings, but coatings with stress. Figure 5.11 shows a change from almost no fringes before coating to many fringes after coating. The coated surface is concave, indicating tensile stress due to the coating. If we count the fringes over some diameter, we can calculate the radius of curvature using Stoney's Equation. The r_s in this case is near enough to infinity so that we can ignore it.

Figure 5.12 shows a microscope cover glass approximately 0.16-mm thick and 22 mm square. These cover glasses are widely available at less than ten

Optical Coatings

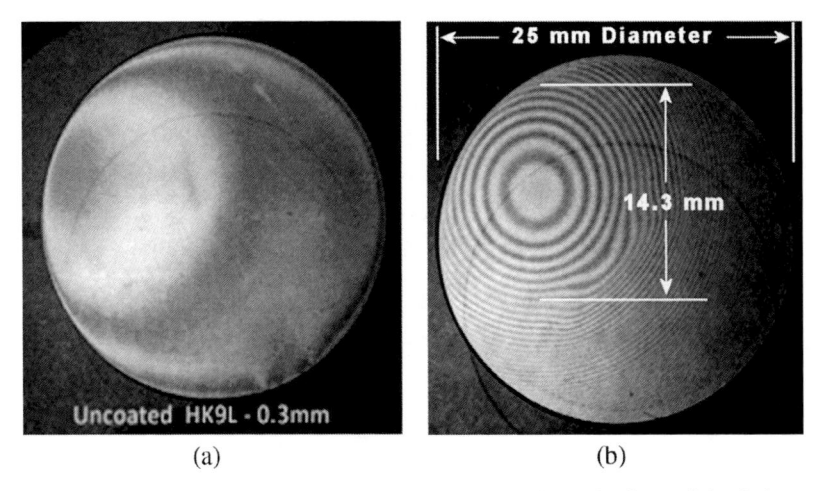

Figure 5.11 (a) Interference fringes from an uncoated, optically polished, borosilicate H-K9L optical glass of 0.3-mm-thickness on an optical test flat illuminated by a compact fluorescent light. (b) The same test piece after coating with MgF$_2$ at 300°C.

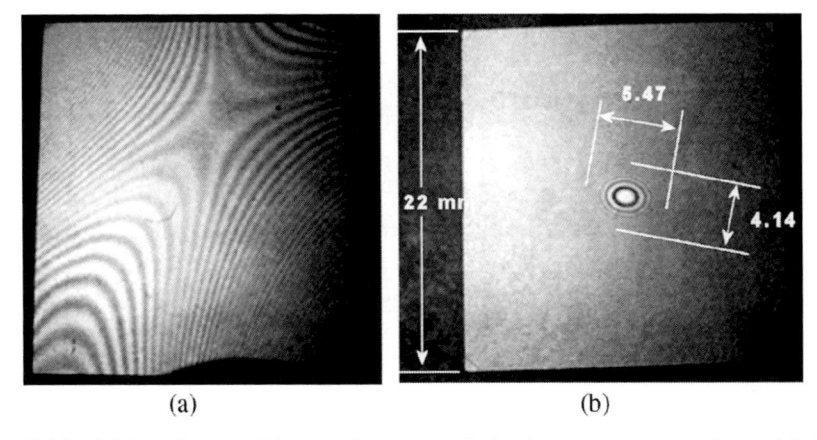

Figure 5.12 (a) Interference fringes of an uncoated microscope cover glass of 0.16 mm thickness on an optical test flat illuminated under compact fluorescent light. (b) The same test glass after coating with approximately 550 nm of MgF$_2$ at 300°C for comparison with Fig. 5.11.

cents apiece! This saves a lot of fabrication cost in making samples like that shown in the left of Figure 5.11. The uncoated piece, as seen here on an optical test flat and observed with an illuminant of only a common (inexpensive) compact fluorescent light, shows an irregular ("not-too-flat" saddle-shaped) surface. However, when coated, the tensile stress pulls the cover glass into such a concave radius that it is regular enough to measure.

The resulting fringes on cover glass test pieces are usually elliptical as seen in Figs. 5.12(b) and 5.11. The recommended measurement procedure is to measure the diameter of the 7th fringe on both the large and small axes of the ellipse, as shown in Fig. 5.12(b). That is easily done on a properly scaled photo of the fringe image. With a compact fluorescent lamp for a light source and the cover glass on an optical flat, the 7th fringe is about the limit of what can be clearly discerned. This would also not be a problem with a monochromatic source. A different number of fringes could be used as long as the calculations are adjusted appropriately.

In the current example, the effective wavelength for the fringes is taken as 550 nm (at the peak of visible light; so that the height of one fringe in air above the test flat is 275 nm or 2.75×10^{-4} mm (per fringe). 70 GPa is used here as a representative Young's modulus for glass and 0.20 as a representative Poisson's ratio. The radius of the stressed sample is approximated by: $r = x^2/2y$, where x is the average radius of the ring pattern to the 7th fringe in mm and y is the height of 7 fringes at 550 nm (which is $7 \times 2.75 \times 10^{-4}$ mm = 0.001925 mm). Therefore, the stress in the film can be calculated from these measurements as:

Stress in $\sigma_f \approx 289/x^2$ MPa, where x is the radius of the 7th fringe in the pattern in mm.

In the case of Fig. 5.12(b), where the average diameter of the fringe pattern is 4.94 mm and the x is therefore 2.47 mm, the stress is calculated as is 47.4 MPa. For another example, Fig. 5.13, where the coating was deposited using broad beam Ion Assisted Deposition (IAD), the average diameter of the 7th fringe is 15.62 mm, and the stress has been reduced by an order of magnitude to 4.74 MPa! The IAD is done during the deposition process and densifies the film to reduce the tensile stress common to evaporated MgF_2.

Figure 5.13 shows the interference fringes of a coated microscope cover glass of 0.16-mm thickness (as used in Figs. 5.11 and 5.12). This image is after coating with ~550 nm of MgF_2 at 300°C and 3 Drive Amps at 113 Drive

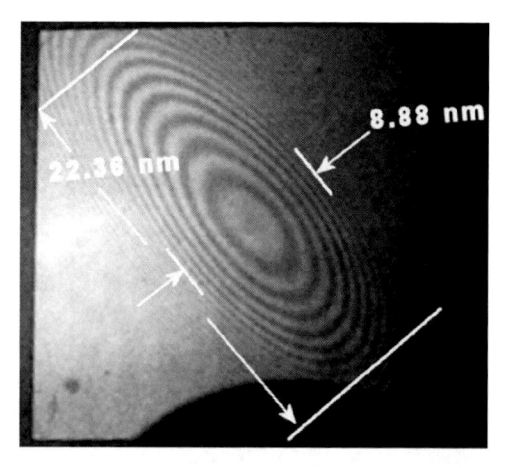

Figure 5.13 Shown are the interference fringes of a coated microscope cover glass of 0.16-mm thickness (as used in Figs. 5.11 and 5.12).

Volts of broad-beam ion-assisted deposition. Much lower stress is noted, compared to samples in Figs. 5.12 and 5.13.

If the only difference between the round test piece of polished H-K9L in Fig. 5.11 and the cover glass in Fig. 5.12 were the difference in thickness from 0.30 to 0.16 mm, then the diameter of the 7th fringe in Fig. 5.11 would be expected to be about 17.4 mm in diameter. The actual diameter in Fig. 5.11 is 14.3 mm or 82% of what was expected. Considering that the details of the glasses may differ somewhat; this implies that the reproducibility of this measurement may be on the order of ± 10-20%, which is adequate to pursue minimizing stress in a process of interest. Samples shown in Figs. 5.12 and 5.13 used "No. 1" microscope cover slides (0.13 to 0.16 mm thick).

5.2.5 Masking for uniformity

There has been an evolution in the 21st century of the masking techniques for uniformity of coating thicknesses over large coating areas. Masks can be used to reduce the amount of material which arrives at selected regions of the substrates. Figure 5.14 illustrates a plan view of a chamber with a calotte (flat or domed) with leaf-shaped masks, as has been commonly used in the past. The sources are near the baseplate of the chamber, while the calotte is near the top of the chamber. Because the sources do not often have identical material distribution patterns, the masks over each source are not identical. There is some effect of both masks on the uniformity from one source, but the mask farthest from the source usually has a much smaller effect. Placing the mask in the position directly over its source has the most effect for the smallest mask because of its proximity to the source. However, this is obstructing the best material which arrives at the substrate with near-normal incidence.

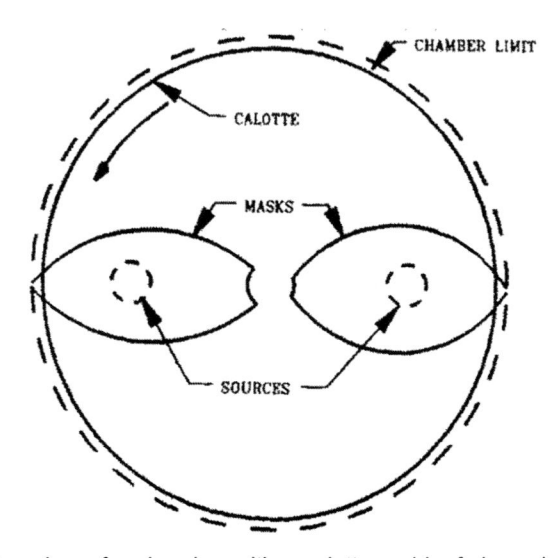

Figure 5.14 Plan view of a chamber with a calotte and leaf-shaped uniformity masks.

Another approach is to put the mask opposite the source for the material to be masked. This masks that undesirable material which would be at large angles to the substrate normal. This avoids masking the most desirable material depositing over the source at near-normal incidence because of its less dense growth properties. In this position, the mask would also be less sensitive and would probably have a larger area. The difficulty with this approach occurs when more than one material must be masked with different masks; the masks would then need to be moved out of the path when not being used for their intended materials.

The application of source position and masking to a domed calotte and drums or cylinders operates on the same principles as the above and differs only in detail.

The 21st century has seen the use of a single mask midway around the circle between the sources and as far from the sources as practical, as illustrated in Fig. 5.15. This mask could be trimmed by adjusting the edge nearest the source of concern. This is the preferred approach because it masks the materials away from the best near-normal material over the sources.

5.2.6 Monitoring and control

The purpose of optical thin film monitoring is to enable the control of the properties, primarily optical thickness, of a layer being deposited to within adequate tolerances for the film to perform its intended function. When one considers controlling the optical thickness of a layer by its optical characteristics, there is a broad range of options. The simplest of these is using the eye to observe when a metallic film becomes opaque as in

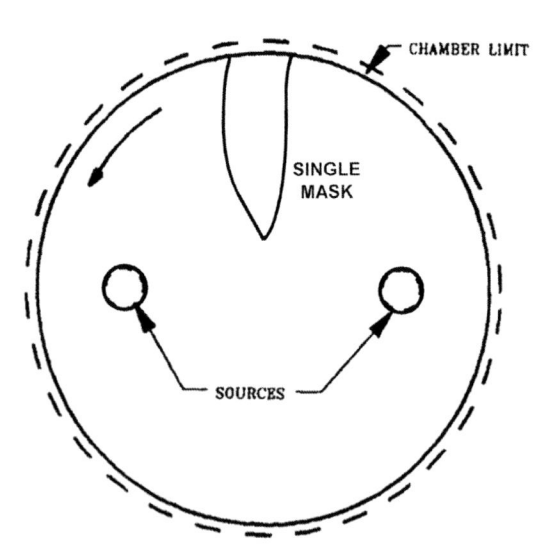

Figure 5.15 Single mask to trim two sources, and also in a position to remove high-angle-of-incidence material which gives a lower quality deposition.

aluminizing a mirror, or observing when a single-layer AR coating has the right color to terminate deposition. This was the primary means of controlling thickness in the 1940s through the 1960s. The QCM became available only in the 1960s and became practical in the early 1970s.

A photoelectric sensor can be used to replace the human eye. Polster [6] was the first to publish this in 1952. This could observe the reflection or transmission of the part being coated at a single wavelength or over many wavelengths (broad-band optical monitoring (BBOM)) in more recent years (1979) as reported by Vidal et al. [26]. It is now possible to use lasers to have a monochromatic monitoring source, but most monitors have isolated one wavelength at a time using a NBP filter or monochromator. Spectral ellipsometers using broadband sources and spectrometer detection are also common now.

With monochromatic monitoring, a layer might be terminated at a certain %R or %T, or when the signal "turns a corner," or cut at some %R or %T after a turning point (TP) has been passed.

Prior to the advent of computer-controlled monitors, the operator was the one who decided when to terminate the layer based on his or her observation of the above occurrences. Many systems can now delegate these decisions to a computer program. When the layer termination was operator-dependent, more than a few layers was difficult, but with 21st century computer control, hundreds of layers are fairly commonplace. Many monitoring strategies are discussed by Willey [14] and their relative merits and applications.

A significant step forward for the 21st century was reported by Zöller et al. [27,28] at Leybold Optics on systems where the monitor is at the center of a ring of deliverable parts in a large rotating calotte as illustrated in Fig. 5.16. The monitor is synchronized to the calotte's rotation so that it only observes one spot on only one of the monitored parts as it comes through that position on each rotation.

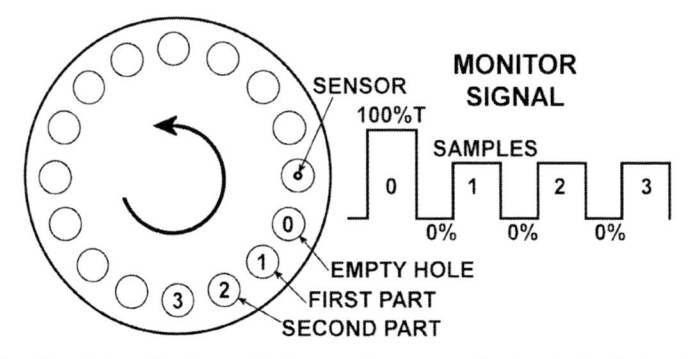

Figure 5.16 The intermittent monitoring system used by Leybold Optics in its Helios system, and by others.

This system has all of the advantages of a true double-beam spectrophotometer. In rapid succession, it samples the 100% and 0% transmittance of the monitor and then the transmittance of the part being coated. Therefore, a drift of source, detector, or environment is instantly calibrated out of consideration.

The relative quality of the parts around the same radius will depend upon whether they receive the same amount of depositing material as the monitored part. This will mostly depend on a constant deposition rate and the effects of when the shutter is closed and the speed of calotte rotation. The relative quality of the parts in other areas of the calotte will depend on the above factors and also the uniformity of the depositing materials in a radial direction from the center of the calotte. This is generally controlled by the choice of deposition source positions and additional masks to block some material from radii where there would be an excess.

The experimental and production examples reported seem to show that the intermittent approach is a significant improvement. It is truly direct monitoring on the monitored part, and if conditions are quite stable for the other areas of the calotte, those areas should also produce satisfactory parts.

Figure 5.17 shows the side view of the Leybold Optics layout for their Helios system, and Fig. 5.18 is a photo of such a system.

Gencoa [29] recently introduced its OPTIX plasma emission monitor (PEM). This analyzes the spectrum of a plasma of the sputtering gas in the chamber to determine the species and relative amount of a gas similar to the output of a residual gas analyzer (RGA). A key advantage of the OPTIX is that it can operate at pressures up to 1 mBar without differential pumping as would be required by an RGA. This is four orders of magnitude higher pressure than the RGA can handle, and it can be very helpful for the analysis of sputtering and other processes.

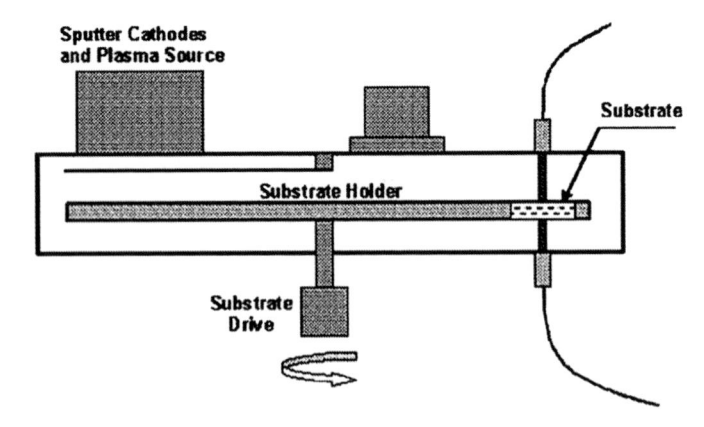

Figure 5.17 Side view of the layout of the Helios system by Buhler Leybold Optics [27]; figure is taken from their earlier web site.

Figure 5.18 Photo of the Helios system by Leybold Optics as in Figs. 5.16 and 5.17 from their earlier web site.

The spectral data is taken into the computer and can then be processed in many ways to display what is in the residual gas.

The OPTIX can also be used as a He leak detector, and here is the comparison of its detection sensitivity limits with the RGA and a dedicated He leak detector: the OPTIX limit is 10^{-6} mBar l/s, the RGA is 10^{-7} mBar l/s, and the Dedicated is 10^{-9} mBar l/s. If the OPTIX is always on the chamber, this could be an added benefit.

In 2019, Gencoa introduced the adaptation of the Optix to control reactive sputtering processes such as Al_2O_3 from aluminum metal with O_2 gas, etc. The control of reactive processes has been refined in the 21st century by sputter voltage feedback to a fast gas-flow controller, but the residual gas feedback of the Optix may prove even more effective.

5.2.6.1 Broad-band optical monitors

Vidal, et al. [30] in 1979 reported the definitive work with broadband optical monitoring and the use of computer re-optimization of the remaining layers to correct for errors detected.

Where the performance over a relatively broad band is important, it is desirable to have some means to sense what is happening in the whole band (per Vidal et al.) or at least what is predicted to be the impact on the whole band of a more narrowly sensed error. If the controllability of the layers involved is sufficient, it may be unnecessary to utilize these more sophisticated monitors.

It can be said in summary about error compensation, that it is appropriate to "nail down" the performance where it is most important, by monitoring it at that wavelength. Without the aid of computer re-optimization, the NBP filter and a broad class of edge filters can be monitored at one wavelength and will have certain "natural" error compensation [14]. This is why NBP filters have been possible even though the state of the technology might not have seemed to be accurate enough to give an acceptable result. If a broad-band coating must be controlled, the impact of each detected error must be considered on the broad band, and subsequent layers can be adjusted on the basis of computer "compensation."

An instance was encountered some years ago where a BBOM was being touted as the way to monitor a NBP filter. It is this author's opinion that this approach is outright wrong, for the reasons mentioned above! Use a broad-band monitor for broad-band requirements, and use a narrow-band monitor for narrow-band requirements!

The first BBOM was the human eye. Banning [5] described the use of the eye as a BBOM in 1947.

Vidal et al. [26,31] typically compared the actual spectrum at a given point in the deposition with the expected/desired spectrum over the band. The difference between the present spectrum and the desired spectrum at the end of the layer (cut point) was computed very frequently to give a merit (demerit) value. The termination of each layer is the sum of the magnitude of errors between the target transmittance versus wavelength for that layer and the actual transmittances at a given point of the deposition. If a perfect match were achieved, the demerit would go to zero. As a function of film thickness when the cut point was approached, the merit value generated a curve approaching a "turning point." If that point were past, the layer would be getting thicker than desired. This curve would then look much like a single-wavelength monitor for terminating layers at the turning points, but it would represent the weighted merit of all the wavelengths in the band of interest.

When errors are present (and noise), the TP will not go to zero demerit, only a minimum. Manually determining this point is similar to single-wavelength TP monitoring, but possibly more sensitive, since it is not sinusoidal.

Vidal and Pelletier [31] first pointed out the potential to calculate the actual index, dispersion, and homogeneity of a layer. Flory et al. [32] then actually computed the index versus thickness and thereby homogeneity by using the values of the reflectance or transmittance at the spectral peaks and valleys. The ability to determine index, packing density, homogeneity, and thickness during the process is a major contribution of BBOM. This is as much to the credit of computing power as to the BBOM itself. Li Li and Yi-hsun Yen [33] repetitively (with increasing thickness) solved for the four variables a, b, c, and d, where d is the physical thickness and the others are for the dispersion formula $n_i = a + b/\lambda^2 + c/\lambda^4$. Sullivan and Dobrowolski [34] carried this further to use the effective media approximation (EMA). Here, the equivalent of the *a*, *b*, and *c* variables above were used for the densest form of the index, and that index was then "diluted" by a mixing fraction of voids according to the EMA to fit the calculated index and dispersion. This reduced the solution to two variables, the thickness and mixing fraction or density of the film. The information gained by such computations should describe the film much closer to reality than is possible by any single-wavelength approach. This more accurate knowledge of the film up to any point in the deposition, should allow the best opportunity to make whatever corrections are possible to the remaining layers in order to get as close as possible to the target result for the total coating.

In 2000, Sullivan et al. [35] reported on the implementation of a system on a commercial scale which used an improved BBOM. The deposition rate of the earlier rf-magnetron sputtering system was deemed insufficient for commercial use along with other limitations of the smaller prototype system. Dual ac-magnetron sputtering sources were used in the new system which gave a fivefold increase in deposition rate. The earlier BBOM with ∼3-nm resolution was replaced by one with ∼0.5-nm resolution over 400–900 nm to overcome limitations for films with fine spectral structure. Strategies for dealing with layers whose monitoring was either insensitive or too sensitive were also discussed an demonstrated by experimental results.

The LZH group published a paper by Lappschies et al. [36] about the use of a BBOM for rugate filter deposition. Little was said about the BBOM. The deposition was by ion-beam sputtering of two targets of SiO_2 and TiO_2 which were abutted to each other and movable across the axis of the beam to vary the mix of the two materials as they were sputtered. Although it was not discussed in the paper, it appears that the BBOM offers the best opportunity to control the rugate index versus thickness which is needed. The BBOM can simultaneously measure the thickness and the index versus wavelength and thereby control the position of the targets to correct for errors in the mix. This author is unaware of any other rugate deposition and control system which could be as straightforward and stable as that.

Wilbrandt et al. [37] demonstrated the use of a BBOM to calibrate a QCM. In addition to calibrating a crystal controller, another possible use of the broad-band monitor could be to electronically select a single wavelength to monitor wherein the monitor has no moving parts as in a monochromator.

In summary, it would seem logical that the greatest benefit of a BBOM would be for broadband coatings such as beamsplitters, color correction filters, and broadband AR coatings. It is hard to visualize how a BBOM would benefit the production of a DWDM or other NBP filter, which is the ultimate narrow-band task. It is usually possible to design a single wavelength monitoring process which will automatically compensate for reasonable errors at and near the monitoring wavelength. When the parameters related to the other wavelengths than the one being monitored are sufficiently stable, the results of such single wavelength monitoring can be satisfactory over a broad band. This has some similarity to the use of the intermittent monitor like the Helios to control a whole calotte of parts by monitoring only at one of the parts. However, in the case of the BBOM, since the monitoring wavelength is the whole band, it should be possible to extend this compensation to the broad band rather than just the smaller band around a single wavelength.

It must be kept in mind, however, that the various monitors tend to view and control only one small area. A chamber full of parts depends on the properties of the coating to be uniform from the monitored area to all of the parts being coated. It has also been seen that the stability of the deposition rate can be critical because it affects not only the monitoring but also the index of the deposition to a greater or lesser degree depending on chemical reactions in the process.

5.2.6.2 Data logging

With the ever increasing computer power, the last few decades have seen the inclusion of data logging on most optical coating systems. This can be a great asset in trouble-shooting and process refinement. Having a record by time of pressures and all of the other parameters of a process allows a check of the "health" of a system and aids in the understanding of the behavior of each aspect of the process.

5.2.7 Processes

5.2.7.1 Atomic layer deposition

ALD first appeared in the 1960s under the name "molecular layering" in the Soviet Union. It appeared again as "atomic layer epitaxy" (ALE) in Finland in the 1970s. ALE was first applied on a large scale primarily in ZnS thin films for display panels. In the 21st century, ALD processes have been refined for a great many materials, and many previous limitations have been overcome. The application of ALD to optical thin films and semiconductors has become practical, and it is worthy of examination as a possibility for any new product or application.

Two particular benefits of ALD make it attractive: its exact thickness control and the fact that it can be a conformal film to essentially any shape. ALD has even demonstrated its ability to produce DWDM NBP filters for fiber optics communications, which is one of the most demanding coatings in the industry.

Pfeiffer et al. [38] demonstrated the practical application of SiO_2, Al_2O_3, TiO_2 and Ta_2O_5 to AR coatings which were conformal on highly curved surfaces and the thickness was controlled by only counting layer cycles for each material.

Ghazaryan et al. [39] created very low index SLARs which could be optimized over the 193 to 1200 nm range using ALD. They deposited a few atomic layers of SiO$_2$ and then a few of Al$_2$O$_3$ and then repeated the pair to get the desired thickness. The coating was then chemically etched to remove the Al$_2$O$_3$. What remained was a porous structure of SiO$_2$ whose effective index depended on the ratio of the number of SiO$_2$ to Al$_2$O$_3$ atomic layers.

Pfeiffer et al. [40] carried the above findings further to make a hybrid AR coating design, combining an 8-layer Al$_2$O$_3$/TiO$_2$/SiO$_2$ interference multilayer stack by ALD with a low-index nano-porous SiO$_2$ top-layer by ALD and wet etching. This showed the excellent performance expected by the employment of a last low index layer of index about 1.15.

Paul et al. [41] applied the above ALD processes to a 5-layer AR coating on PMMA plastic. Here the process challenges were to obtain good adhesion and crack resistance when the coating is subjected to environmental conditions. The resulting processes were successful, and the benefits of conformity and thickness control were preserved.

It is clear that ALD should be seriously examined as an option when any new optical thin-film coating production project is being considered.

5.2.7.2 The H$_2$O problem

Titania or titanium dioxide has been commonly used since the availability of e-guns for the evaporation of high melting point refractory oxides. Although its application is old, new and improved life has been given to TiO$_2$ by the recent work of Wayne Sainty from Australia.

Sainty [42] published an insightful paper on the deposition of TiO$_2$ at low pressures with IAD. Figure 5.19 shows the modified Balzers BAK 760 with his ST55 ion source and a Meissner trap. The Meissner trap more than doubles the pumping speed for water vapor, which is normally the primary gas to be pumped. This is in order to get to a low pressure for the process which minimizes the adverse effects of water vapor on the film growth.

Figure 5.19 Sainty's chamber setup to get high index with IAD TiO$_2$.

Figure 5.20 shows screenshots of his RGA with and without the operation of the Meissner trap, showing the process oxygen for the IAD and the order of magnitude reduction in the water vapor.

Figure 5.21 shows the results of the index of the TiO_2 versus the ion flux of the IAD. There is a broad region of ion flux over which optimum results were obtained.

Figure 5.22 is an interesting result when compared with the discussion in Willey [14] as relates to the work of Pulker et al. [43]. Pulker showed a large difference in results depending on the starting material, but Sainty shows

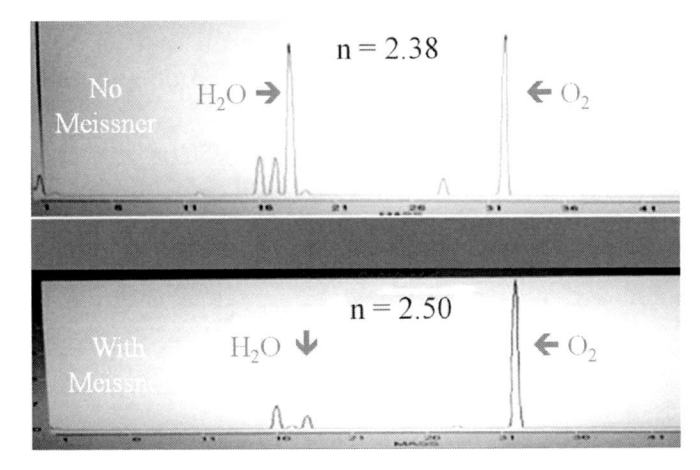

Figure 5.20 Screenshot of RGA showing the difference in water vapor pressure with and without the operation of the Meissner trap.

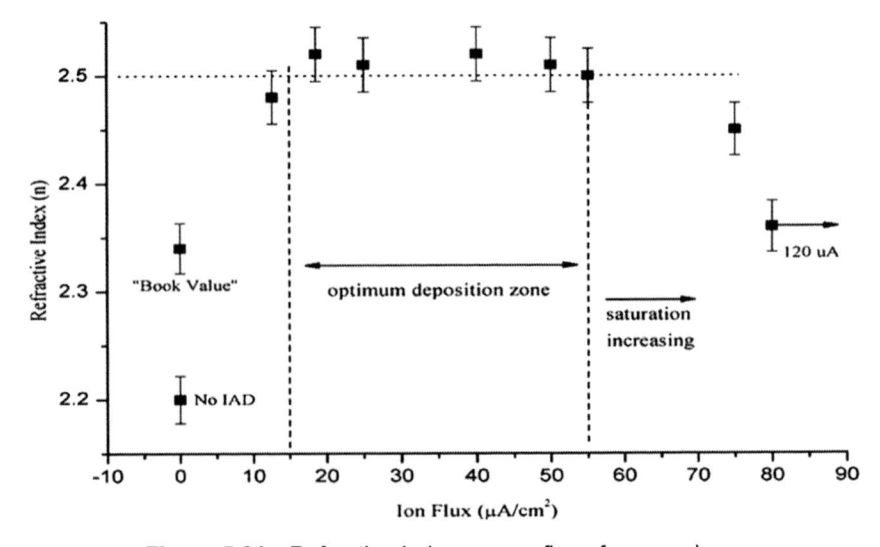

Figure 5.21 Refractive index versus flux of oxygen ions.

Start Material	Initial Outgassing	Deposition Pressure	Refractive Index
TiO2 (dioxide)	Slight & momentary	~3.0 x10^{-5} mbar	2.52 +/- 0.02
TiO (monoxide)	Negligible	~2.5 x10^{-5} mbar	2.52 +/- 0.02
Ti (metal)	Zero	~1.0 x10^{-5} mbar	2.53 +/- 0.02

Figure 5.22 Resulting index and pressure depending on starting material.

essentially no measurable difference in the final index of refraction, indicating a fully densified and stable result due to the reduction of chamber pressure.

5.2.7.3 Reactive sputtering

Sputtering is another field whose history goes back into the 19th century, but has blossomed in the 21st century. It was little used for optical coatings until the later 20th century. Reactive DC Magnetron Sputtering increased the rates at which materials could be sputtered. If dielectric coatings were produced, they would form an insulating layer on everything, and the anode became electrically hidden, causing what is called the "disappearing anode" problem. Just before the 21st century, two solutions to this problem were developed which employed reversing the polarity of the sputtering at a "medium" frequency. The electrons which would build up on the insulating coating needed to be removed before they reached a concentration which would arc through the insulator. One approach was to reverse the polarity for a proportionately shorter time because the electrons (being of much lower mass than atoms) would rapidly move to ground as compared to the massive charged atoms. This was the principal of the Sparkle unit from advanced energy. The other approach was to have dual magnetrons which were alternately acting as cathode and anode at a medium frequency to discharge electron buildup. Figure 5.23 illustrates this, which seems to be the most common and satisfactory form of application at the present time.

When it is desired to sputter-deposit a dielectric such as Al_2O_3, the usual approach is to sputter a metal target such as aluminum and react it with the oxygen gas as it deposits on the substrate. The difficulty with this is that the metal target tends to react with the oxygen in the chamber to form a dielectric insulator on the target, as in Fig. 5.24, or "poison" the target. This same thing can and does happen with other reactive sputtering processes using metal targets such as Ti and Zr to sputter TiN of ZrN_x. The Sparkle or Dual Magnetron approaches can be used to solve these problems.

Mattox [44] reviewed the history of reactive sputtering and mentions that Veszi coined the term in 1953. There has been quite a bit of work done in recent decades to gain greater control over reactive sputtering and achieve

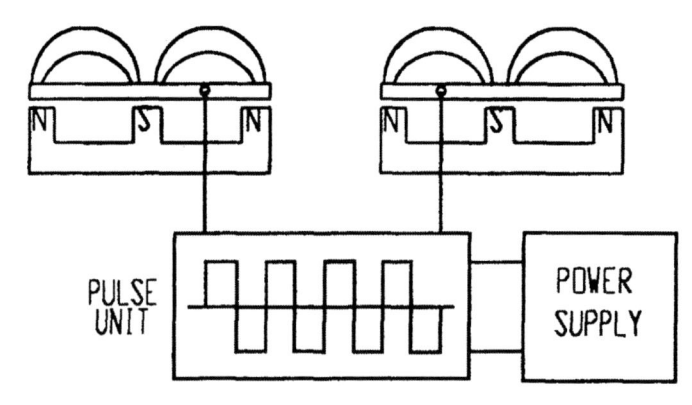

Figure 5.23 Dual Magnetron Sputtering, alternated at a medium frequency.

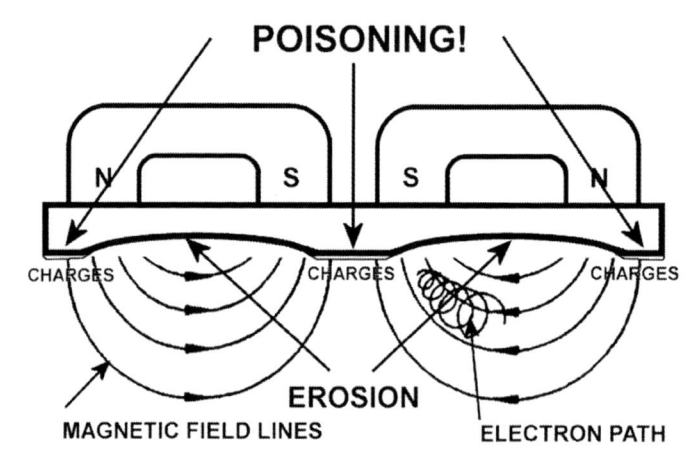

Figure 5.24 Poisoning of a DC Magnetron target in a reactive gas.

higher deposition rates. Rossnagel [45] has provided an overview and review of the technology. Figure 5.25 is after the fashion of Rossnagel's presentation.

Figure 5.25 is conceptual only; the "corner" of a real process would not generally be as sharp nor would the edges be very straight. However, this does represent the general behavior of many metals when being reactively sputtered to make an oxide or nitride of that metal. Let us assume that the target is aluminum, the reactive gas is oxygen, and we wish to sputter Al_2O_3. We might start by sputtering the metal with Ar as the sputtering gas with no oxygen. Conditions would then be at the left edges of Fig. 5.25 (sub-graphs a, b, and c). The deposition rate is high, the chamber pressure is low, and the discharge voltage is high. As the flow of reactive gas increases (along the dashed paths in Fig. 5.25b), the deposit is being reactively oxidized to something less than stoichiometric aluminum oxide. The areas of the target which are less actively sputtered will start to oxidize and perhaps arc because of electron build-up on the insulating surface. It will be assumed for the moment that the arcing is

Optical Coatings   181

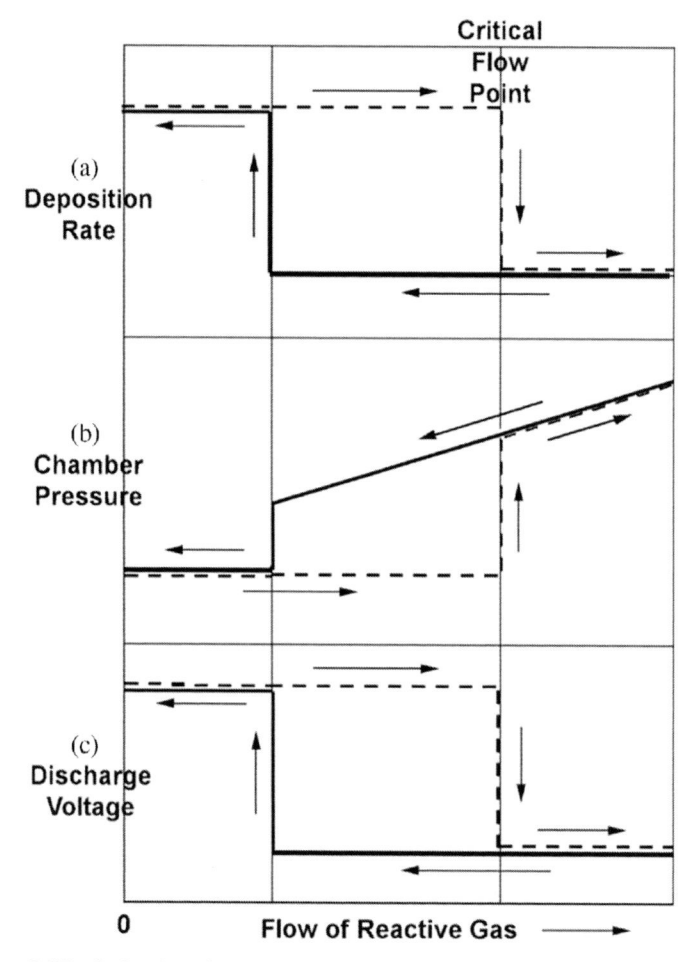

Figure 5.25 Behavior of reactive magnetron sputtering after Rossnagel [45].

held at bay by appropriate electronic means so that we can ignore arcing for this discussion. As the oxygen flow is increased from zero up to some flow rate, the chamber pressure is relatively constant because the metal is "gettering" or reacting with all of the oxygen. Then, at a critical flow point, the gas flow is too much and the target starts to "poison" or glaze over almost entirely with oxide on the surface of the metal target. At this point, much less metal is sputtered, less gas is reacted with the sputtered metal, and the chamber pressure rises abruptly while the deposition rate and discharge voltage drop precipitously. The process is no longer in a metallic sputtering mode, but it is in a dielectric/insulator sputtering mode which is much slower. Just before this critical point, the oxide deposited is stoichiometric and the deposition rate is almost as high as the metal-only rate. If we try to regain the high rate by reducing the reactive gas flow, we find that we are returning to low flow rates on the solid lines of Fig. 5.25. Once the excess oxide forms on the target, the process runs away and cannot be readily recovered in the case shown. The chamber pressure has to drop a long way by reducing the oxygen flow before the oxide glazing on the target is sputtered away and the metallic mode resumes. This hysteresis effect is seen in Figs. 5.25 and 5.26. If a scheme could be devised to hold the conditions just before the run-away point where the deposited film is just the desired stoichiometric Al_2O_3, we would have a robust and high-rate oxide sputtering process. Figure 5.26 shows the results of a real titanium nitride reactive sputtering process similar to Fig. 5.25b; the plot is from Carter et al. [46]. In recent decades, there have been successful approaches to solving this problem.

The challenge is how to sense and control the partial pressure of the reactive gas. Sproul [47] mentions five sensor possibilities, "which include a mass spectrometer, an optical gas controller (OGC), an optical emission spectrometer (OES), a spinning rotor gauge (SRG), and a change in the cathode voltage. There are advantages and disadvantages to each of these methods." He goes into some detail about each, but it seems that OGC, OES, and cathode voltage have become most common. The OPTIX system mentioned above is such an OES. The response time of the sensor and the gas flow control must be fast enough to control the reactions. A piezoelectric gas valve is apparently fast enough.

Schill et al. [48] described the OGC which uses optical signals generated from the gases entering the gauge to both provide a measure of the partial pressure of two gases and to then automatically control the partial pressure of these two gases. The OGC does not look at light generated from the sputtering process itself, but instead, when the process gases enter the OGC sensor head, they are activated by electron impact. When the activated gas species relax, they emit light, and that passes through filters and is detected by a photomultiplier. An OGC looks at two gases at any given time (such as Ar and O_2), but for most reactive sputtering processes, control is only needed for two gases.

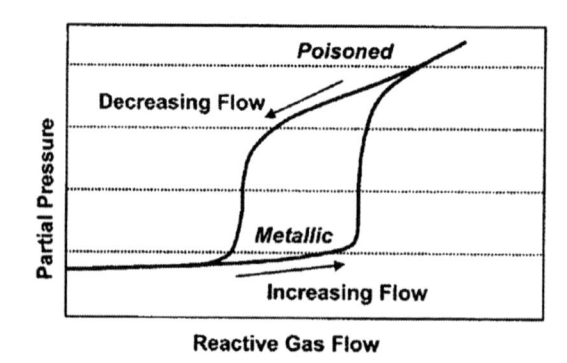

Figure 5.26 Real process results adapted from Carter et al. [46].

Sproul [47] says that, "The OGC has several advantages. It operates at sputtering pressures, it does not have to be differentially pumped, and it does not suffer from pump failures. The sensor head for the OGC is very small, and it can be located very close to the reactive sputtering zone. The OGC is very stable, and it provides accurate control of the process over a long period of time."

Optical emission spectroscopy (OES) can be used to control reactive sputtering processes. OES uses the optical signal of the sputtering plasma to generate a control signal at an appropriate wavelength. The strength of this reactive gas signal can change with time as the target erodes, so measures need to be taken to compensate for that.

Another possible source for a control signal for reactive sputtering is cathode voltage. The overall composition of the sputtering atmosphere changes as reactive gas is added to the chamber. This leads to a change in the impedance of the electrical circuit and the cathode voltage. Changes can be detected quickly, but other factors such as the erosion of the target will change the cathode voltage, and the best operating point changes with time. Additional sophistication is needed to make sure that the system operates at the right point.

Carter et al. [49] describe a system controlled by cathode voltage. They said that: a largely metallic aluminum surface sputters at a much higher voltage than a surface that has been converted to a dielectric in an oxygen rich environment. This target poisoning is seen in their characteristic hysteresis curve of in Fig. 5.27. The target voltage is plotted against oxygen flow as constant power is regulated to the cathode. As the flow into the reaction chamber increases, the target condition and operating voltage change very little as long as the sputtered aluminum production is adequate to getter the majority of the reactive gas. As the introduction rate of oxygen begins to

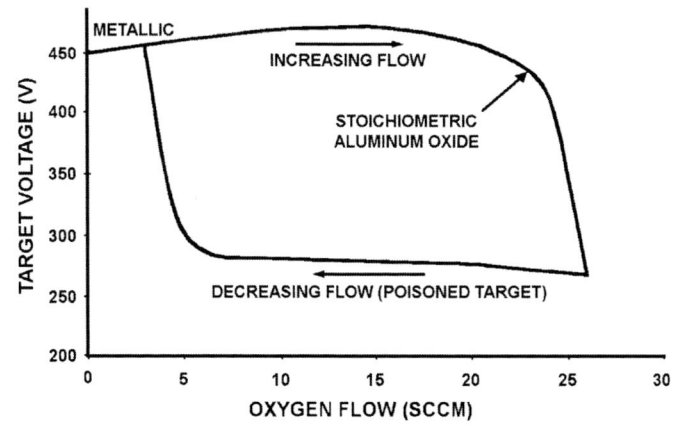

Figure 5.27 Target voltage versus gas flow under control by target voltage. (Adapted from Ref. [49])

exceed the liberation rate of aluminum, the process quickly becomes unstable and the target poisons, or converts to a dielectric surface condition. The poisoned aluminum target is characterized by behavior more similar to that of the oxide, causing the discharge voltage to drop dramatically and consequently the deposition rate. The optimum condition for maximum rate and good quality dielectric deposit occurs where the target begins to transition from a metallic to an oxidized state at or near the knee of this curve.

5.2.8 Index versus thickness

There is increasing interest in the design of films with thicknesses on the order of 10 nm and less for a variety of applications of nano particles, plasmonics, quantum dots, solar reflectors, black mirrors, etc. The indices of refraction (n and k) for the effective media of such coatings depend on the materials with which such "layers" interface and the specific process parameters used to produce those films. The structures may typically be nucleating island structures and may also be continuous films. A key factor is that the n- and k-values vary with thickness until some thickness is obtained, usually greater than 20 nm. Heretofore, films have not been designed where the index variation with thickness is allowed within the design process. Software is now available wherein the index at a given thickness is computed at each iteration of the design optimization process. This allows more realistic design results utilizing the full representation of the behavior of the layers in question; the resulting coatings, when produced, should be in better agreement with the designs. Including the n and k versus wavelength and thickness in the design process is referred to here as Double Dispersion. Optical thin film design has typically used the simplifying assumption that real films are homogeneous throughout their thickness, although this has been known to be only an approximation. Layers thinner than 10 nm in optical thin film designs have commonly been avoided until recent times, probably due to the uncertainty of the index of such layers as a function of thickness. This could be due to the effects of nucleation and other deposition and film growth properties.

Foteinopolou et al. [50] recently introduced a feature issue in Optical Materials Express on "Beyond Thin Films: Photonics with Ultrathin and Atomically Thin Materials," which emphasizes the broad current interest in these very thin films in both pre-percolated and continuous forms. Papers in that issue that relate to the present work include the following: Milewska et al. [51] discuss simple methods to control the self-organization of gold atoms where surface diffusion of gold atoms can be suppressed and also their high surface mobility can be harnessed to fabricate large-area substrates. Liu et al. [52] review Van der Waals or 2D materials. Baburin et al. [53] discuss the pursuit of depositing the silver "dream" film. Wang et al. [54] discuss second harmonic generation spectroscopy on two-dimensional materials. Taghinejad et al. [55] review the semiconductor-like heterostructures produced on this nanoscale. Roberts et al. [56] demonstrate the use of very thin TiN films in Fabry-Pérot resonators to produce selective absorbers for solar energy. Secondo et al. [57] show how to obtain reliable modeling of ultrathin alternative plasmonic materials using spectroscopic ellipsometry. Knopf et al. [58] demonstrate the integration of atomically thin layers of transition metal dichalcogenides into high-Q, monolithic Bragg-cavities. All of these papers deal with films on the order of 10 nm thick, and some are dealing with films whose index variation with thickness must be taken into account in the coating design processes such as we are discussing here.

As discussed above, when the film thickness is monitored by a QCM, after a few nanometers have registered, the actual islands are usually several times (to an order of magnitude) higher than the QCM reading, because the material droplets are "piled-up" and not spread evenly over the surface. With many materials, the percolation occurs at about 10–20 nm as read on a QCM. If the conditions promote wetting as reported by Formica et al. [59] using a very thin precoat of copper (Cu), the percolation might occur at only a few nm of QCM reading.

The net effect of the described behavior of the nucleating atoms in a new layer is that the indices of refraction are quite different as a function of the thickness for most layer materials until thicknesses of several tens of nm are reached. These very thin layers are used in modern applications such as solar control coatings, black mirrors, metamaterials, and plasmonic structures, and they must be designed. To realistically model such systems, it is necessary to know how the n and k of these layers vary with thickness, and to incorporate that knowledge into the thin film design process. For example, a solar control coating might have three silver (Ag) layers on the order of 10 nm thick which are separated by much thicker dielectric layers, plus another very thin layer for some additional absorptance as described by Medwick et al. [60]. Similarly, a black mirror might have two or three layers of chromium (Cr) or other materials which are less than 10 nm thick.

Stenzel and Macleod [61] have provided a very extensive tutorial on the technology related to this subject. They pointed out that the approach at that time was to acknowledge that the index of very thin films did vary with thickness and therefore needed to be characterized at each thickness of interest. The design approach was to pick a specific thickness which had been characterized and to hold that thickness constant while varying the other layers of the design. They state that: "The thickness of a metal island film must not be varied during a design procedure. A metal island film with a given thickness and given effective optical constants has to be tackled as a fixed building block which can be introduced into an interference stack, but should not be modified during the synthesis and refinement." They also later state: "A limitation that is missing from normal coating design is the strict constraint on the thickness of the composites and this is something that we must accept." This section deals with relieving this constraint and the expansion of the capability of the design software to allow the metal island or other variable index film to vary in thickness during the design process in a realistic way.

As will be shown, the n and k versus thickness can be a strong function of the materials and processes used on both sides of the variable thickness layers. The characterization layers need to be first deposited on a substrate or layer of the preceding material, then the variable index layer, and then a capping layer of the following material. All of these depositions are to be done by the processes to be used in application.

Various modeling schemes such as polynomials and Bruggeman theory, as discussed in Hummel and Guenther [62], were considered to fit n and k versus wavelength and thickness to the measurements [63]. However, it was decided that the real curves of a variety of materials are too varied in shape for these models to be practical. Therefore, the chosen approach is to use the real measured n and k at the deposited thicknesses and interpolate the n and k for any thicknesses in between these actual measurements. Normally, 3 to 9 test thicknesses have been deemed adequate. For each different material combination and process, a named table is built with the number of thickness runs, the thicknesses, and length of the n and k data versus wavelength. The n and k data is stored as the usual index files in the FilmStar software and the material, process, and thickness are coded in the file with an identifying name. In the first version of the software, for any one material and process, the files of wavelengths and the n and k data needed to be the same range, spacing, and length in order to allow easy interpolation. Since this work, FilmStar has simplified the process (BASIC not required) through a new THK variable in User-Defined Index Functions. The Index Functions evaluator directly supports interpolation and no longer requires that n and k files have the same wavelength range and spacing.

In the normal optical thin-film design process, when a variable index versus thickness layer is encountered, the layer's identifying code is passed to a FilmStar BASIC program which generates a temporary n and k table to be used for that layer at that thickness for that one time. The optimization process otherwise proceeds normally. As mentioned in the previous paragraph, this process has been simplified in a newer FilmStar revision.

Figure 5.28 illustrates how radically the n and k can vary with thickness in silver films deposited on Fused Silica (FS). At 2.4 through 9.4 nm thickness, island formation is apparent and plasmonic/polariton behavior seems to come into play. In the region beyond 9.4 nm thickness, there seems to be some percolation, whereas 20 nm shows a move towards a continuous film, and 30 nm seems to behave more like bulk silver.

Figure 5.29 plots the n and k versus thickness and wavelength for two different designs and material interfaces; Figs. 5.29(a) and (c) are for silver

Optical Coatings

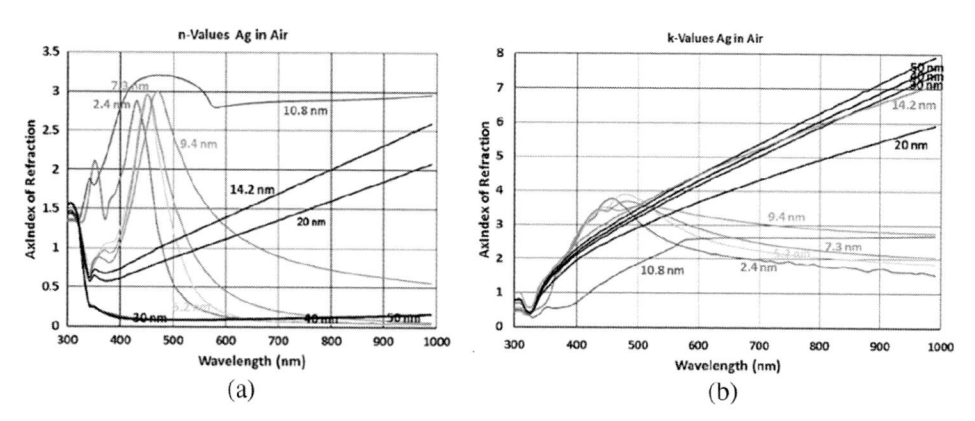

Figure 5.28 (a) Index (n) vs. wavelength for various thicknesses of Ag deposited on FS in air. (b) Index (k) vs. wavelength for various thicknesses of Ag deposited on FS in air.

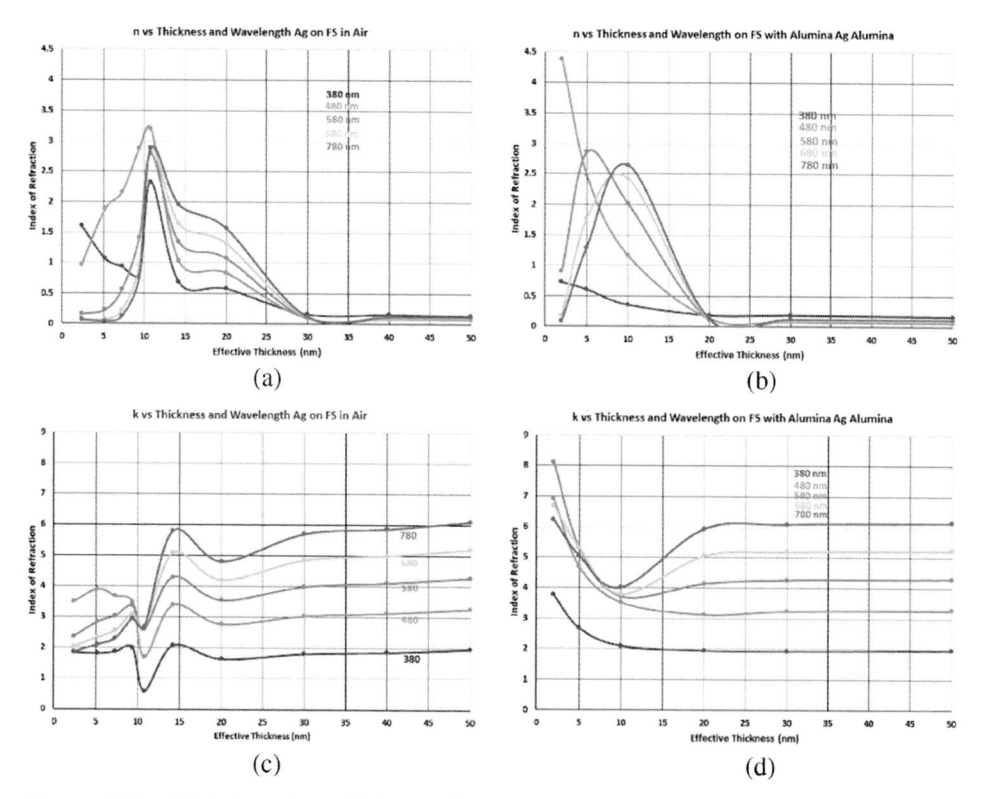

Figure 5.29 (a) Index (n) vs. thickness for various wavelengths of Ag deposited on FS in air. (b) Index (n) vs. thickness for various wavelengths of Ag deposited on Alumina and capped with alumina. (c) Index (k) vs. thickness for various wavelengths of Ag deposited on FS in air. (d) Index (k) vs. thickness for various wavelengths of Ag deposited on Alumina and capped with alumina.

deposited on FS in air, and Figs. 5.29(b) and (d) for FS substrates coated with 7 nm of alumina (via ALD), then the silver layer, and then that is capped with 7 nm of alumina (via ALD). These plots illustrate the influence of both of the materials which interface with the silver.

It is clear that some (if not all) materials and processes show a strong variation in indices with thickness for very thin layers, up through the point of percolation, that is: before bulk properties are established. This can also be true for some materials and processes after percolation. Very thin layers are used now more frequently than ever for solar energy control, black mirrors, metamaterial applications, plasmonic applications, etc., where the design tools described herein are needed to properly apply the real variation of indices with thickness.

5.3 Conclusions

It has always been true that we stand on the shoulders of our predecessors to look toward and work for the future. The pioneers of the 19th and 20th centuries laid the foundations for the progress that has been made in the 21st century. Many discoveries and inventions are found before their application is needed or capable of being implemented. Much of what has made the progress of the last two decades is actually based on the work of the previous two centuries. The development of the digital computer in the mid-20th century has allowed the advancement of optical thin film technology as it has also benefitted most aspects of our lives today. Digital computers allowed the computations and simulations needed for design in our field, and it allowed the measurement, control, automation, and data logging of our equipment and processes for the process development and production of those designs.

The word "nano" has become ubiquitous in the last two decades, sometimes to excess. However, it cannot be denied that work on this scale has made great advances in many fields. In the field of optical thin films, it has greatly expanded our knowledge and understanding of material properties and processes. We can now even assemble atoms on a nano-level to make meta-materials which have properties which have not been available in nature, such as hardness, durability, index of refraction, etc.

The expanded understanding of the energy interactions that occur in atoms that are growing thin films has allowed the application and control by not only heat but also energetic ions and atoms with plasma, sputtering, and IAD. Reactive sputtering has become well refined so that most materials can be economically deposited by sputtering, and ALD has matured to the point of being a viable process for optical coatings. Layers which are an order of magnitude thinner than previously practical can now be modeled, designed, and produced. Stress and the effects of water vapor in processes are better understood and thereby more manageable than they were before the turn of the century. Layer thickness and index control are significantly better monitored and controlled than previously.

The practice of optical thin film production has come from being a "black art" in the 20th century to a true science in the 21st century.

References

[1] J. Strong, "Evaporation technique for Aluminum," *Physical Review* **43**(6), 498 (1933).

[2] W. Geffken, "Deutsches Reichs Pat. No. 716153 Interferenzlichtfilter," (1939). German Patent 716153.

[3] H. A. Macleod, "The history of optical coatings," in *50 Years of Vacuum Coating Technology and the Growth of the Society of Vacuum Coaters, chapter 16*, ed. by Donald M. Mattox and Vivienne Harwood Mattox, Society of Vacuum Coaters (2007).

[4] M. Auwarter, "The historical development of thin film physics and technology," in *Handbook of optical properties: thin films for optical coatings*, ed. by Hummel, Rolf E and Guenther, Karl H, CRC Press (1995).

[5] M. Banning, "Practical methods of making and using multilayer filters," *JOSA* **37**(10), 792–797 (1947).

[6] H. D. Polster, "A symmetrical all-dielectric interference filter," *JOSA* **42**(1), 21–24 (1952).

[7] C. I. Bright, "Review of transparent conductive oxides (TCO)," in *50 Years of Vacuum Coating Technology and the Growth of the Society of Vacuum Coaters, chapter 16*, ed. by Donald M. Mattox and Vivienne Harwood Mattox, Society of Vacuum Coaters (2007).

[8] K. H. Guenther, "Microstructure of vapor-deposited optical coatings," *Applied Optics* **23**(21), 3806–3816 (1984).

[9] M.-C. Liu, C.-C. Lee, M. Kaneko, K. Nakahira, and Y. Takano, "Microstructure of magnesium fluoride films deposited by boat evaporation at 193 nm," *Applied optics* **45**(28), 7319–7324 (2006).

[10] E. Chason and P. R. Guduru, "Tutorial: Understanding residual stress in polycrystalline thin films through real-time measurements and physical models," *Journal of Applied Physics* **119**(19), 191101 (2016).

[11] E. Chason, M. Karlson, J. Colin, D. Magnfält, K. Sarakinos, and G. Abadias, "A kinetic model for stress generation in thin films grown from energetic vapor fluxes," *Journal of Applied Physics* **119**(14), 145307 (2016).

[12] R. R. Willey, *Practical Design of Optical Thin Films*, Lulu.com (2018).

[13] M. Friz and F. Waibel, "Coating materials," in *Optical interference coatings*, 105–130, Springer (2003).

[14] R. R. Willey, *Practical production of optical thin films*, Lulu.com (2019).

[15] A. Herpin, "Calcul du pouvoir réflecteur d'un système stratifié quelconque," *Comptes rendus hebdomadaires des séances de l'Académie des sciences* **225**(3), 182–183 (1947).

[16] R. R. Willey and R. Shakoury, "Stable, durable, low-absorbing, low-scattering mgf2 films without heat or added fluorine," in *Advances in Optical Thin Films VI*, **10691**, 106910C, International Society for Optics and Photonics (2018).

[17] U. Schulz, N. Gratzke, S. Wolleb, F. Scheinpflug, F. Rickelt, T. Seifert, and P. Munzert, "Ultraviolet-transparent low-index layers for antireflective coatings," *Applied Optics* **59**(5), A58–A62 (2020).

[18] U. Schulz, F. Rickelt, H. Ludwig, P. Munzert, and N. Kaiser, "Gradient index antireflection coatings on glass containing plasma-etched organic layers," *Optical Materials Express* **5**(6), 1259–1265 (2015).

[19] U. Schulz. private communication.

[20] k-Space Associates, Inc. https://www.k-space.com/product/mos/.

[21] A. E. Ennos, "Stresses developed in optical film coatings," *Applied optics* **5**(1), 51–61 (1966).

[22] E. Quesnel, B. Rolland, V. Muffato, D. Labroche, and J. Robic, "The ion beam sputtering: a good way to improve the mechanical properties of fluoride coatings," in *Proceedings of the annual technical conference-Society of Vacuum Coaters*, 293–300, Society of Vacuum Coaters (1997).

[23] W. Lechner, G. Strauss, and H. Pulker, "Correlation between optical and mechanical properties of ion plated ta 2o 5 films," in *Proceedings of the annual technical conference-Society of Vacuum Coaters*, 287–290, Society of Vacuum Coaters (1998).

[24] G. G. Stoney, "The tension of metallic films deposited by electrolysis," *Proceedings of the Royal Society of London. Series A, Containing Papers of a Mathematical and Physical Character* **82**(553), 172–175 (1909).

[25] M. Ohring, "Materials science of thin films 2nd ed.," (2002).

[26] B. Vidal, A. Fornier, and E. Pelletier, "Wideband optical monitoring of nonquarterwave multilayer filters," *Applied optics* **18**(22), 3851–3856 (1979).

[27] A. Zoeller, M. Boos, H. Hagedorn, W. Klug, and C. Schmitt, "High accurate in-situ optical thickness monitoring," in *Optical Interference Coatings*, Optical Society of America (2004).

[28] A. Zoeller, M. Boos, R. Goetzelmann, H. Hagedorn, and W. Klug, "Substantial progress in optical monitoring by intermittent measurement technique," in *Advances in Optical Thin Films II*, **5963**, 59630D, International Society for Optics and Photonics (2005).

[29] GENCOA http://www.gencoa.com/optix/.

[30] B. Vidal and E. Pelletier, "Nonquarterwave multilayer filters: optical monitoring with a minicomputer allowing correction of thickness errors," *Applied Optics* **18**(22), 3857–3862 (1979).

[31] B. Vidal, A. Fornier, and E. Pelletier, "Optical monitoring of nonquarterwave multilayer filters," *Applied Optics* **17**(7), 1038–1047 (1978).

[32] F. Flory, B. Schmitt, E. Pelletier, and H. Macleod, "Interpretation of wide band scans of growing optical thin films in terms of layer microstructure," in *Thin Film Technologies I*, **401**, 109–116, International Society for Optics and Photonics (1983).

[33] L. Li and Y.-H. Yen, "Wideband monitoring and measuring system for optical coatings," *Applied optics* **28**(14), 2889–2894 (1989).

[34] B. T. Sullivan and J. Dobrowolski, "Deposition error compensation for optical multilayer coatings. i. theoretical description," *Applied optics* **31**(19), 3821–3835 (1992).

[35] B. T. Sullivan, G. A. Clarke, T. Akiyama, N. Osborne, M. Ranger, J. Dobrowolski, L. Howe, A. Matsumoto, Y. Song, and K. Kikuchi, "High-rate automated deposition system for the manufacture of complex multilayer coatings," *Applied Optics* **39**(1), 157–167 (2000).

[36] M. Lappschies, B. Görtz, and D. Ristau, "Application of optical broadband monitoring to quasi-rugate filters by ion-beam sputtering," *Applied optics* **45**(7), 1502–1506 (2006).

[37] S. Wilbrandt, O. Stenzel, N. Kaiser, M. Trubetskov, and A. Tikhonravov, "On-line re-engineering of interference coatings," in *Optical Interference Coatings*, WC10, Optical Society of America (2007).

[38] K. Pfeiffer, U. Schulz, A. Tünnermann, and A. Szeghalmi, "Antireflection coatings for strongly curved glass lenses by atomic layer deposition," *Coatings* **7**(8), 118 (2017).

[39] L. Ghazaryan, Y. Sekman, S. Schröder, C. Mühlig, I. Stevanovic, R. Botha, M. Aghaee, M. Creatore, A. Tünnermann, and A. Szeghalmi, "On the properties of nanoporous sio2 films for single layer antireflection coating," *Advanced Engineering Materials* **21**(6), 1801229 (2019).

[40] K. Pfeiffer, L. Ghazaryan, U. Schulz, and A. Szeghalmi, "Wide-angle broadband antireflection coatings prepared by atomic layer deposition," *ACS applied materials & interfaces* **11**(24), 21887–21894 (2019).

[41] P. Paul, K. Pfeiffer, and A. Szeghalmi, "Antireflection coating on pmma substrates by atomic layer deposition," *Coatings* **10**(1), 64 (2020).

[42] W. G. Sainty, "Deposition of titanium dioxide thin films by low-pressure ion-assisted deposition," in *Proceedings of the annual technical conference-Society of Vacuum Coaters*, Society of Vacuum Coaters (2016).

[43] H. K. Pulker, G. Paesold, and E. Ritter, "Refractive indices of tio 2 films produced by reactive evaporation of various titanium–oxygen phases," *Applied optics* **15**(12), 2986–2991 (1976).

[44] D. Mattox, "The historical development of controlled ion-assisted and plasma-assisted pvd processes," in *Society of Vacuum Coaters 40 th Annual Technical Conference*, 109–118 (1997).

[45] S. Rossnagel, "Thin film deposition with physical vapor deposition and related technologies," *Journal of Vacuum Science & Technology A: Vacuum, Surfaces, and Films* **21**(5), S74–S87 (2003).

[46] W. Sproul, D. Christie, and D. Carter, "47th annual technical conference proceedings," in *Society of Vacuum Coaters*, 96 (2004).

[47] W. D. Sproul, "Control of a reactive sputtering process for large systems," in *Society of Vacuum Coaters. 36 th Annual Technical Conference*, 504–509 (1993).

[48] S. Schill, C. Gogol, and R. Mueller, "A new technique for controlling partial pressures in reactive processes," in *Society of Vacuum Coaters Annual Technical Conference*, (1989).

[49] D. Carter, H. Walde, G. McDonough, and G. Roche, "Parameter optimization in pulsed dc reactive sputter deposition of aluminum oxide," in *Proceedings of the Annual Technical Conference-Society of Vacuum Coaters*, 570–577 (2002).

[50] S. Foteinopoulou, N. C. Panoiu, V. M. Shalaev, and G. S. Subramania, "Feature issue introduction: Beyond thin films: photonics with ultrathin and atomically thin materials," *Optical Materials Express* **9**(5), 2427–2436 (2019).

[51] A. Milewska, A. S. Ingason, O. E. Sigurjonsson, and K. Leosson, "Herding cats: managing gold atoms on common transparent dielectrics," *Optical Materials Express* **9**(1), 112–119 (2019).

[52] C.-H. Liu, J. Zheng, Y. Chen, T. Fryett, and A. Majumdar, "Van der waals materials integrated nanophotonic devices," *Optical materials express* **9**(2), 384–399 (2019).

[53] A. S. Baburin, A. M. Merzlikin, A. V. Baryshev, I. A. Ryzhikov, Y. V. Panfilov, and I. A. Rodionov, "Silver-based plasmonics: golden material platform and application challenges," *Optical Materials Express* **9**(2), 611–642 (2019).

[54] Y. Wang, J. Xiao, S. Yang, Y. Wang, and X. Zhang, "Second harmonic generation spectroscopy on two-dimensional materials," *Optical Materials Express* **9**(3), 1136–1149 (2019).

[55] H. Taghinejad, A. A. Eftekhar, and A. Adibi, "Lateral and vertical heterostructures in two-dimensional transition-metal dichalcogenides," *Optical Materials Express* **9**(4), 1590–1607 (2019).

[56] A. S. Roberts, M. Chirumamilla, D. Wang, L. An, K. Pedersen, N. A. Mortensen, and S. I. Bozhevolnyi, "Ultra-thin titanium nitride films for refractory spectral selectivity," *Optical Materials Express* **8**(12), 3717–3728 (2018).

[57] R. Secondo, D. Fomra, N. Izyumskaya, V. Avrutin, J. Hilfiker, A. Martin, Ü. Özgür, and N. Kinsey, "Reliable modeling of ultrathin alternative plasmonic materials using spectroscopic ellipsometry," *Optical Materials Express* **9**(2), 760–770 (2019).

[58] H. Knopf, N. Lundt, T. Bucher, S. Höfling, S. Tongay, T. Taniguchi, K. Watanabe, I. Staude, U. Schulz, C. Schneider, et al., "Integration of atomically thin layers of transition metal dichalcogenides into high-q, monolithic bragg-cavities: an experimental platform for the enhancement of the optical interaction in 2d-materials," *Optical Materials Express* **9**(2), 598–610 (2019).

[59] N. Formica, D. S. Ghosh, A. Carrilero, T. L. Chen, R. E. Simpson, and V. Pruneri, "Ultrastable and atomically smooth ultrathin silver films grown on a copper seed layer," *ACS Applied Materials & Interfaces* **5**(8), 3048–3053 (2013).

[60] P. A. Medwick, A. V. Wagner, P. J. Fisher, and A. D. Polcyn, "Nanoplasmonic ("sub-critical") silver as optically absorptive layers in solar-control glasses," in *Society of Vacuum Coaters. 60th Annual Technical Conference Proceedings*, (2017).

[61] O. Stenzel and A. Macleod, "Metal-dielectric composite optical coatings: underlying physics, main models, characterization, design and application aspects," *Advanced Optical Technologies* **1**(6), 463–481 (2012).

[62] R. E. Hummel and K. H. Guenther, *Handbook of optical properties: thin films for optical coatings*, vol. **1**, CRC Press (1995).

[63] R. Willey, A. Valavicius, A. Lamouri, and K. Patel, "Methods for determining and modelling indices of refraction versus thickness of silver, ITO, and chromium," in *Society of Vacuum Coaters. 62nd Annual Technical Conference Proceedings*, (2019).

Ronald R. Willey has over 50 years of experience as an individual contributor and director of optical systems and optical coating development, design, and production. He graduated from the Massachusetts Institute of Technology, having majored in optical instrumentation. He also earned a Masters degree in computer science from the Florida Institute of Technology. He has taught optics at the Florida Institute of Technology and the University of Wisconsin. He has taught optical thin film design, development, and production at the Society of Photo-optical Instrumentation Engineers, the Society of Vacuum Coaters, Florida Institute of Technology, and the University of Wisconsin. He is versed in optical and thin film design and fabrication, and he is an expert in optical tolerancing and practical optical instrument development. He has served as an expert forensic witness for optical systems and coating litigations. He teaches short courses internationally. He is the author of many technical papers and five books. He is a Fellow of SPIE and the Optical Society of America.

Chapter 6
Infrared Optical Systems

Adam Phenis
AMP Optics LLC, 13308 Midland Rd., Unit 1304, Poway, CA 92074, USA

Jason Mudge
Golden Gate Light Optimization, LLC, 2912 Diamond St., Ste. 347, San Francisco, CA 94131, USA

6.1 Introduction

From 2000 to 2020, infrared (IR) optical systems have experienced a strong transition from primarily a military technology and application to an increase in commercial and consumer products. This transition has enabled or driven many component advances primarily related to detectors and sources. This chapter discusses two basic aspects of this transition: (1) new applications and (2) component developments. The two categories are, effectively, intertwined; the former often drives the latter or said another way "necessity is the mother of invention." Component developments include advancement in detectors, materials, fabrication techniques, and metrology, to name just a few. A major trend in infrared systems is significant miniaturization of previously bulky systems reducing size, weight, power consumption, and cost (SWaP-C) over the past two decades. It is this factor which paved the way into a large variety of new industrial and commercial applications.

The infrared spectrum division into bands can vary between industries and personal views but tend to be defined from three different perspectives: (1) light propagation through air, i.e., atmospheric transmission; (2) emission source, e.g., lasers, thermal, etc.; and (3) detector materials. In this work, we are taking the latter approach, which is consistent with the approach chosen for ANSI/OEOSC OP1.007 – Spectral bands [1]. Currently, there is interest in an international version of this standard, but it has not been fully resolved.

The IR spectrum ranging from 0.75 μm to 30 μm is divided into near-infrared (NIR), short-wave infrared (SWIR), mid-wave infrared (MWIR), long-wave infrared (LWIR) and very-long-wave infrared (VLWIR). NIR light extends from 0.75 to 1.1 μm and has been widely used for imaging and sensing of digital products and night vision imaging in security cameras. SWIR light, considered to range from 1.1 to 3 μm, is regularly employed in agricultural and industrial inspection and covers the mainstream fiber optical communication wavelength range of about 1.53 to 1.57 μm. Due to transparent windows in the NIR and SWIR in biological tissue, imaging spectroscopy is also used for medical imaging. MWIR light ranges from 3 to 5 μm, lies in the transition zone from reflective radiation to thermal radiation, and is widely used in defense, industrial inspection, and various remote sensing applications. LWIR light is from approximately 5 to 14 μm, measures the thermal radiation of objects, and is commonly utilized for defense, medical imaging, industrial and environmental monitoring. It is important to note that the typical definition of LWIR is around 8 to 14 μm due to the atmosphere having little to no transmission in the 5- to 8-μm region – light absorption of H_2O and CO_2 in the atmosphere. This band is being standardized to cover this atmospheric absorption window for continuity and applications where atmospheric transmission is not a factor or at least has minimal effect on system performance. Additionally, there is a VLWIR region that ranges from 14 to 30 μm with applications in spectroscopy, astronomy and long-range missile detection. All these bands are primarily based on a combination of detector technologies and atmospheric transmission as summarized in Table 6.1 along with typical optical materials.

In the past two decades, NIR optical systems have seen significant development. Common NIR systems include oximeters, heart-rate monitors, proximity sensors, and biometric systems based on fingerprint, face, iris, or vein recognition. These applications have seen explosive growth in consumer

Table 6.1 IR spectral bands, common optical materials, and typical detectors [1].

Spectral Band	Wavelength Range (μm)	Common Optical Materials	Typical Detectors	Prevalent Sources
NIR	0.75–1.1	Visible glasses	Si, InGaAs, HgCdTe	Incandescent lamps, LED, laser diodes, fiber lasers, DPSS
SWIR	1.1–3.0	Visible glasses, Si, ZnSe, ZnS, Chalcogenides, CaF_2, BaF_2	InGaAs, HgCdTe	LED, laser diodes, DPSS, black bodies
MWIR	3.0–5.0	Si, Ge, ZnSe, ZnS, CaF_2, BaF_2, Chalcogenides	InSb, HgCdTe, PbSe	Lasers, QCLs, black bodies
LWIR	5.0–14.0	Ge, ZnSe, ZnS, CaF_2, BaF_2, Chalcogenides	HgCdTe, Bolometer, Pyroelectric, Thermopiles	Lasers (CO_2), QCLs, black bodies
VLWIR	14.0–30.0	NaCl, KCl, Kbr, CdTe, KRS-5, CsBr, CsI, IR Plastics	Doped Si, Bolometer, Pyroelectric, Thermopiles	QCLs, black bodies, monochromator

electronics and wearable medical devices. Additionally, new applications such as commercial 3D sensing have been driven by LiDAR/LADAR advancements and increased computational access.

Due to relatively high spatial resolution and spectral signatures, SWIR and MWIR systems have been further explored in agricultural and industrial inspection, with applications including plant and produce inspection, smoke-see-through monitoring, biological and medical imaging, art conservation and forgery detection, remote sensing and reconnaissance, to name a few. Thermal imaging applications for the consumer market include automotive and imaging thermography, which have encountered a much lower barrier to access due to the component developments. While adoption of new techniques in the medical industry can be slow relative to the consumer market, imaging thermography has shown promise in body-temperature monitoring as well as the diagnosis of neuropathy due to diabetes, and vascular disorders [2,3]. Other commercial thermal applications, such as fire safety, building efficiency investigation, and welding defect detection have benefited from component advances, specifically from mass production of uncooled thermal imagers, e.g., microbolometers. Furthermore, active thermography is broadly used in non-destructive testing for aerospace or industrial inspection. Additionally, gas detection has seen significant advances with the commercial production of tunable quantum cascade lasers (QCLs) produced from a variety of manufacturers. Imaging polarimetry has allowed dehazing due to atmospheric scatter and increased the resolution capability of long-range infrared imaging [4].

Lastly, with the advances in freeform fabrication and improved metrology techniques, infrared optical systems are achieving increased optical performance with wider field of view, lower f-number and lower distortion optical systems all with lower SWaP-C [5]. In general, advancements in photonics tend to follow Moore's law with a 20-year lag [6].

6.2 New Applications

During the period of 2000 to 2020, with tremendous infrared technology development and substantial reduction in size and cost, several previously expensive defense IR systems have been modified and developed for industrial and consumer use. As a result, IR systems have been extended into new application areas for use by the general public.

6.2.1 Industrial imaging

There are a significant number of potential uses of thermal imaging such as the ability to view surroundings in an inferno within a building. This allows firefighters to move seamlessly through a burning building and save lives. Due to environmental issues that affect us all, a building's thermal efficiency has become a high priority. These new technologies have the potential to provide significant cost savings for up-and-coming businesses and, simultaneously, conserve world resources to prevent further significant erosion of our planet. Another important aspect is to automate or improve manufacturing in general. As an example, welding processes can be monitored by being able to image the glow of an ongoing weld and viewing in the IR naturally cuts out the dominant ultraviolet light. This allows an IR imaging system to be used as feedback of the current state of the weld, allowing the welding process to progress efficiently and, more importantly, to lower the risk of defects. In the next couple sections, some specific applications are discussed in further detail.

6.2.1.1 Automotive and/or autonomous driving

As the world moves in the direction of autonomy, e.g., driverless automobiles, active depth sensors such as range-finders and LiDARs of all flavors (incoherent and coherent) are becoming desirable allowing robots to move through the environment seamlessly. Additionally, imaging microbolometers are desirable for passive night vision for enhanced driver safety.

LiDAR One method of viewing the world is through a 3-D sensor also known as a LiDAR, which is an active optical sensor. Generally, there are two main categories of LiDARs: (1) incoherent and (2) coherent [7]. In the last several years, these sensors have become quite popular due to the quest for full autonomous driving. Additionally, augmented reality (AR) and virtual reality (VR) have become a new method for interacting with computers requiring range detection. Typically, these sensors are developed in the NIR or SWIR since the active source (typically a laser) should not be detected by the human eye because it would be an irritant [8]. One advantage of the NIR is that it can use a silicon-based detector material, which is quite inexpensive with a drawback that the quantum efficiency (QE) is lower than at visible wavelengths. On the other hand, InGaAs (indium gallium arsenide) detectors work well in the SWIR and have quite a high QE across most of the NIR, but their cost is significantly higher, and they carry more dark-current shot noise. These detectors are often used in the S-band (or 1550 nm) or even the O-band (1310 nm) [8].

Passive night vision Automotive thermal imaging dates back to 2000 when GM introduced a microbolometer thermal imager on the Cadillac Deville [9]. Thermal imaging provides passive night vision, which has an advantage over active systems such as LiDAR due to their need for active illumination sources. This makes thermal imaging night vision relatively inexpensive, effective and compact (low SWaP). Additionally, it has the natural ability to discern between living people and animals versus inanimate objects, which could play a significant role in the decision making for autonomous vehicles. Furthermore, automotive thermal imagers can be designed for long range viewing, allowing human reaction to potential dangers within a reasonable amount of time. In 2020, there are only a handful of luxury cars, e.g., Rolls-Royce, that are available with thermal imaging, but this may change in the near future as the price point continues to drop [10].

6.2.1.2 Medical and imaging thermography

With two windows [11] in the NIR (650–950 nm) and SWIR (1000–1350 nm) in biological tissue, where the penetration depth of light is a maximum, light in these regions have a long history in medical applications. Spectroscopy, diffuse optics, fluorescence imaging, optical coherence technology (OCT) have been widely used for ophthalmic imaging, dermatological diagnosis, blood-glucose-level monitoring, heart-rate monitoring, oximetry, brain-computer interface research, as well as others.

In the LWIR, imaging thermography is extensively used for temperature monitoring. Body-temperature monitoring has become an area of interest during the troubling times of the COVID-19 pandemic–the first in over 100 years. Body-temperature measurement is indispensable in monitoring and attempting to contain the spread of the virus by assuming those with an elevated temperature are more likely to have the coronavirus. This is critical to minimize the rapidly spreading COVID-19 and its variants. Current single-point, non-contact thermometers typically use thermopiles (many thermocouples in series) as the detector, and for a full image, bolometers detector arrays are more typical. Thermography has also been used as a diagnostic tool in oncology, allergic diseases, angiology, plastic surgery, rheumatology, diabetes neuropathy, and vascular disorders [12].

6.2.2 Defense

Infrared systems are extensively used for various defense applications and defense arenas. Particular IR sensor application are night vision, surveillance and reconnaissance, seeker systems to guide intercept missiles autonomously to their targets, and more. Such surveillance and reconnaissance use both visible and IR sensors. Examples of this are Predator, Global Hawk, and Landsat 7 [13]. Predator, a UAV, can handle a variety of payloads, but in particular it can have two types of staring arrays which can be employed: a 512×512 platinum silicide (PtSi) detector and a 256×256 indium antimonide (InSb). These detectors operate in the MWIR and must be cooled using a Stirling cycle cooler [13]. Additionally, there are two types of rangefinders both of which operate in the SWIR. The first operates at an eye-safe 1.54-μm wavelength using an Erbium glass laser with an operating range of 15,000 m, and the other is a 1.06-μm Nd:YAG laser with a range of 9995 m. The Global Hawk (UAV) again has both visible and IR imaging sensor as payloads. The IR imaging sensor payload operates from 3.6–5.0 μm and with a NIIRS of 5.5 at a 45-degree elevation angle at a 28-km range. Lastly, Landsat 7 is a surveillance and reconnaissance earth-viewing satellite. Its sensor payload is a hyperspectral imager covering the visible and into the SWIR. There are 7 bands plus a panchromatic band. This imaging data is used by a variety of US government agencies [13].

The Space-Based Infrared System (SBIRS) is a US IR imaging missile warning system and is being replaced by the Next-Gen Overhead Persistent Infrared (NG-OPIR) satellites system to enhance detection [14]. This improves reporting of intercontinental ballistic missile launches, submarine-launched ballistic missile launches, and tactical ballistic missile launches.

There are other non-US satellites system such as SPOT 2, 4 and 5 in both the visible and NIR bands and was developed in France along with Belgium and Sweden. Additionally, some older Russian satellites sell image data mostly in the visible systems and are extensively used for defense applications.

6.2.3 Science

The Clouds and the Earth's Radiant Energy System, or CERES, is a radiometer for Earth radiation budget measurements developed out of Langley Research Center (named in honor of Dr. Samuel P. Langley, who is credited with the invention of the microbolometer). Monitoring the earth's health via radiation has been one of the longest consecutive measurements in National Aeronautics and Space Administration's (NASA) history (some 40+ years). The instrument CERES uses an uncooled microbolometer and is one quite large pixel that is mechanically scanned across the earth [15]. One of the more interesting features of this particular bolometer is the special coating of gold-black, which has the ability to absorb light energy from 200 nm to 50 μm [16]. Follow-on missions are proposed using a non-scanning system and a focal plane array to reduce SWaP and cost while improving spatial resolution. Thermopiles also have excellent properties for radiometers and significant room for improvement before they reach their theoretical performance limit more so than microbolometers. The Radiation Budget Instrument (RBI) was a NASA mission that proposed to monitor earth's radiation and planned on using thermopiles as its detector [17]. Thermopiles tend to be less responsive relative to bolometers, which indicates thermopile tend to have a larger effective dynamic range.

Multispectral and hyperspectral infrared remote sensing systems are of great importance to monitor global warming and climate change. The majority of remote sensing hyperspectral imagers cover a range commonly referred to as VNIR-SWIR (visible, NIR and SWIR) with data products targeting vegetation and agriculture, geology and soils, land use, urban, water resources and disaster topics and applications primarily including monitoring, classification, mapping, properties, exploration, quality, bathymetry, prevention and post-crisis [18].

6.3 Component Development

Component advancement has enabled access for many applications through cost reduction. Dividing them up into two natural groups: (1) sensors and (2) sources. In this instance, sensors consist of both the detector and the optics – a camera is a subset of a sensor. Both groups have had significant improvements over the last two decades.

6.3.1 Infrared sensing

Within sensing, there consists the detector and the optics which collect the light or emission. Both are discussed in this section with references for further review.

6.3.1.1 Detectors

As previously discussed, IR technology has evolved from primarily a military technology to include commercial and consumer applications. Affordable uncooled thermal detectors have significantly extended the use of infrared systems, since cooling detectors to cryogenic temperatures is cost prohibitive for mass-production applications.

IR detectors and focal plane arrays fall into two branches, those that are sensitive to photons (quantum) and those that are sensitive to temperature (thermal detectors). In photon or quantum detectors, the radiation is absorbed by the material, and electrons are generated where the photon transfer curve applies [19]. This process is illustrated in Fig. 6.1 [20] as well as in [21]. The electrical output signal results from the changed electronic energy distribution. Such sensitive materials are summarized in Table 6.1. Commercially available IR sensors focus on the two main atmospheric transmission windows – the MWIR band of 3–5 μm and 8–14 μm in the LWIR band. Note that silicon (Si) is still useful in the NIR even with the diminished quantum efficiency (QE). A key advantage of HgCdTe over InSb in the MWIR is that it can be tuned to particular wavebands with a change in operating temperate and the relative composition makeup. Unfortunately, one main disadvantage of HgCdTe and InSb are that they need be cooled to cryogenic temperatures to prevent the well from being filled with electron carriers which are easily

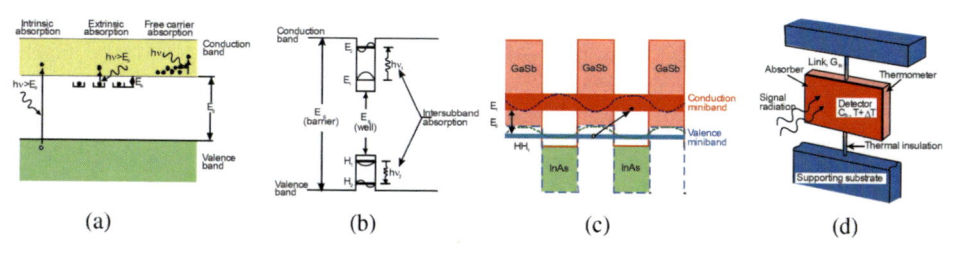

Figure 6.1 Optical excitation in: (a) bulk semiconductors, (b) quantum wells, (c) type-II InAs/GaSb superlattices, and (d) thermal detectors [20].

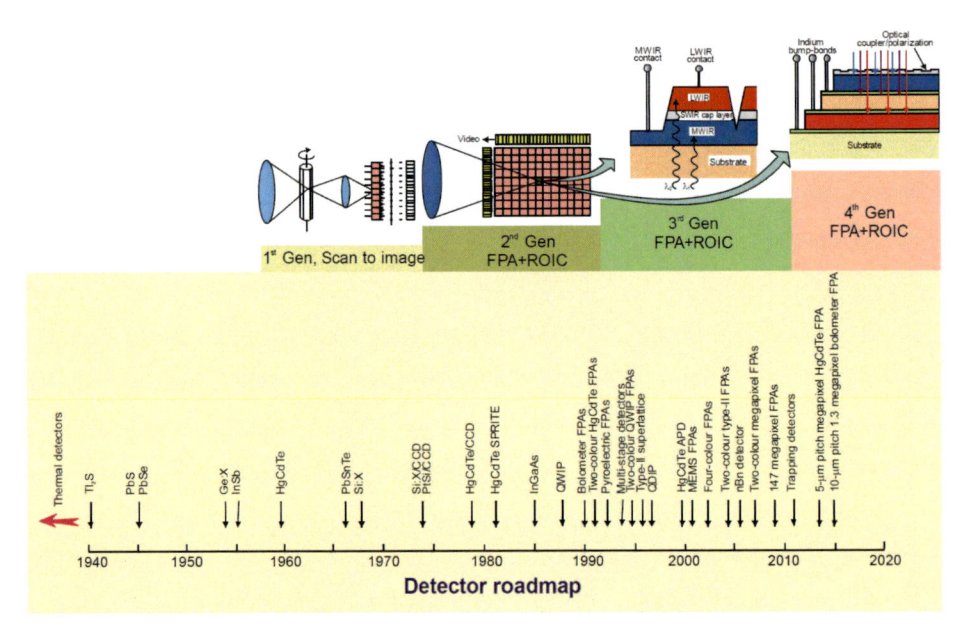

Figure 6.2 History of the development of infrared detectors and systems. Four generations of systems can be considered for principal military and civilian applications [20].

generated by random room temperature thermal generation and recombination [22] as well as significant background housing emission making them somewhat impractical for commercial use. The development of IR detectors is illustrated in Fig. 6.2. In the second branch are thermal detectors where the incident radiation is absorbed by the material, which changes a physical electrical property of the material and is used to generate an electrical output. These detectors, such as microbolometers, measure differences in energy (or temperature) meaning they can perform without cooling. However, maintaining the detector's immediate surrounding temperature (sensor housing and optics) relatively constant or at a fixed temperature is desired to accurately detect the signal. Being able to measure differential energy is effectively what prevents these detectors from being saturated by the background (assuming it is relatively constant with respect to the signal). This is key in that keeping optics very cold is a cost driver, which makes room-temperature detectors more commercially viable and accessible to the general public. These facts have driven significant technical improvements in thermal detectors and is the focus of this section.

Figure 6.3 summarizes the evolution of HgCdTe and α-Si microbolometers. This trend to smaller pixels and larger array pixel count is bringing these detector technologies in line with silicon-based detectors and semiconductor fabrication processes. The issue of how small a pixel should be to still be useful has been summarized in [23] where the authors concluded for LWIR systems that minimum pixel size for usefulness is 5 μm and 3 μm in the MWIR.

Infrared Optical Systems

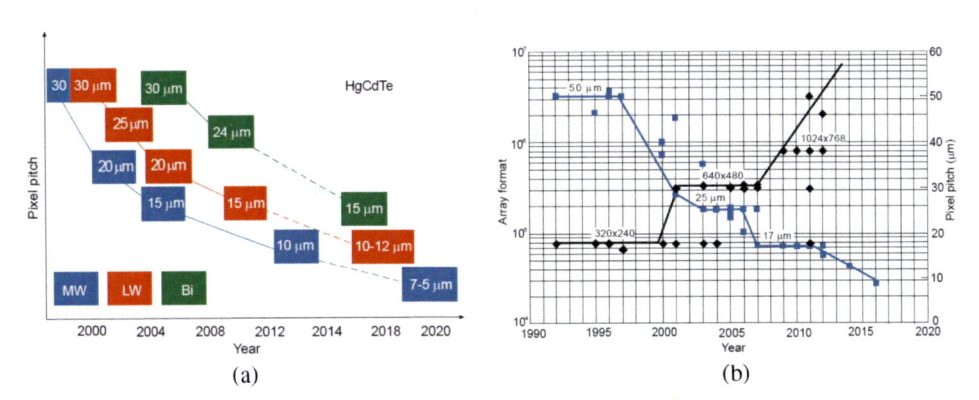

Figure 6.3 Pixel pitch for (a) HgCdTe photodiodes and for (b) a-Si microbolometers have continued to decrease due to technological advancements [20].

A comparison of uncooled or room temperature detectors is shown in Table 6.2. It tabulates the relative benefits and drawback of each of the uncooled IR detectors. It is interesting to point out that bolometers and pyroelectric detectors are "high" on the list for improvement and thermopiles are medium. Again, note that bolometers and thermopiles are "excellent" for use in radiometers. There is a very clear and mathematically succinct publication by Foote, which compares bolometers and thermopiles. He discusses the manufacturability of each and compares their relative performance given a background or instrument housing operation temperature for these two uncooled and unchopped thermal detectors [24].

Bolometers The first thermal or uncooled detector discussed is the bolometer or microbolometer, which has seen explosive growth from 2000–2020 with a reduction in pixel size and increased array size as shown in Fig. 6.3. They are produced in larger volumes than all the other IR detector technologies combined [20]. Pixel sizes have gone from sizes on the order of 30 μm to 50 μm prior to 2000 down to 12 μm and even 10 μm in readily commercially available arrays with primary manufacturing efforts focused on reduction in pixel size/pitch and an ever-increasing array size. Figure 6.4 shows the trade space on pixel size, noise and difficulty providing a pictorial of the challenges faced. Microbolometers are primarily vanadium oxide (VO_x) or amorphous silicon (α-Si). The current state is that VO_x is the most used technology due in part to VO_x being the lowest production cost. The state of the art commercially available microbolometers have reached HD format (1920 × 1200 pixels) [26]. Figure 6.5 shows the structure from various commercial microbolometer manufacturers. These structures show some of the mechanisms used to absorb the incoming light and range from thin films to umbrella-like structures to optimize absorption. Per Table 6.2 in the previous section, bolometers are regularly used in radiometers for scientific research.

Table 6.2 Comparison between uncooled thermal detectors [25].

	Monolithic resistive bolometer	Monolithic pyroelectric detector	Hybrid pyroelectric detector	Hybrid ferroelectric detector	Monolithic thermoelectric detector
Responsivity	High	High	High	High	High
Bias required	Yes	No	No	Yes	No
DC response	Yes	No	No	No	Yes
Chopper required	No	Yes	Yes	Yes	No
Dynamic range linearity	High	Unknown	Low	Low	High
Radiometric capability	Excellent	Unknown	Poor	Poor	Excellent
Array availability	Yes	No	No	Yes	Yes
Production status	Medium	No	No	High	Medium
Possibility of performance improvement	High	High	Low	Low	Medium

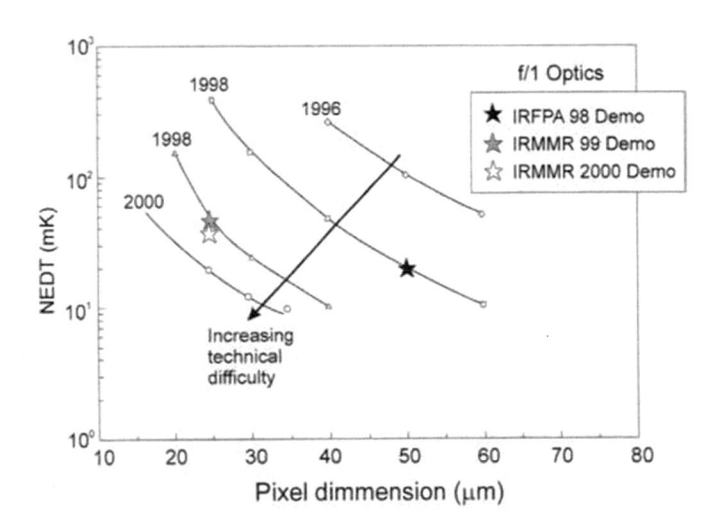

Figure 6.4 The trends in noise and pixel dimensions and difficulty [27].

Thermopiles Thermopiles use the "Seebeck effect", named after the inventor. They are one of the oldest IR detectors. They are essentially a collection of connected thermocouples – many thermocouples in series increases the signal voltage potential. Because they do not require cryogenic cooling, they are another potential detector for consumer products. One main attraction of thermopiles in measuring temperature remotely is that their voltage output is very linear with temperature. Thanks to advances in CMOS and lithographic fabrication processes, thermopile's on-chip circuitry technology has encountered mass production through miniaturization. Many of these processes work well with silicon as the substrate and deposited thin film metals for the thermoelectric materials. While thermopiles are not as sensitive as bolometers and pyroelectric detectors, their reliability and good cost to performance ratio

Infrared Optical Systems 205

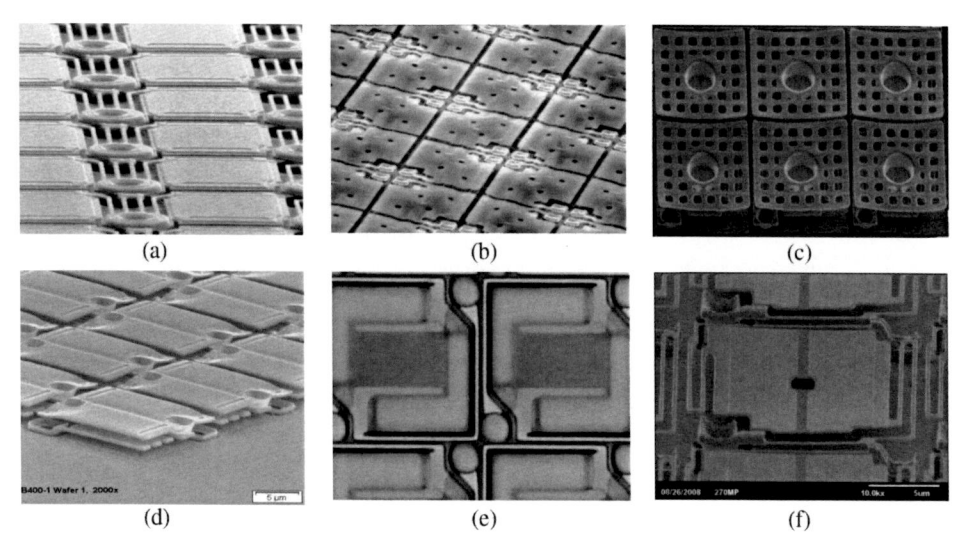

Figure 6.5 Commercial bolometer design: (a) VO$_x$ bolometer from BAE, (b) α-Si bolometer from ULIS, (c) VO$_x$ umbrella design bolometer from DRS, (d) VO$_x$ bolometer from Raytheon, (e) VO$_x$ bolometer from SCD, and (f) α-Si/α-SiGe bolometer from L-3 Communications [20].

allows them to displace these other detectors in many applications. This is seen in the explosive growth of the handheld, single-point, non-contact thermometers. As with bolometers, thermopiles are regularly used in radiometers, a particular advantage of thermopiles is that they do not require a measure of current or, more generally, a bias applied as shown in Table 6.2 [28].

Pyroelectric detectors Of the three uncooled detectors discussed, pyroelectric detectors are the youngest – proposed in the late 1950s by Chynoweth [29] with applications arriving in the 1960s [30,31]. However, this effect has been observed for centuries, first being described by Theophrastus in 315 BCE [28]. This pyroelectric effect is found most commonly in ferroelectric materials, which can generate a surface electric polarization (charge). With a temperature change of the material, there is a brief charge polarization in the materials that can be measured as a voltage. However, this polarization only lasts a short time, meaning it is a more useful detector in an AC signal implementation (chopper wheel). Recent published papers have shown that fundamental limits of performance can be reached with semiconductor fabrication processes.

6.3.1.2 Materials

While materials have been covered in Chapter 2 focusing on visible glasses that can generally perform well out to about 1.7 μm, it is important to note that advancement in IR materials have enabled considerable IR optical sensors capabilities. This is the focus of this particular section. Reference [32] is an excellent reference for IR materials pre-2000.

Optical properties Advancements in chalcogenides and IR Gradient Index (IR-GRIN) materials have allowed IR sensors to cover a broader wavelength spectrum with good optical performance. The tailoring in developmental IR glasses for specific dispersion and thermal characteristics have further opened up the optical design space to allow designs that are achromatic over a broad spectrum and athermal over an increased temperature range, some combinations even enabling lenses that are achromatic over multiple spectral bands [33].

Chalcogenides are materials containing one or more of the chalcogen elements (sulfur, selenium, or tellurium). Chalcogenide glasses are vitreous materials made with one or more of these chalcogen elements and can present unique properties including highly nonlinear refractive indices and very broad wavelength transmission ranges. Some of the material combinations can exhibit transmission up to 25 μm when tellurium is combined. A plot of these materials' transmission curves is shown in Fig. 6.6, and a table of commercially available chalcogenide IR glasses is shown in Table 6.3.

The other material category that has advanced significantly in the last two decades is GRIN, which has extended to the IR wavelengths. These materials have the capability to compositionally tune optical and mechanical properties that result in planar photonic systems. The particular tunable glass properties include spectral window, refractive index, dispersion, dn/dT, thermal and chemical stability, coefficient of thermal expansion (CTE), to name a few. This results in an optical element that can be combined with more traditional IR crystals and chalcogenides to achieve performance requirements for very diverse applications. IR-GRIN materials have the potential to significantly reduce SWaP and cost of optical elements while additionally enhancing performance. These materials are still at the research and development phase,

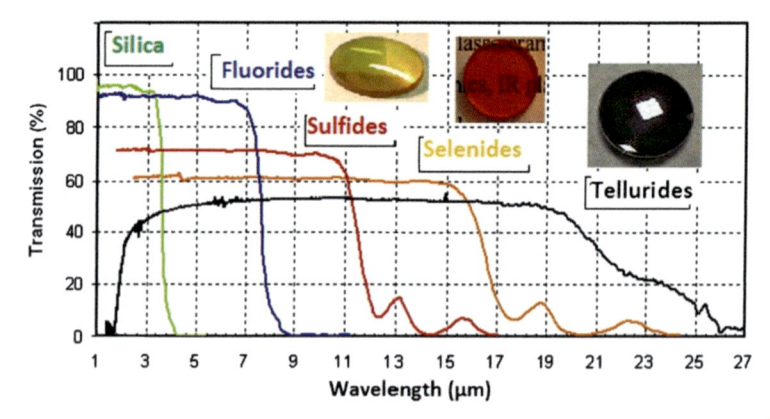

Figure 6.6 Chalcogen transmission. Reproduced from [34]. Copyright © 2017 Elsevier Masson SAS. All rights reserved.

Table 6.3 Refractive index (n), Abbe number, and thermo-optic coefficients (dn/dT) of currently commercially available chalcogenide glasses. All data at room temperature [35,36,37,38,39,40,41,42,43,44].

	n(λ)		Abbe Number		dn/dT
Glass (Commercial name)	4 μm	10 μm	3–5 μm	8–12 μm	(10^{-6}/K)
As$_2$S$_3$ (IRG7 / IRG27)	2.4169	2.3873	159	48	−3.2 (at 5 μm), −3.7 (at 10 μm)
Ge22As20Se58 (GASIR®1)	2.5100	2.4944	196	120	55 (at 10.66 μm)
Ge33AS12Se55 (AMTIR 1, IRG2 / IRG22)	2.5146	2.4981	202	109	86 (at 3.4 μm), 72 (at 10 μm)
Ge20Sb15Se65 (GASIR®2)	–	2.5842	–	101	58 (at 10.6 μm)
GeSbSe (AMTIR 3)	2.6216	2.6027	159	110	98 (at 3 μm), 91 (at 10 μm)
GeSbSe (GASIR®3)	2.6287	2.6105	170	115	53 (at 10.6 μm)
Ge20As12Se33Tes5 (IG3 / IRG23)	2.8034	2.7870	153	164	130 (at 4 μm), 145 (at 10.6 μm)
Ge10As40Se50 (IRG4 / IRG24)	2.6222	2.6090	30	175	21.1 (at 5 μm), 20.3 (at 10 μm)
GE28Sb12Se60 (IRG / IRG25)	2.6219	2.6030	173	108	62 (at 5 μm), 61.1 (at 10 μm)
As40Se60 (IRG6 / IRG26, MG-463)	2.7947	2.7781	169	160	33.4 (at 5 μm), 32.1 (at 10 μm)
GASIR®5	2.793	2.777	169	160	32 (at 10.6 μm)

but they are a promising future commercial material [45]. Design with IR-GRIN materials have been shown to reduce the number of optical elements as discussed in the presentation by Gibson et al. [46] where the surfaces can give you the optical power and the material can give you the color correction. This approach is analogous to achromatic IR singlets that utilize diffractive surfaces to create the inverse dispersion profile of the substrate material without the compromises encountered through the use of diffractive surfaces.

An effort was started in 2012 and is currently underway to standardize IR materials by a group comprised of the Optics and Electro-Optics Standards Council (OEOSC) and The International Society for Optics and Photonics (SPIE). The standardization of these IR optical materials is intended to be similar to those that exist for visible glasses as called out in ISO10110-18. The goal is to obtain grades and quality metrics for IR materials that designers, fabricators and users can consistently rely on to obtain optical elements that meet their needs. The two groups, OEOSC and SPIE, have two distinct roles. OEOSC "administers the development of American domestic standards in optics (ASC OP) and American participation in International Organization for Standardization or ISO technical committee for optics and photonics (ISO TC172)" [47], and their main support for this effort is the documentation and authoring of standards to be published at the ANSI and ISO standards levels. The SPIE portion of the group is to advise the ISO group with measurement information and to develop and vet characterization techniques. To date, National Institute of Standards and Technology (NIST) has supported this effort with high quality refractive index measurements of Germanium [48]. Other materials will be measured to not only provide the most accurate standard measurements of these materials but also give a deeper understanding of the sampled homogeneity of the refractive index throughout the material boules and batches. Additional information on this effort can be found in meeting documents related to OEOSC TF6 IR Materials [47]. At the international level, ISO standards are under development that address refractive index measurements (ISO/NWIP 6760-1, ISO/NWIP 6760-2, ISO/CD 21395-2, and ISO/DIS 21395-1), CaF_2 in the IR (ISO/DIS 22576:2020), chalcogenides (ISO/NWIP 17411), and IR spectral bands (ISO/NWIP 8424) for standardizing material properties. NWIP8424 is expected to be an ISO adopted version of ANSI/OEOSC OP1.007.

Mechanical properties Two mechanical materials have seen significant technical development since the year 2000: (1) Negative Thermal Expansion Alloys as discovered at Texas A&M in 2010 and commercialized by ALLVAR have opened up the athermalized optical design space; and (2) fine-structure aluminum manufactured by RSP Technologies allowing uniform material properties and visible image quality surface finishes used in reflective aluminum mirrors.

The first material is Negative Thermal Expansion (NTE) alloy technology. NTE technology was first discovered at Texas A&M University in 2010 and is only a decade old. The core technology relies on the union of known alloy chemistries and known alloy processes, which combined in a unique way produce a tailored macroscopic thermal expansion. During early alloy development, Cu, Fe, Ni, Co, and Ti-based alloys exhibiting negative thermal expansion were identified through work funded by the National Science Foundation (NSF). The company ALLVAR, a Texas A&M University spin off, was founded in 2014 to focus on scaling the production of these alloys. The first commercial products, Alloy-16 and then Alloy-30, were made available in 2017. The offerings of Ti-based Alloy-16 were thin sheets, plates and strips with −16 ppm/°C CTE at room temperature [49]. Through further material and process development of Alloy-30, the CTE magnitude was almost doubled to −30 ppm/°C CTE at room temperature and production shifted to rod, tube, and bar due to the demand for these form factors from the optics industry. NASA funded projects continue to develop the technology for use in space optics applications. The three key areas of development are: (1) NTE tubes for infrared optical athermalization, (2) zero CTE struts for reflective optics, and (3) NTE washers and spacers for constant force fasteners. Much of the current material development work is focused on producing alloys with greater NTE magnitudes, a more linear strain versus temperature response, and higher operating temperatures. Additionally, a material property database including mechanical, thermal conductivity, micro-yield, micro-creep, etc., is being built. Database development is a necessary step toward gaining wide acceptance of this young family of alloys.

While some ceramics, intermetallics and polymers, such as zirconium tungstate (ZrW_2O_8), lead titanate ($PbTiO_3$), carbon fibers, Fe_3Pt, and stretched rubber, can exhibit room-temperature NTE [50], ALLVAR Alloys are the first bulk metals to exhibit NTE. The brittle nature of ceramics and intermetallics, along with the low elastic modulus and mechanical strength of polymers, limit their use as structural members in mechanical systems. The ability to compensate for the natural Positive Thermal Expansion (PTE) of other structural materials using a strong and ductile metal offers a new athermal design paradigm.

Measures of thermal expansion must be discussed to better compare the thermal expansion properties of ALLVAR Alloys to traditional metals. Most engineering approaches use a single linear coefficient of thermal expansion (CTE), α, to model thermal displacements. While this works well for systems with loose thermal performance requirements, it does not capture the complexity of the true strain versus temperature response that can impact high performance optical systems. Figure 6.7a displays the strain versus temperature curve for various metals commonly used in optical designs while Figure 6.7b displays the instantaneous or tangent CTE defined as the first derivative of the curves in Figure 6.7a. The dashed lines correspond to publicly available data through the NIST Index of Material Properties [51] while the dotted lines correspond to data supplied courtesy of ALLVAR [52,53]. Aluminum 6061 (Al6061), A286 Stainless Steel (A286), Titanium 6Al-4V (Ti64), and Invar 36 (Invar36) all exhibit fairly constant positive instantaneous CTE values between −75°C and +80°C. However, large changes in instantaneous CTE are observed below −75°C. ALLVAR Alloy-30 exhibits more CTE curvature between −75°C and +80°C and a more

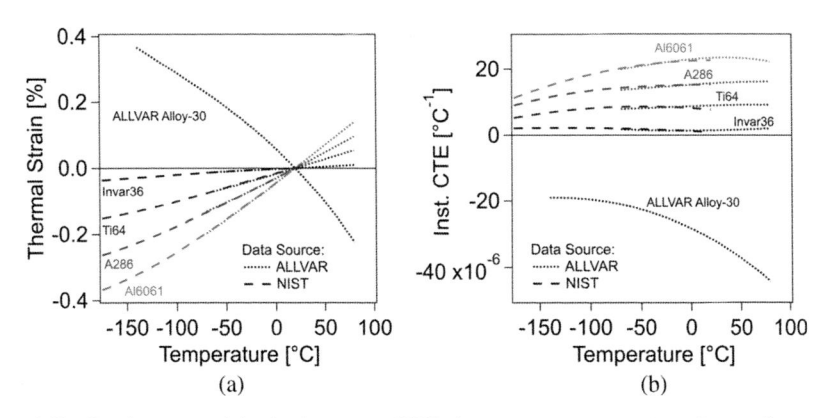

Figure 6.7 Strain, a, and instantaneous CTE, b, versus temperature for various metals courtesy of NIST and ALLVAR [51,52,53].

constant CTE below −75°C. These changes in instantaneous CTE are due to the higher-order nonlinear response observed in the thermal strain. The instantaneous CTE changes are most drastic for the largest CTE magnitude metals with ALLVAR Alloy-30 and Aluminum 6061 exhibiting larger CTE fluctuations than Titanium 6Al-4V or Invar 36.

To compare the highest CTE magnitude metals to high-CTE polymers commonly used in fixed focus infrared optics, Figure 6.8 displays the instantaneous CTE for Delrin®, Nylon 6, Aluminum 6061, and ALLVAR Alloy-30. While ALLVAR Alloy-30 exhibits more instantaneous CTE curvature than Aluminum 6061, it is much less than Delrin's® or Nylon 6's CTE curvature. Therefore, as an optic's operational temperature range increases and/or stability requirements become more stringent, the nonlinearity in the thermal strain should be taken into consideration beginning with Delrin 100 followed by Nylon 6, ALLVAR Alloy-30, Aluminum 6061, A286, and ending with Ti64.

Novel NTE ALLVAR Alloys will potentially have a huge impact on the optics community, but the magnitude and benefits of this technology are still being evaluated. Some obvious gains for fixed-focus infrared optics include reduced diameter and length by replacing switch-back athermalizing assemblies made with high CTE and low CTE components as displayed in Fig. 6.8 Specifically, a length savings of 40% was calculated by replacing the commonly used Aluminum 6061 low CTE and Delrin® high CTE switch-back assembly with a simple ALLVAR Alloy 30 tube in systems with −30 ppm/°C thermal focus shift. Interestingly, this 40% length savings was maintained for systems with thermal focus shift values below −30 ppm/°C by replacing the switch-back's Aluminum 6061 component with ALLVAR Alloy-30 and keeping Delrin® as the high CTE material. A surpriseing result from the same study was that a switch-back design made with Aluminum 6061 as the high-CTE material and ALLVAR Alloy-30 as the low-CTE material

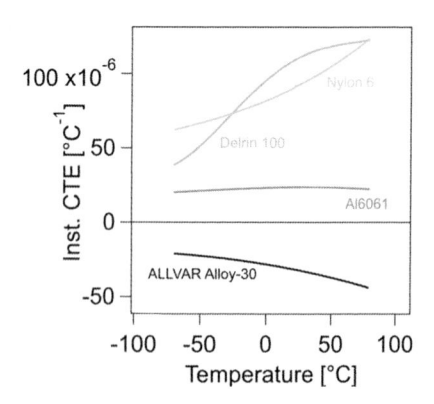

Figure 6.8 Instantaneous CTE versus temperature for various materials courtesy of ALLVAR [52,53].

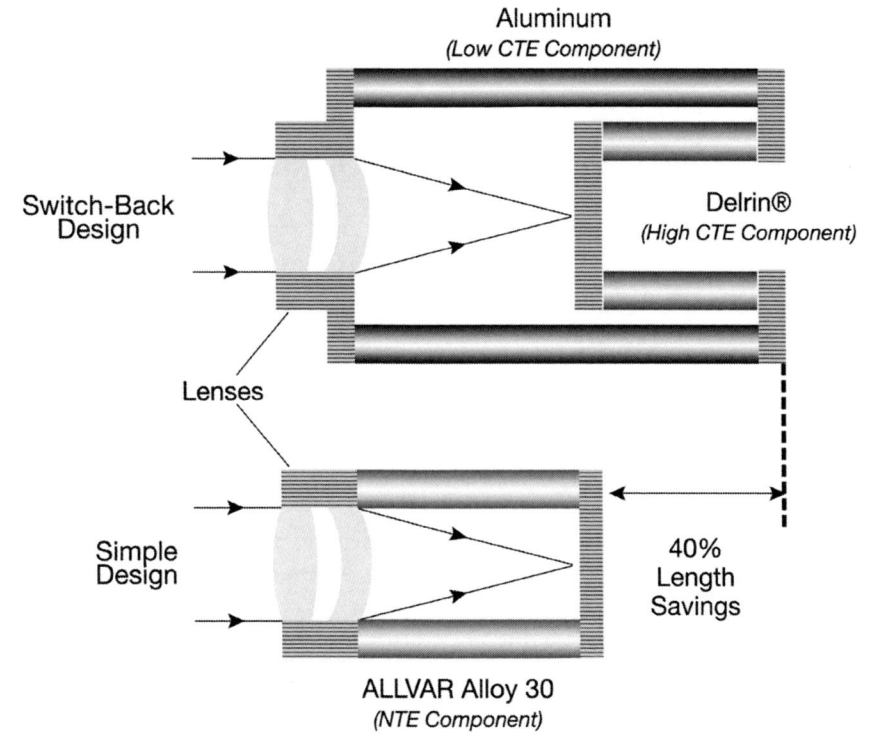

Figure 6.9 ALLVAR Alloys provides passive athermalization for IR optics. (Adapted from [54].)

produced a length savings down to −110 ppm/°C thermal focus shift when compared to the Aluminum and Delrin® switch-back.

This ability to make an all-metal athermalized switch-back design offers several potential advantages in addition to the length and diameter savings discussed above. For example, eliminating polymer components like Delrin® and Nylon 6 would improve an athermalizing assembly's mechanical and environmental stability by removing coefficient of moisture expansion effects, increasing strain versus temperature and instantaneous CTE linearity, and improving rigidity and impact resistance. Additionally, the tighter achievable machining tolerances and the drastically reduced batch to batch CTE variability of metals compared to polymers could improve manufacturing yield. This rapidly developing application space for NTE alloys will grow as more optical and mechanical designers and engineers add this emerging technology to their library of available materials.

The second mechanical material that has seen significant application advancement for optical systems is RSA aluminum made by RSP Technology which has been around since the 1990s (RSP stands for Rapid Solidification Process, and RSA stands for Rapid Solidified Aluminum). However, this material only began to see significant applications in optics starting in 2006.

Figure 6.10 Baffle mirror made from SPDT RSA-6061; RMS = 2nm (TNO, The Netherlands) [55].

These unique materials allow single-point diamond-turned aluminum mirrors with surface finishes that are acceptable for visible optical systems. For IR optical systems, the surface roughness that can be achieved (10 Å) can have significant impact on the optical performance from a scattered and stray light perspective. These materials are currently making an impact in high-end optical systems by allowing deterministic surface shaping and finishing of conventional and non-conventional (freeform) optical surfaces. This area has been further advanced through freeform optics and the metrology required to produce them.

RSP Technology produces and develops super alloys through a melt spinning production method which forms the basis for unique materials. The process reaches a cooling speed of more than 1M degrees/second. Because of this ultrarapid cooling technology, the liquid metal solidifies creating a super-strong alloy with a very fine homogeneous microstructure. This process is known as Rapid Solidification Process (RSP) [56]. The very fine microstructure allows production of mirrors or molds with a factor of >2 smoother surface finish and can allow the surface to a good finish without a coating [57]. The roughness of diamond turned RSA-6061 routinely is around 10 Å and can be even less with post-polishing.

The amorphous microstructure of RSA-9xx (RSA-905, RSA-902) is a further step for an even better surface finish, as well as improved properties (E-mod; CTE). Diamond machining results in typical surface finish of approximately 25 Å, but polishing techniques (such as Magnetorheological Finishing (MRF)) bring this further down to <5 Å. In addition, these amorphous alloys are non-heat treated, which offers further improvement of shape stability.

As a point of contrast, conventional Al-6061 achieves rms roughness >40 Å, while nickel-plated Al-6061 can achieve rms roughness to less than

Infrared Optical Systems 213

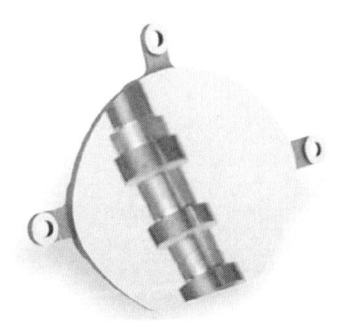

Figure 6.11 Head-up display mirror made from SPDT RSA-905; RMS = 1-2 nm (Sumipro, The Netherlands) [55].

Figure 6.12 Mirror substrate RSA-443 + polished NiP plating; RMS =1nm (Fraunhofer IOF, Germany) [55].

15 Å, but bimetallic effects between the plating and the aluminum substrate can have significant impact on surface figure error with temperature change.

When nickel-plating is chosen for the mirror surface, there is an RSP alloy which can be used as a substrate material, eliminating the bimetallic effect. Here AlSi40% (RSA-443) has exactly the same CTE as the nickel-plating eliminating this effect. In addition, such substrate materials offer high stiffness and low density.

A comparison of microstructure for conventional aluminum and RSP is shown in Fig. 6.13.

RSP fabrication steps are as follows and illustrated in Fig. 6.14: (1) the alloy needs to be prepared using melting and different alloying elements, (2) the melt is poured through a small nozzle onto a rotating copper wheel, creating a rapidly solidified ribbon, (3) this ribbon is chopped to flakes and (4) collected in a vessel, (5) the flakes are degassed and (6) subjected to hot

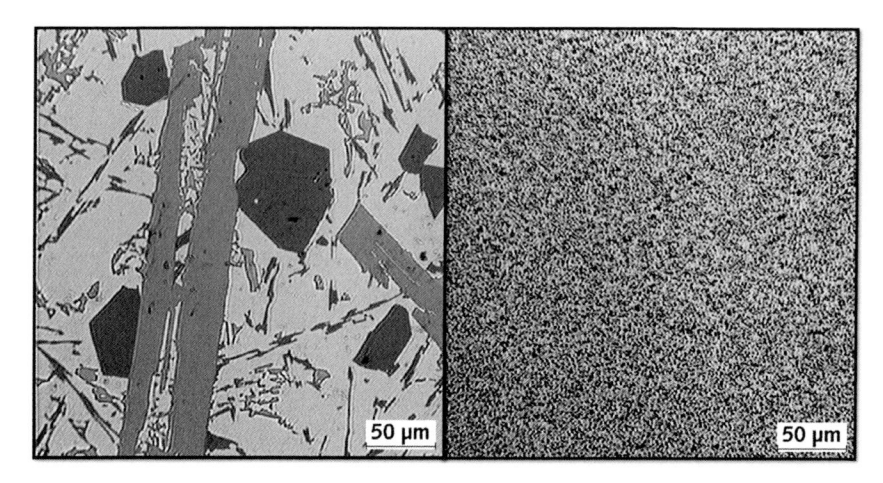

Figure 6.13 Left - Conventional aluminum microstructure. Right - RSP microstructure [55].

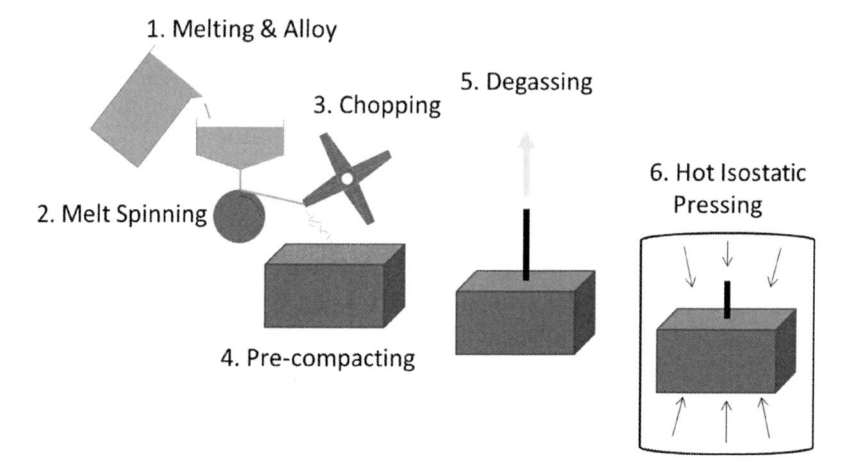

Figure 6.14 Processing steps in Rapid Solidification Processing [56].

isostatic pressure (HIP) processing to create a consolidated material. The resulting billets are up to 1 m in diameter, and the billets can be extruded or forged to different dimensions, as needed.

6.3.2 Infrared sources

As sensors (optics and detectors) improve, so do the sources to either test the sensors or act as a part of the sensor, i.e., active sensor.

6.3.2.1 Scene projector systems

"The development of very-large format infrared detector arrays has challenged the IR scene projector (IRSP) community to develop larger-format infrared emitter arrays." [58] IRSPs have seen the majority of their use in the IR sensor and seeker system performance calibration, and these systems appear to be driving the IRSP development. Advances in the development of these systems are increases in resolution with development systems getting to greater than a 2,000 × 2,000 emitter format with scene generation maximum temperatures reaching the 1,500-K equivalent temperatures in the MWIR band, corresponding to a 2.47 W/cm^2-sr radiance and running at framerates on the order of 500 Hz. The individual pixels are actually relatively small (48 μm) emitters able to achieve greater than 1,000-K temperature rise in less than 5 ms.

There has also been significant development in IR LED (light-emitting diode) based IRSP systems primarily focused on the MWIR band. These systems are utilizing 15-stage, Interband Cascade Light Emitting Diodes (ICLED) with a designed peak emission of 3.4 μm at 77-K operation. The arrays being developed implement IR LED pitches ranging from 24–32 μm with work showing a promising path to get to a 2,048 × 2,048 format running at a frame rate greater than 240 Hz. The work thus far shows an MWIR apparent temperature of greater than 900 K. While the IR LEDs demonstrated have to run at 77 K, there appears to be a path to get to higher operational temperature with similar device structures [59].

6.3.2.2 Quantum cascade lasers (QCLs)

Quantum cascade lasers, with their wavelength tuning ability, have undergone tremendous development and enabled many applications related to spectroscopy. They have elevated remote stand-off active spectroscopy for many applications including gas sensing and environmental monitoring. Besides the suitable wavelength range, QCLs usually feature a relatively narrow linewidth and good wavelength tunability, making them very suitable for many applications [60].

QCLs were first demonstrated at Bell Labs in 1994 by J. Faist and F. Capasso [61]. Stable, continuous operation was achieved in 2002 [62,63] and they demonstrated significant emission ranging from the SWIR [64] all the way into the THz range [65] Instead of interband transitions (conduction band to valence band), QCLs rely on intersubband transitions between the sublevels of quantum wells with a much smaller photon energy. In order to more efficiently utilize the electric power and generate a higher gain, a cascade of such transitions is used. QCLs are typically three-level lasers where electrons tunnel from the lowest level of one quantum well to the upper laser level of the next quantum well, contributing three laser photons.

While continuously operating, room-temperature devices are normally limited to output power levels in the lower milliwatt region, hundreds of milliwatts are easily possible with liquid nitrogen cooling. Even at room temperature, watt-level peak powers can be achieved using short pump pulses.

6.4 Acknowledgments

Parts of this research were supported by the optics consulting companies AMP Optics, LLC and Golden Gate Light Optimization, LLC. The authors would like to thank James Monroe of ALLVAR and Roger Senden of RSP Technology for their communications and support of this effort. Appreciation to A. R. for the Polish translation of [12]. In memory of Eugene Cross, "a true gentleman."

References

[1] American National Standards Institute/Optics and Electro-Optics Standards Council, "ANSI/OEOSC OP1.007-2020 – for optics and electro-optical instruments – optical elements and assemblies – infrared spectral bands," (2020).

[2] B. Lahiri, S. Bagavathiappan, T. Jayakumar, and J. Philip, "Medical applications of infrared thermography: a review," *Infrared Physics & Technology* **55**(4), 221–235 (2012).

[3] S.-R. Tsai and M. R. Hamblin, "Biological effects and medical applications of infrared radiation," *Journal of Photochemistry and Photobiology B: Biology* **170**, 197–207 (2017).

[4] J. Mudge and M. Virgen, "Real time polarimetric dehazing," *Applied optics* **52**(9), 1932–1938 (2013).

[5] T. P. Johnson, J. Sasian, and L. G. Cook, "Optical design using image distortion for orthorectification," *Applied Optics* **59**(22), G175–G184 (2020).

[6] T. Day, "The future of quantum cascade laser technology," *AZoMaterials* (2020).

[7] J. W. Goodman, *Speckle phenomena in optics: theory and applications*, Roberts and Company Publishers (2007).

[8] J. Mudge, "Wavelength selection approach for an incoherent optical detection sensor (lidar)," *Applied Optics* **59**(33), 10396–10405 (2020).

[9] "Cadillac DeVille DTS: Infrared camera gives DTS a new look," 2000. Autoweek. https://www.autoweek.com/news/a2127071/cadillac-deville-dts-infrared-camera-gives-dts-new-look/.

[10] "5 current car models with in-built night vision," (2019). Lanmodo. https://www.lanmodo.com/lanmodo-night-vision-system-insights-into-night-vision-gadgets/5-current-car-models-with-in-built-night-vision.html.

[11] A. M. Smith, M. C. Mancini, and S. Nie, "Second window for in vivo imaging," *Nature nanotechnology* **4**(11), 710–711 (2009).

[12] D. Mikulska, "Contemporary applications of infrared imaging in medical diagnostics," in *Annales Academiae Medicae Stetinensis*, **52**(1), 35–9 (2006).

[13] J. C. Leachtenauer and R. G. Driggers, *Surveillance and reconnaissance imaging systems: modeling and performance prediction*, Artech House (2001).

[14] "Lockheed receives up to $4.9 billion for Next-Gen OPIR satellites," (2021). Air Force Magazine. https://www.airforcemag.com/lockheed-receives-up-to-4-9-billion-for-next-gen-opir-satellites/.

[15] G. L. Smith, G. L. Peterson, R. B. Lee III, and B. R. Barkstrom, "Optical design of the CERES telescope," in *Earth Observing Systems VI*, **4483**, 269–278, International Society for Optics and Photonics (2002).

[16] N. B. Munir, J. Mahan, and K. J. Priestley, "First-principle model for the directional spectral absorptivity of gold-black in the near infrared," *JOSA A* **36**(10), 1675–1689 (2019).

[17] G. Mariani, M. Kenyon, J. Pearson, and W. Holmes, "Far-infrared room-temperature focal plane modules for radiation budget instrument," in *2016 41st International Conference on Infrared, Millimeter, and Terahertz waves (IRMMW-THz)*, 1–2, IEEE (2016).

[18] J. Transon, R. d'Andrimont, A. Maugnard, and P. Defourny, "Survey of hyperspectral earth observation applications from space in the sentinel-2 context," *Remote Sensing* **10**(2), 157 (2018).

[19] J. R. Janesick, *Photon Transfer DN→λ*, SPIE Press (2007).

[20] A. Rogalski, "Next decade in infrared detectors," in *Electro-Optical and Infrared Systems: Technology and Applications XIV*, **10433**, 104330L, International Society for Optics and Photonics (2017).

[21] J. D. Vincent, S. Hodges, J. Vampola, M. Stegall, and G. Pierce, *Fundamentals of Infrared and Visible Detector Operation and Testing*, John Wiley & Sons (2015).

[22] A. Rogalski and K. Chrzanowski, "Infrared devices and techniques," *Optoelectronics Review* **10**(2), 111–136 (2002).

[23] R. G. Driggers, R. H. Vollmerhausen, J. P. Reynolds, J. D. Fanning, and G. C. Holst, "Infrared detector size: how low should you go?," *Optical Engineering* **51**(6), 063202 (2012).

[24] M. C. Foote, "Temperature stabilization requirements for unchopped thermal detectors," in *Infrared Technology and Applications XXV*, **3698**, 344–350, International Society for Optics and Photonics (1999).

[25] P. W. Kruse, *Uncooled thermal imaging: arrays, systems, and applications*, vol. **51**, SPIE press (2001).

[26] "Athena 1920," (2019). BAE Systems. https://www.baesystems.com/en-us/download-en-us/20190319150448/1434628459234.pdf.

[27] A. Rogalski, "Comparison of photon and thermal detector performance," in *Handbook of Infra-red Detection Technologies*, 5–81, Elsevier (2002).

[28] A. Rogalski, *Infrared and terahertz detectors*, CRC Press (2019).

[29] A. Chynoweth, "Dynamic method for measuring the pyroelectric effect with special reference to barium titanate," *Journal of applied physics* **27**(1), 78–84 (1956).

[30] J. Cooper, "Minimum detectable power of a pyroelectric thermal receiver," *Review of Scientific Instruments* **33**(1), 92–95 (1962).

[31] E. Putley, "The pyroelectric detector," in *Semiconductors and Semimetals*, **5**, 259–285, Elsevier (1970).

[32] D. C. Harris, *Materials for infrared windows and domes: properties and performance*, vol. **158**, SPIE press (1999).

[33] S. Bayya, D. Gibson, V. Nguyen, G. Beadie, J. Sanghera, and M. Kotov, "Expanded ir glass map for multispectral optics designs," in *Advanced Optics for Defense Applications: UV through LWIR*, **9822**, 98220N, International Society for Optics and Photonics (2016).

[34] L. Calvez, "Chalcogenide glasses and glass-ceramics: Transparent materials in the infrared for dual applications," *Comptes Rendus Physique* **18**(5-6), 314–322 (2017).

[35] J. D. Musgraves, J. Hu, and L. Calvez, *Springer Handbook of Glass*, Springer Nature (2019).

[36] SCHOTT, "Infrared Chalcogenide Glass IRG 22," (2018). https://www.us.schott.com/d/advanced_optics/8240cca9-245b-41ec-8049-fd4c39279756/1.14/schott-infrared-chalcogenide-glasses-irg-22-english-us-11052017.pdf.

[37] SCHOTT, "Infrared Chalcogenide Glass IRG 24," (2018). https://www.us.schott.com/d/advanced_optics/53066181-5783-4eed-936b-313e8a6f67f3/1.13/schott-infrared-chalcogenide-glasses-irg-24-english-us-10042017.pdf.

[38] SCHOTT, "Infrared Chalcogenide Glass IRG 25," (2018). https://www.us.schott.com/d/advanced_optics/d91edc7a-ce5a-4d0a-a772-02278ec382c0/1.13/schott-infrared-chalcogenide-glasses-irg-25-english-us-10042017.pdf.

[39] SCHOTT, "Infrared Chalcogenide Glass IRG 26," (2018). https://www.us.schott.com/d/advanced_optics/866702a2-39e9-41a5-ac2b-c28e10f0fbff/1.13/schott-infrared-chalcogenide-glasses-irg-26-english-us-10042017.pdf.

[40] SCHOTT, "Infrared Chalcogenide Glass IRG 27," (2019). https://www.us.schott.com/d/advanced_optics/056534c4-9480-4cbe-a984-5a0d7afad333/1.0/schott-infrared-chalcogenide-glasses-irg-27-english-us-11042017.pdf.

[41] Umicore GASIR®1 https://eom.umicore.com/storage/eom/gasir1-for-infrared-optics-old.pdf.

[42] Umicore GASIR®5 https://eom.umicore.com/storage/eom/gasir5-for-infrared-optics-old.pdf.

[43] Redwave Glass, LLC MG-463 http://www.redwaveglass.com/.

[44] B. Gleason, "Designing optical properties in infrared glass," (2015).

[45] K. A. Richardson, M. Kang, L. Sisken, A. Yadav, S. Novak, A. Lepicard, I. Martin, H. Francois-Saint-Cyr, C. M. Schwarz, T. S. Mayer, et al., "Advances in infrared gradient refractive index (grin) materials: a review," *Optical Engineering* **59**(11), 112602 (2020).

[46] D. Gibson, S. Bayya, V. Nguyen, J. Sanghera, M. Kotov, C. McClain, J. Deegan, G. Lindberg, B. Unger, and J. Vizgaitis, "Ir grin optics: design and fabrication," in *Advanced Optics for Defense Applications: UV through LWIR II*, **10181**, 101810B, International Society for Optics and Photonics (2017).

[47] OEOSC. http://oeosc.org/.

[48] J. H. Burnett, E. C. Benck, S. G. Kaplan, E. Stover, and A. Phenis, "Index of refraction of germanium," *Applied optics* **59**(13), 3985–3991 (2020).

[49] J. A. Monroe, J. S. McAllister, D. S. Content, J. Zgarba, X. Huerta, and I. Karaman, "Negative thermal expansion allvar alloys for telescopes," in *Advances in Optical and Mechanical Technologies for Telescopes and Instrumentation III*, **10706**, 107060R, International Society for Optics and Photonics (2018).

[50] K. Takenaka, "Negative thermal expansion materials: technological key for control of thermal expansion," *Science and technology of advanced materials* (2012).

[51] NIST. https://trc.nist.gov/cryogenics/materials/materialproperties.htm.

[52] Private communications with James Monroe of ALLVAR ALLOYS.

[53] Data collection funded by NASA SBIR Phase II Contract 80NSSC19C0176 Ultra-Stable ALLVAR Alloy Strut Development for Space Telescopes, 2019.

[54] J. A. Monroe, J. S. McAllister, D. S. Content, and J. Zgarba, "Negative thermal expansion allvar alloys for smaller optics," in *Optical Architectures for Displays and Sensing in Augmented, Virtual, and Mixed Reality (AR, VR, MR)*, **11310**, 1131013, International Society for Optics and Photonics (2020).

[55] Private communications with Roger Senden of RSP Technology, Netherlands.

[56] R. ter Horst, M. de Haan, G. Gubbels, R. Senden, B. van Venrooy, and A. Hoogstrate, "Diamond turning and polishing tests on new rsp aluminum alloys," in *Modern Technologies in Space-and Ground-based Telescopes and Instrumentation II*, **8450**, 84502M, International Society for Optics and Photonics (2012).

[57] RSP. http://www.rsp-technology.com/.

[58] S. McHugh, G. Franks, and J. LaVeigne, "High-temperature mirage xl (lfra) irsp system development," in *Infrared Imaging Systems: Design, Analysis, Modeling, and Testing XXVIII*, **10178**, 1017809, International Society for Optics and Photonics (2017).

[59] D. T. Norton Jr, J. LaVeigne, G. Franks, S. McHugh, T. Vengel, J. Oleson, M. MacDougal, and D. Westerfeld, "Development of a high-definition ir led scene projector," in *Infrared Imaging Systems: Design,*

Analysis, Modeling, and Testing XXVII, **9820**, 98200X, International Society for Optics and Photonics (2016).

[60] R. Paschotta, *Field guide to laser pulse generation*, vol. **14**, SPIE press Bellingham (2008).

[61] J. Faist, F. Capasso, D. L. Sivco, C. Sirtori, A. L. Hutchinson, and A. Y. Cho, "Quantum cascade laser," *Science* **264**(5158), 553–556 (1994).

[62] M. Beck, D. Hofstetter, T. Aellen, J. Faist, U. Oesterle, M. Ilegems, E. Gini, and H. Melchior, "Continuous wave operation of a mid-infrared semiconductor laser at room temperature," *Science* **295**(5553), 301–305 (2002).

[63] "QCL primer: History, characteristics, applications." https://www.photonics.com/Articles/QCL_Primer_History_Characteristics_Applications/a58343#:~:text=Suris%20proposed%20the%20idea%20of,to%20confine%20electrons%20%E2%80%94%20quantum%20wells.&text=However%2C%20it%20 wasn't%20until, electron%20transitions%20within%20quantum%20wells.

[64] O. Cathabard, R. Teissier, J. Devenson, J. Moreno, and A. Baranov, "Quantum cascade lasers emitting near 2.6 μm," *Applied Physics Letters* **96**(14), 141110 (2010).

[65] C. Walther, M. Fischer, G. Scalari, R. Terazzi, N. Hoyler, and J. Faist, "Quantum cascade lasers operating from 1.2 to 1.6 thz," *Applied Physics Letters* **91**(13), 131122 (2007).

Adam Phenis is the CEO of AMP Optics, LLC with nearly two decades of optical design and analysis experience throughout the optical spectrum (UV through VLWIR). He graduated from the University of California, Davis with a B.S. in Optical Science and Engineering and M.S. from The University of Arizona's James C. Wyant College of Optical Sciences. His expertise lies in making optical systems across a wide variety of applications perform. He has specific experience in the IR with respect to designing and building these optical systems and getting into the fine details to ensure that the systems that he works on are successful. He started his career at Lockheed Martin's Advanced Technology Center in Palo Alto, CA and has worked at various companies on optical systems ranging from IR, free space optical communications, defense, aerospace, and semiconductor. He currently serves as the task force leader of ASC OP TF6 Infrared Materials and SPIE IR Materials Working Group, a Director of OEOSC, and is the ISO SC3/WG2 Convener. Adam is an SPIE Senior Member (2018).

Jason Mudge is a principal at Golden Gate Light Optimization, LLC an optical science consulting company located in San Francisco, CA – the heart of Silicon Valley. He has attended Foothill College (AS), University of California, Davis (BS & PhD), Stanford University (MS) and The University of Arizona's James C. Wyant College of Optical Sciences (MS). Jason began his career Lockheed Martin's Advanced Technology Center in Palo Alto, CA. His expertise lies in optical sensor-level performance simulation, analysis and data processing. Dr. Mudge has over 25 publications on topics ranging from imaging polarimetry to LiDAR to image quality and currently holds 4 utility patents. He is an SPIE Senior Member (2021).

Chapter 7
Polymer Optics

Robert Parada, Jr., Douglas Axtell, and Dan Morgan
Syntec Optics, 515 Lee Rd, Rochester, NY 14606, USA

7.1 Introduction

The field of polymer optics has expanded at an explosive rate during the period between 2000 and 2020. Reasons for this increased prominence include reduced cost, decreased weight, expanded geometry options, and simplified assembly. Numerous publications focus on the benefits of polymer optics in depth [1,2,3,4,5,6,7,8]. Since such detail is beyond the scope of this work, the summary in Table 7.1 is provided to give the uninitiated reader a sense for the advantages that polymer solutions offer over traditional glass optics.

The use of polymer optical components and systems is now several decades old. Their penetration into commercial and consumer products [9], medical instrumentation for surgical and diagnostic testing [10,11,12,13,14,15,16], and enhanced vision systems and armaments for defense and aerospace platforms [7,17,18,19,20,21] has grown significantly in the preceding two decades. Some recent representative examples are shown in Fig. 7.1. The diversity of applications continues to expand; the final part of this section discusses several of the emerging techniques and applications that form the frontier of polymer optics.

7.2 Polymer Materials

The workhorse materials fabricating polymer optics are polycarbonate (PC), polystyrene (PS), and acrylics such as polymethyl methacrylate (PMMA). In addition to these standard polymers, cyclic olefin polymers and copolymers (COP, COC) and polyethylenimine (PEI) are now in common use. More recent polymers now finding application in optics include high-refractive index options such as the Osaka OKP family (O-PET) and EP5000 (PC), and low-birefringence variants of existing polymers such as ZEONEX F52R (COP) and TOPAS (COC) – the latter also has advantageous transmittance

Table 7.1 Representative advantages and disadvantages of polymer optics.

Characteristic	Performance Relative to Glass Optics
Cost Reduction	The typical unit cost of a polymer optical system can be at least one order of magnitude lower than that of a corresponding glass optical system produced using traditional grind-and-polish techniques. While some specialized polymer materials are relatively more expensive, manufacturing method/time is a significant driver of overall cost.
Weight Reduction	Due to their lower specific gravity (or density), optics constructed from common optical polymers typically have a 2.5x to 3x weight advantage compared to similar glass optics.
Form and Size	Polymer optics frequently have geometries that are difficult to achieve using glasses. Even-symmetry aspheres are common, with less symmetric toroidal and freeform surfaces gaining in use. Multi-element lenslet arrays are in widespread use, as are more delicate micro-structures such as diffractive elements. These options, which can greatly impact performance while reducing the number of required elements in a system, are more difficult to achieve in glass elements. While singlets with outer diameters (OD's) between 25–150 mm are most common, components as small as a few millimeters in diameter are now manufacturable in high volume.
Spectral Range	Polymer materials have a narrower spectral passband than glasses. A representative range of 450–900 nm is commonly cited. At the ends of this range, polymers absorb radiation. It is less common to use polymers for ultraviolet (UV) applications, and their use for applications beyond the near-infrared (NIR) region is likewise limited due to several absorption windows. New polymer materials are expanding the utility of polymer optics in both these spectral regions.
System Integration	The ability to manufacture alignment and mounting features directly into a polymer optical element can significantly reduce the complexity and time required to assemble a multi-element optic. This benefits overall cost and unit-to-unit consistency. In contrast, such features are usually manufactured as separate optomechanical elements in glass optics.
Unit-to-unit Repeatability	The use of high-precision Computer Numerical Control (CNC) machining and replicative molding methods allows polymer optics to achieve relatively narrow distributions in individual parameters as compared to glass optics. This can aid the designer in the tolerancing of designs, to desensitize them for expected unit-to-unit variability.
Environmental Performance	Polymer optics have mixed performance relative to glass optics in this category. Polymers like polycarbonate have relatively high fracture resistance, making them common choices for applications where safety is a factor; however, polymers are more prone to cosmetic damage due to their relative softness. Acrylics and some other polymers have proven to be biocompatible, making them suitable for implants; generally speaking, they have lower chemical resistance which limits their cleaning options and use in toxic environments. Thermal performance is another relative disadvantage of polymer optics: their Coefficient of Thermal Expansion (CTE) is 3x to 10x higher than typical glasses which can require the implementation of more complex athermalization and/or autofocus to avoid performance degradation over wide operating temperature ranges.
Birefringence	The inherent processes used to create polymer optics - whether precision machined from a bulk substrate or injection molded from liquified resin - results in a degree of internal stress in the final optic. While such inhomogeneities can be mitigated through proper annealing and process control, it cannot be eliminated completely in most cases. This can impact the performance of applications that are sensitive to polarization.

characteristics. Starting in the early 2000s, the COC/COP polymers offered birefringence and light transmission performance equal to PMMA, giving optical designers a polymer substitute for BK7 with a closer refractive index match than PMMA (1.53 versus 1.49) and thus reducing lens curvature and

(a)　　　　　　　　　　　(b)　　　　　　　　　　　(c)

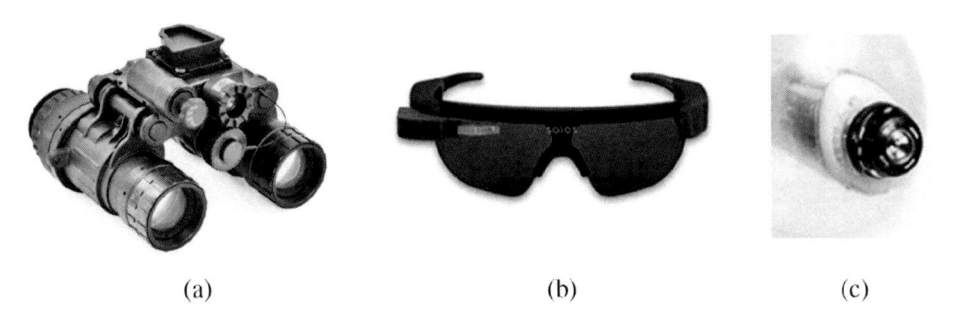

Figure 7.1 Examples of diverse products employing polymer optics. (a) Binocular nightvision goggles use polymer elements to reduce weight [22]. (b) Augmented reality glasses use polymer components to bend the light path [23]. (c) Disposable colonoscope scanners use polymer parts to reduce cost [24].

mass. The Osaka polyester copolymers (OKP family) offer lower birefringence than COC and PMMA with increased refractivity (1.65 versus 1.49) at the tradeoff of a higher material cost. Numerous sources are available that cite the basic optical, mechanical, and environmental properties of these and other optical-grade polymers [3,25,26,27,28,29,30]. Table 7.2 covers common properties. Figures 7.2 and 7.3 provide supplemental information.

Were one simply to focus on refractive properties, application of polymers in optical systems might seem to be rather limited. The range spans roughly from 1.49 to 1.66, which is considerably narrower than glass and crystalline materials. Nevertheless, this range is sufficient to design high-performance imaging and illumination systems for many applications. The relatively low density of polymers makes them attractive for applications where weight is a factor. While the lower refractive index of polymers means their geometries will be more curved than equivalent glass elements, this increase in element volume is offset by lower specific gravity. Another key distinction between glasses and polymers is the much lower transition temperature of polymers. While replicative molding processes have now been developed for certain glass types, polymers exhibit exceptionally good moldability traits which makes them well suited for large-volume needs in which multiple parts are simultaneously produced using a replicative molding process. This greatly reduces the unit cost of polymer elements. On the converse side, polymers have appreciably higher sensitivities to temperature changes, both in terms of their volume (CTE) and their optical power (dn/dt). This requires special attention during the optical design process, to minimize performance impacts when broad operating temperature ranges are required. The spectral transmission and absorption characteristics of polymers must also be considered when assessing their merit for a specific application. Polymers have good transmissivity across the visible and near-infrared spectra. Beyond this window, absorption effects may limit polymer utility. Polymer optics tend to have higher birefringence (internal stress) than similar elements made from

Table 7.2 Common properties of optical-grade polymer materials. Values collected from various online sources. Be sure to confirm precise values from the chosen supplier.

Material Type	PC	PC	PS	PMMA	COC	COP	COP	PEI	O-PET	GLASS
Manufacturer & Code	Lexan HF1110	MGC EP5000	STYRON P675W	Plexiglass VLD100	Topas 5013L	ZEONEX E48R	ZEONEX F52R	Ultem 1010	Osaka OKP4	Schott N-BK7
Glass Code	585.299	636.239	590.309	492.572	533.570	531.518	535.560	658.180	607.270	517.640
Specific Gravity (g/cm^3)	1.20 (ASTM D792)	1.24	1.04 (ASTM D792)	1.19	1.02 (ISO 1183)	1.01 (ASTM D792)	1.01 (ASTM D792)	1.27 (ASTM D792)	1.22	2.51
Glass Transition Temperature (T_g, °C)	126	145	100	92	134	139	156	217	121	557
Coefficient of Thermal Expansion (CTE) ($\times 10^{-6}$/°C)	66–70	66	60–80	67	60–70	60–70	60–70	45–55	74	7.1
Temperature Coefficient of Refractive Index (dn/dt) ($\times 10^{-5}$/°C)	−12 to −14	−12 to −14	−12	−8.5	−10	−8	−8	−13	−13	3
Transmission (3.174 mm thickness)	85–91%	89%	87–92%	91%	92%	92%	92%	36–82% (golden)	85–92%	91%
Water Absorption	0.1% (ASTM D570, 23°C, 24hr)	0.07% (JIS K7209, 24hr)	0.2%	0.30% (ASTM D570, 24hr)	0.01% (ISO 62, at Saturation, 23°C)	<0.01% (ASTM D570, at Equilibrium)	<0.01% (ASTM D570, at Equilibrium)	0.25% (ASTM D570, 24hr)	0.44%	<<0.01%
Birefringence	20	5	50	20	20	20	10	No Data	49	2
Hardness	M70	M70	M90	M94-97	M89	M89	M89	M109	No Data	570 Knoop
Relative Cost	$$	No Data	$$	$	$$	$$$	$$$	$$$	$$$$$	$$$

Polymer Optics 227

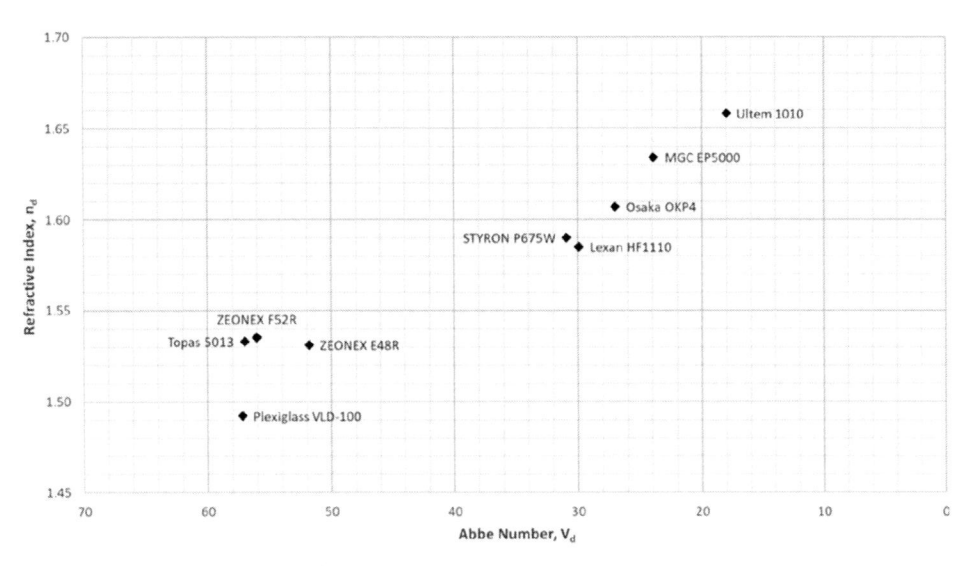

Figure 7.2 Glass map of selected optical-grade polymers.

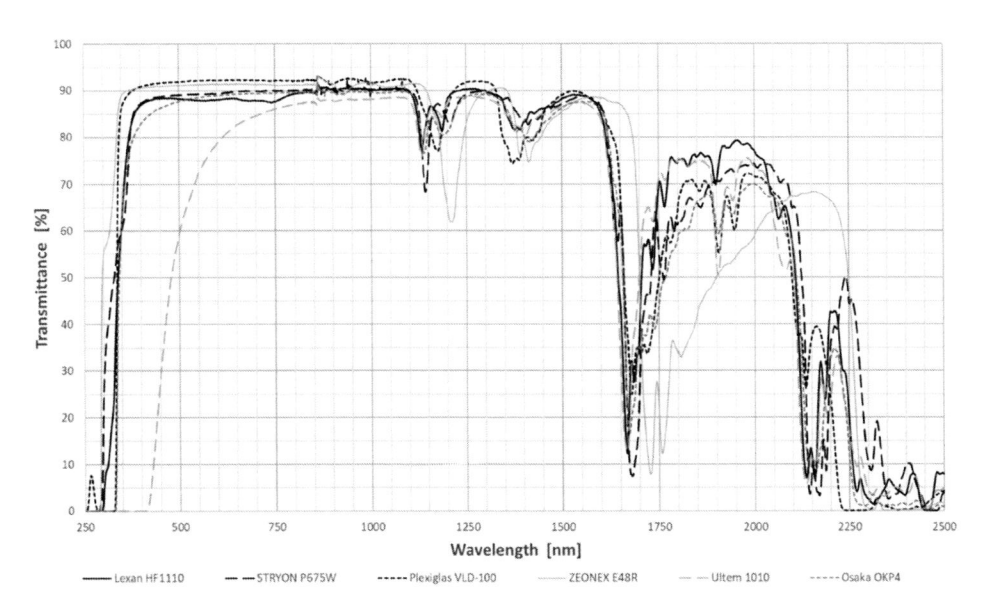

Figure 7.3 Percent transmittance of selected optical-grade polymers. The data shown were measured on a PerkinElmer Lambda 1050 spectrophotometer using 2-mm-thick molded witness pieces and includes surface reflectance losses.

glass. While this effect can be mitigated to a certain extent, it can impact optical performance if unaddressed. Polymers also have lower hardness than many glasses, making them less susceptible to fracturing at the cost of increased cosmetic defects; protective coatings are therefore worthy of consideration.

Trends in these many characteristics may be seen by focusing on the more recent additions to the stable of optical grade polymers: MGC EP5000, Topas 5013L, ZEONEX F52R, Ultem 1010, and Osaka OKP4. OKP4, EP5000, and Ultem 1010 have increased the range of refractivity above 1.60, thereby expanding design space. Ultem 1010 has the additional benefit of increased hardness, making the polymer suitable for harsh operating environments experienced by defense and aerospace applications – with the obvious tradeoff that the polymer has a gold hue making it less applicable for high-accuracy color imaging. Topas 5013L and ZEONEX F52R provide significantly lower birefringence, which is important for imaging applications – particularly those involving the use of polarized radiation sources.

Other important frontiers for polymer development involve broadening the range of spectral transmissivity, manufacturing composite materials, and reducing temperature sensitivity. Variants of common polymers such as Solvay Solef® 1010, Rowland Technologies Solatuf®, AcryLite A100 and Teflon Tefzel™ HT-2183 now provide increased transmissivity in the ultraviolet (UV) spectrum [31,32] between 300–400 nm; polymers that transmit more in the infrared (IR) spectrum likewise are under development [33,34]. Advances also are being made in the production of gradient index materials [35,36,37,38,39,40,41,42]; their enablement for commercial use will require development of an ecosystem of design and manufacturing processes and tools. The CTE of polymers remain an order of magnitude greater than common glasses, leading to performance deterioration across wide operating temperature ranges; this disadvantage deserves further study [43,44,45,46].

Silicone optics are also worthy of mention. Silicone offers an advantage where extreme ranges of temperature and or UV exposure rule out most other moldable polymers. Optical silicones are available from multiple large suppliers and are 90–95% transmissive from 250–1700 nm. Most optical silicones work to about 200C which is 100C higher than PMMA. Optical silicones have been environmentally tested for accelerated UV exposure and have withstood 30-year continuous exposure with no significant color change. Optical silicone has a relatively low refractive index (1.33 to 1.55) compared to other polymers. The lowest-index silicones mitigate the need for broadband antireflection coatings required by most high refractive index polymers and glass types, further reducing cost. Silicone also is not affected by environmental exposure and can be molded directly against a LED or camera module, sealing the system as well as being the first lens or only lens in the system.

7.3 Optical Design Considerations

In this section, we focus on the aspects of the optical design process that are most central to the use of polymers. While many of the common texts on instructing about optical design [47,48,49,50,51,52,53,54] include some mention of polymer elements, there are few whose material is dedicated to the subject. Of notable exception are two works published by Schaub [1,2]. In addition to covering basic design principles, these works provide an overview of the benefits and challenges of using polymer components in an optical system.

A key difference in the design process when using polymer materials is the relative absence of detailed information about the characteristics of these materials as compared to common glass types. Optical design software packages [55,56,57,58,59,60] come supplied with a limited number of polymer types. It is the responsibility of the designer to seek out the design parameters associated with less common polymers. The designer often must directly discuss these details with the material vendor and their selected manufacturing partner to ensure accurate parameters are being used in the final design iteration. Refractive index values are available at the 0.001 level of precision, and only at discrete wavelengths, which is coarser than desired for designs that have tight manufacturing tolerances. In the most sensitive cases, the designer may partner with a manufacturer to explicitly measure these parameters using refractometers and ellipsometers. Even the very process of manufacturing may subtly influence these traits. For example, the use of a high-refractive-index polymer such as OKP-1 requires an understanding of whether a small-volume manufacturing method such as diamond machining is to be used or whether a mass-production method such as injection molding will be employed.

One of the most significant benefits to the use of polymers in an optical design is the ability to incorporate relatively complex surface geometries. The incorporation of even-symmetry aspheres is now commonplace and has the benefit of mitigating aberrations while reducing the number of elements (and therefore cost, size, and weight) in the design [61,62]. The direct manufacture of diffractive elements on a polymer optic can assist with aberration control – albeit with the need to pay close attention to the impact on light transmission efficiency. The optic assembly process benefits from the ability to incorporate custom shapes and geometries. It is commonplace for the design of a polymer element to include custom flanges and/or interlocking features that reduce the number of required optomechanical components and simplify the system-level alignment process. Figure 7.4 shows an example of such mechanical features.

Polymers have CTE values significantly higher (3x to 10x) than those of common glasses, leading to greater changes in optical performance as the operating temperature range increases. Greater care may be required to athermalize an optical design employing polymer components. The physical performance of the finished optic may be affected by changes in stress in a bonded polymer-glass doublet. To alleviate some of these performance degradations, it can be prudent to include an annealing step at the appropriate

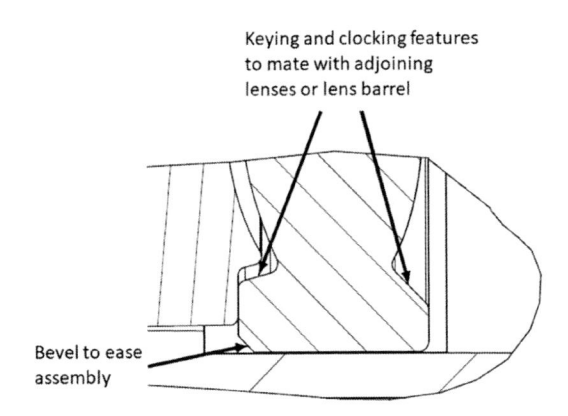

Figure 7.4 Example of a polymer optic with built-in optomechanical keying and clocking features molded into the polymer lens element.

point of the manufacturing process. This reduces the magnitude of internal stresses and inhomogeneities in the final optic. This will be discussed further in a later section; however, it is important for the designer to understand the specification of such steps may be required in their documented design. Further, an autofocus system could be employed to mitigate polymer optical system performance degradation with temperature changes.

One highly active frontier in optical design that involves polymer elements is the creation of freeform optical systems. The departure of these designs from common symmetries in most optical systems requires the development of new tools and macros for use with existing optical design software packages, and reliable, high-accuracy methods for characterizing their physical form and optical performance [63,64,65]. The use of gradient index materials likewise necessitates new infrastructure to simulate performance, manufacture, and measure optics composed of them [35,36,37,38,39,40,41,42].

7.4 Small-Volume Manufacturing

The standard manufacturing technique for small-volume polymer optics is diamond machining, also called single-point diamond turning (SPDT). SPDT is an ultra-precision lathing operation that uses a diamond-tipped tool to carve a desired geometry into a polymer blank. The sophistication of the hardware and software driving this manufacturing method have advanced considerably over the past decade [66,67,68,69,70,71]. The availability of machines with fifth-axis capability and associated control software have enabled the manufacture of novel asymmetric optics, such as microlens arrays and freeform prisms. These shapes can be achieved by driving the equipment in raster mode at nanometer-level accuracy.

The manufacture of a diamond-turned optic depends on several factors. Optical prescriptions with steep slopes and recessions require the design of custom tip geometries, increasing the cost and lead time to complete the work [72,73,74]. Such geometries can also entail complex spindle driving; the simplest optics require two-axis motion (so-called 'x' and 'z'), whereas peaked and asymmetric prescriptions can require off-axis tool positioning and non-rotational blank movement [75,76,77,78,79]. The machine tool path is simulated in advance of the manufacturing cycle to confirm the proper tool and blank positioning strategy have been selected. As designs have required more and more precise surfaces – both in terms of geometry and surface roughness – the design of custom fixturing has become key to the success of a given program [80,81,82,83]. Fixture development is therefore often a closely guarded, proprietary aspect of diamond machining.

The polymer material itself has an important impact on the machining process. Almost all polymers require some degree of preparation to produce an optical quality component. At a minimum, this includes a stress relaxation cycle with an appropriate annealing profile. In many cases, the fabrication of a near-net-shaped blank is also warranted. This reduces the cycle time for the final diamond machining, the costliest part of the manufacturing process. In addition, it can have a significant impact on the final accuracy and smoothness of the optic: the use of a high-precision CNC machine (multi-axis mill or lathe) can reduce the diamond-turning stage to a skimming operation in which minimal material needs to be removed from the blank to achieve the desired prescription. The type of polymer also plays a role in the manufacturing process. For example, most optical designers understand that polycarbonate is a relatively gummy substrate for which it is extremely challenging to achieve acceptable surface roughness. It is common practice to substitute polystyrene for small-volume production needs, since the two materials have remarkably similar optical properties (for which the optical design can be adjusted), and polystyrene does not suffer from the same cutting problems. Experienced diamond-turning technicians develop their own proprietary process for a given polymer covering the tool path, spindle rotation rate, and material feed rate that optimizes the surface quality. In recent years, the availability of higher-refractive-index materials has led to the development of new customized protocols for their diamond machining [84]. Materials such as the OKP family of optical polymers still present challenges to even the most accomplished technicians; the levels of residual internal stress and surface finish can vary greatly depending on component geometry. Ultimately, the skill of the technician is the most important component in a successful machining process.

No single table of specifications can be stated to cover all diamond-machined optics. For the reasons listed above, the achievable specifications depend on a variety of factors that include material, geometry, diamond tool, fixturing, and experience. The figures listed in Table 7.3 are meant to provide a baseline for even-symmetric optics fabricated using common polymers. It is

Table 7.3 Representative manufacturing tolerances for diamond-machined even-symmetric optics. The reader is cautioned that achievable tolerances for any given manufacturing task are a function of multiple aspects of the production process including part design, mold design, material, cavitation, mold construction, and process control/optimization.

Specification	Lower-Cost Tolerances	State-of-the-Art Tolerances
Radius of Curvature	±1–2%	±0.15–1.00%
Irregularity	2–5 fringes	0.5–1.0 fringe
Surface Roughness	100 Å	60 Å
Scratch/Dig	60/40	40/20
Centration/Runout	±0.025 mm	±0.005 mm
Thickness	±0.050 mm	±0.005 mm
Diameter	±0.050 mm	±0.005 mm
Part-to-Part Repeatability	0.5–1.0%	0.03–0.05%

common practice for optical designers to work closely with their selected manufacturing partner to refine a given design to have realistic performance specifications.

The frontiers of diamond-machined optic manufacturing center on the advancement of the equipment used in the process, and on the variety of substrates requiring this manufacturing method. Industry-leading equipment manufacturers now offer fifth-axis workstations [66,67]. As hardware capabilities achieve parity from one manufacturer to the next, the sophistication of the control software will ultimately determine the quality of the optic produced [68,85,86,87,88]. The ability to run multi-day – and in some cases multi-week – cutting cycles in unattended mode will allow higher-complexity freeform optics to be manufactured cost effectively. On the materials side, the introduction of new high-refractive-index polymers [89,90,91,92,93] and the availability of gradient index materials [35,36,37,38,39,40,41,42] will require the development of specialized machining processes which are very likely to remain trade secrets that give manufacturers an edge over their competition.

7.5 Large-Volume Manufacturing

The manufacture of large-volume polymer optics centers on molding techniques. While the replicative molding methods for glass materials covered elsewhere in this text have certain similarities, the molding methods for polymers have their own intricacies.

Once the manufacturing volume for a given program scales significantly, it makes more sense to migrate production to a method that is more cost-effective. The exact definition of 'scales significantly' will depend on the particulars of the optic. Mass-production techniques for symmetric, spherical components may not be cost-justified until the volumes reach several hundred per annum. To the contrary, complex freeform optics may warrant investment in mass-production tooling at volumes closer to ten units, given that their diamond-machining cycles can run multiple days per surface. Other factors that come into play when deciding when to move to mass production include optical performance and lead time. The level of accuracy in surface form and irregularity achievable with diamond-turning techniques is often superior to mass-production methods; this may require an adjustment of the optical prescription with emphasis on looser tolerances. In terms of timing, the completion of a diamond-machining process typically ranges from four weeks for simple optics to twelve weeks or more for complex optics requiring customized tools, fixtures, and blanks. By comparison, the development of tooling for mass-produced optics typically ranges from eight weeks for simple optics to twenty weeks or more for complex optics requiring advanced tooling features and/or higher cavitation.

In most cases, mass production of polymer optics involves an injection molding process [1,94,95,96]. This process entails the liquification of optically pure pellets of a chosen polymer and its injection into a custom mold that results in the desired component geometry. The injection molding of optics differs from the molding of non-optical parts in several ways. In particular, the end goal of the process is to produce high-accuracy, isotropic components rather than driving the injection molding process to minimize cycle time. For this reason, the cycle times and associated cost of molded optics can be significantly higher than similarly sized non-optical parts. The molding tool is constructed with extreme attention given to the components that will produce accuracy in the optical surfaces; these parts of the tool are typically inserts that are diamond-turned to an optimal geometry. The cooling of the tool is similarly given special attention since the uniformity of the cooling process has a strong impact on the residual stress (and therefore uniformity) of the optic. Thermal control of the mold can also affect how the mold fills – ideally either by coating the entire surface of the mold before filling its interior volume or by filling the cavity in one advancing wavefront. Induction heating and/or cooling may be necessary to control the mold temperature as a function of the process state.

The manufacturing process begins with a review of the optical prescription and associated tolerance budget. This review is performed by a senior engineer often having decades of experience in their craft. Iterative conversations are held with the optical designer, to select the optimal approach to building the tool. These conversations cover a broad range of topics, from polymer variants, to injection gate location, to cavitation, to expected lifetime (total number of parts to be produced), to optical performance. They help to determine the proper molding press configuration (screw and barrel selection, required clamping force, etc.). Once a design concept has been established, the manufacturer will often perform mold flow simulations to assess how well the tool will fill with liquified polymer. Advanced simulation software tools such as Moldex3D [97] are now available to assist the tooling engineer with this task, which is an effective way to de-risk a tool development program. After these simulations prove out a design concept, it undergoes a detailed design cycle in which each component of the tool – mold base, optical inserts, cooling lines, etc. – is specified in a solid model.

There are often several types of mold bases that can be used to injection-mold a polymer optic. Choosing the right mold base can reduce cost and cycle time to fabricate the mold. Traditionally, many optics were molded in full-frame molds. To reduce cost, smaller mold frames such as Master Unit Die (MUD), OptiRound and Round Mate frames were innovated. Illustrative examples are shown in Fig. 7.5. The primary advantage of a MUD frame is that it remains clamped in the press once it is aligned with the injection nozzle; the smaller part-specific mold then can be rapidly inserted/pulled out of the MUD frame, thereby reducing the time to align and clamp successive molds in the same press. In the MUD system, the cooling lines are connected to the mold frame, and these must be disconnected from the mold prior to removal just as in the case of the full-frame mold. Round Mate molds were designed to have several sizes featuring a round insert with the cooling integrated into the mold frame. This allows removal of the portions of the mold containing the mold cavity or cavities (both cover and ejector) for rapid changeover to a different optic or service, or cleaning, or adjustment of the optic currently being molded.

Hot tips and hot runner systems are often used to shorten cycle time in molding by reducing the polymer thermal mass and thereby decreasing cooling time. This technology also allows the mold process to be more isothermal which can produce more uniform strain in the lenses and thus more uniform and, ideally lower birefringence.

There are many aspects to consider when developing an optimal molding process for an optic [98]. Optical molding technicians hone their craft over

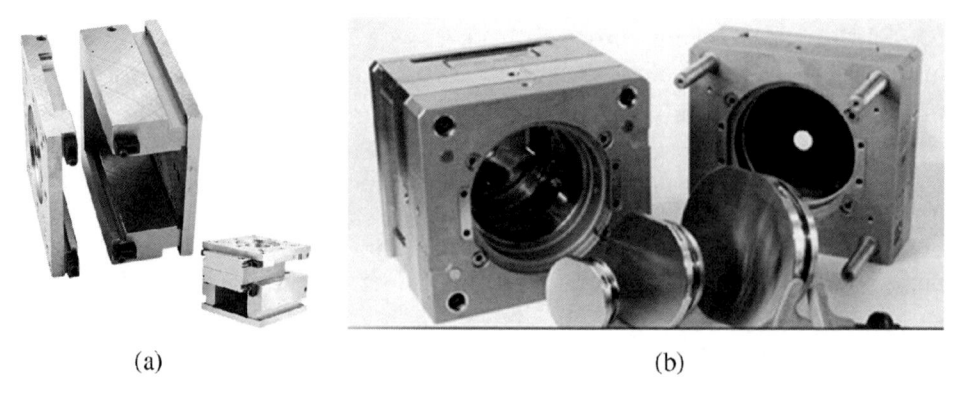

Figure 7.5 Form factors of smaller mold bases: (a) First-generation (DME) MUD Injection Mold Frame & Mold. (b) Round Mate Injection Mold Base & Inserts.

decades of experience. The following paragraphs summarize a few of the many challenging aspects of developing an optimal molding process.

When injection molding an optic, a key concept to keep in mind is that the polymer melt is compressible and elastic. Sometimes during the molding process, the screw bounces back. This happens as the melt decompresses when the pressure is released and is known as cushion. Cushion must be minimized to avoid shot-to-shot variations.

Injection speed is synonymous with shear rate; it can be used in conjunction with screw rotation speed to induce shear, thereby thinning the polymer to achieve low viscosity and fill the mold. While turbulent flow from high injection speeds is usually acceptable in the part sprue and runner, care must be taken to maintain laminar flow while filling the active optic region to avoid knit lines and the flow front leaving flowlines in the optic. These artifacts may only be visible in polarized light. Nevertheless, they can produce retardation gradients having several orders of magnitude, resulting in birefringence that is objectionable in all but the most rudimentary applications.

In conventional (non-optical) injection molding, polymer shrink is often controlled by adding non-polymer components for example glass strands or micro-balloons. This inhomogeneous material is not an option for injection-molded optics, so other methods of controlling shrink are required. One tactic is to increase the pack pressure. Another tactic is to change the sprue and gate geometry to facilitate rapid and continuous fill of the optic cavity. Expert optical molding engineers and technicians collaborate with their metrology teams to incrementally optimize the molding process for a given optic.

As was mentioned for SPDT optics, no single table of specifications can be stated to cover all injection-molded optics. The figures listed in Table 7.4 are meant to provide a baseline for even-symmetric optics fabricated using common polymers. Comparison to Table 7.3 shows that many of the specifications in the table below are looser than for diamond-machined optics. This is an important consideration when scaling production to higher-volume levels; sensitive designs with tight tolerances can sometimes not be scaled for injection molding, resulting in inflated product costs.

Frontiers in optical molding include pushing the limits of part size and geometry, as well as advanced tooling features and molding techniques. It is now commonplace for optics to be molded in sizes from 10–150 mm in diameter. The state of the art in molding micro-optics has advanced to the point where optics 2–3 mm in diameter are now possible [99,100,101,102,103]. Furthermore, while a component thickness-to-diameter ratio of at least 1:10 is preferred, optics having a thickness less than 0.5 mm and a diameter greater than 10 mm have been achieved [104]. Advances in the capabilities of molding presses and tool features are fueling the ability to achieve such challenging form factors.

Table 7.4 Representative manufacturing tolerances for injection-molded even-symmetric optics. The reader is cautioned that achievable tolerances for any given manufacturing task are a function of multiple aspects of the production process including, part design, mold design, material, cavitation, mold construction, and process control/optimization.

Specification	Lower-Cost Tolerances	State-of-the-Art Tolerances
Radius of Curvature	±2–3%	±1–2%
Irregularity	5–7 fringes	2–5 fringe
Surface Roughness	80 Å	40 Å
Scratch/Dig	80/50	60/40
Centration/Runout	±0.075 mm	±0.025 mm
Thickness	±0.125 mm	±0.050 mm
Diameter	±0.125 mm	±0.050 mm
Part-to-Part Repeatability	1–2%	0.5–1.0%

As advances in optic designs push the limits of imaging performance and compactness, the capabilities of injection molding presses and process equipment must continue to evolve to address these frontiers. A new class of micro-injection molding presses are emerging, with shot sizes down to 0.05 g. As the shot weights decrease with tighter tolerances in process control, the injection molding presses are moving from single-screw presses to screw-plunger or screw-piston machines. Some molds and/or presses may use injection-compression with vacuum to improve fidelity, for example, in Fresnel lenses and other diffractive elements. Injection-compression increases the packing pressure by using the hydraulic system of a press to compress a mold filled with molten polymer. Vacuum molding also warrants mention: shooting polymer into a vacuum eliminates the need to infinitely compress trapped gas to accurately reproduce minute features in the optic.

Hot tip systems are used to eliminate the sprue and runner in a mold, thus reducing the distance the molten material must flow in the mold before entering the cavity that forms the optical part. This also eliminates the colder resin that remains in the tip of the barrel from the previous shot ('cold shot'). The difference in density between the colder resin and hotter, fresh resin can produce increased birefringence and flow lines in the optics which can result in optics that fail to meet print specifications.

7.6 Metrology Considerations

A thorough overview of the many techniques and instruments used to measure the physical and optical properties of a manufactured optic is presented elsewhere in this text. This section focuses on how these methods are applied to metrology of components and systems made from polymer materials. Preferred methods do not differ much from those used for glasses, crystals, and metals. For example, surface form validation for polymer aspheres commonly involves the use of a profilometer. What is often distinct from measurement of a similar glass component is the sampling scheme and the use of non-contact options, where possible.

The sampling scheme for polymer optics tends to be based on volume – and thus on manufacturing cost. Expensive diamond-machined components with aspherical surfaces are seldom put through 100% inspection because they typically require contact metrology that can damage the optic, making it unsuitable for use in its final application. Given the extremely high level of repeatability from one unit to the next, a first-last measurement scheme is common. In this approach, the first and last optics produced in a manufacturing run are measured for all critical-to-function components, including those requiring contact methods. High consistency in the measurement sets for these two units helps to confirm the remaining units in the manufacturing run are within specified tolerances. Mass-produced optics have a significantly lower unit cost. Quality impact and metrology cost preclude 100% metrology; it is common to select a sampling frequency by which to ensure an active molding process continues to produce components meeting specifications. In many cases, a sampling rate of 2–3 per mold cavity per hour is representative of the sampling scheme for a mature molding process. The measured units are either discarded or kept as retained units for comparison during subsequent molding runs. In all cases, a professional manufacturer will document a clear metrology plan ('quality plan') that specifies the critical-to-function parameters to be measured, the method and instrumentation to be used for each measurement, and the frequency of each measurement. This plan may be reviewed with the customer, to ensure both parties agree on how the polymer optic is being certified.

The set of parameters requiring certification for a polymer optic can be extensive – and will certainly be application-specific. Optics to be used for imaging applications may require measurement of effective focal length or back focal length – and those involving polarized illumination sources may require special attention to the level of birefringence (internal stress) in the optic. Optics with molded diffractive elements to be used in illumination systems may have custom requirements on transmitted wavefront error and efficiency. In most cases, the validation of form and irregularity, surface roughness, center thickness, mounting feature dimensions, and cosmetic defects are required. For optics that receive an optical thin film coating, additional radiometric certifications are required. In other sections of this chapter, representative manufacturing tolerances for different methods of polymer optic manufacturing are given. In Table 7.5, common metrology methods for different parameters are shown, along with the level of precision each method can achieve. In addition to the intrinsic capabilities of the instrumentation, care must be taken to ensure the environment in which the instruments are used and the skill level of the operator all match (or exceed!) the requirements of the product under test.

Table 7.5 Common methods and capabilities for metrology of polymer optics.

Parameter	Instrumentation & Methodology	Equipment Capabilities
Form & Irregularity	Interferometers (e.g., Zygo Verifire™) for flats and spheres	1/20 λ
	Contact profilometers (e.g., Zeiss Surfcom, Ametek® Form Talysurf PGI) for even-symmetrical aspheres	±0.2 m
	Contact CMM's for aspheres and freeforms; non-contact profilometers (e.g., Panasonic UA3P Ultrahigh Accurate 3-D Profilometer) for even-symmetrical aspheres and freeforms	±10 nm
	Non-contact profilometers (e.g., Ametek® LUPHOScan, OptiPro UltraSurf 5X) for even-symmetrical aspheres and freeforms	±30 nm
Surface Roughness	White light interferometers (e.g., Zygo NewView™ 9000); contact and non-contact profilometers (e.g., Panasonic UA3P Ultrahigh Accurate 3-D Profilometer)	<1 nm
Mid-Spatial Frequency Errors	White light interferometers (e.g., Zygo NewView™ 9000); contact and non-contact profilometers (e.g., Panasonic UA3P Ultrahigh Accurate 3-D Profilometer)	<1 nm
Focal Length	EFL/BFL benches; MTF benches (e.g., TRIOPTICS ImageMaster® HR, Optikos OpTest® Lens Measurement System)	±0.5%
Mechanical Thickness	Drop indicators, laser indicators, and CMM's for center thickness and flange thickness	±0.5 nm
Tilt and Decentration	Air bearing spindle and laser (e.g., Opto-Alignment Technology LAS-P™, TRIOPTICS OptiCentric®)	tilt to 1 arcsec decentration to 0.1 μm
Cosmetics	Certified cosmetic standards/rulers (e.g., Gage-Line Reference Standard) corresponding to accepted specifications (e.g., MIL-STD-13830, ISO 10110, ISO 14997, ANSI OP1.002 Reference Standard)	limited by microscope resolution
Stress / Birefringence	Crossed polarizers (e.g., Strainoptics® PS-100 Polarimeter / Polariscope Systems) for subjective assessment; digital polarimeters (e.g., Axometrics AxoscanTM) for objective assessments	retardance 0.25 nm–6000 nm
Transmitted Wavefront Error	Interferometers (e.g., Zygo Verifire™)	1/20 λ
Coating Performance	Spectroradiometers for absolute transmittance and reflectance (e.g., LAMBDA 1050+ UV/Vis/NIR Spectrophotometer)	±0.1%

Significant advances have been made over the preceding decade in the ability to quantify the amount of residual texture imparted on optics – particularly those produced using precision machining methods. The performance of an optic may be particularly sensitive to textures within certain spatial frequency bands, giving rise to the term 'mid-spatial frequency' artifacts. There are now multiple options for measuring the magnitude of textures in these frequency ranges at high accuracy. Such metrology provides the SPDT technician the feedback needed to correct the issue – whether the cause is tool wear, machine cutting parameters, or environmental factors.

During this same period, the optics industry has been migrating to more quantifiable metrics for cosmetic flaws. This has been a beneficial transition for polymer optic manufacturing. Because polymer materials are more susceptible than glasses to surface defects and inclusions, leading to cosmetically compromised performance. Furthermore, the replicative nature of injection molding means that a scratch on the mold will produce the same cosmetic defect on every optic it molds. Depending on the design of the optic, some surfaces will have a higher sensitivity to cosmetic defects than others. It is helpful to have objective specifications and instrumentation for assessing these defects, such as side-by-side evaluation of digital microscope images to high-precision standards. Examples of cosmetic standards and direct comparison via digital microscopy are shown in Fig. 7.6.

Birefringence is another figure of merit of particular importance in polymer optics. Most optics produced from polymers have some level of internal stress. As polymer optics have proliferated into high-quality imaging applications – particularly those using polarized light sources such as some AR/VR headsets – the need to accurately quantify and mitigate such nonuniformities has increased in importance. Depending on the acceptance criteria for a given application, this assessment can range from a simple comparison to a known good standard to a statistical analysis of the entire clear aperture field using a digital polarimeter. Figure 7.7 shows an example of using a digital polarimeter to assess the internal stress of a polymer lens.

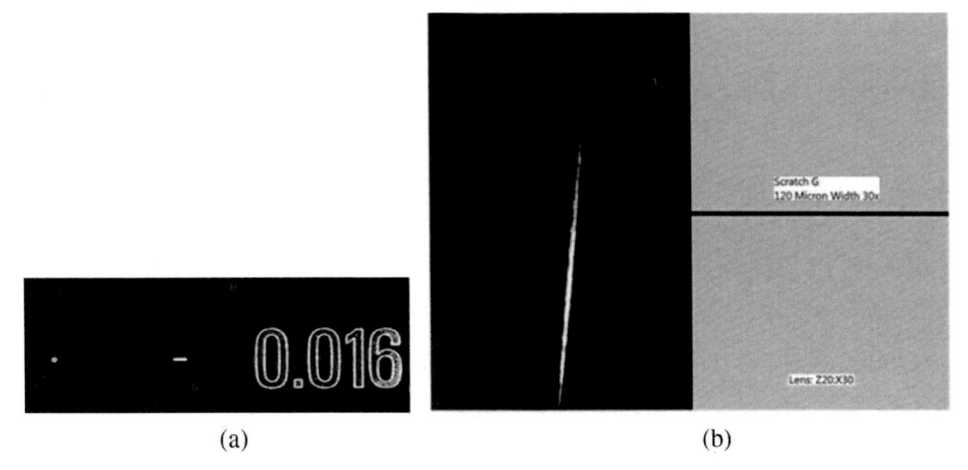

Figure 7.6 Example of modern cosmetic defect assessment in polymer optics. (a) Cosmetic defect on ISO 10110 standard from Gage-Line, Inc. The physical extent of both a 'round' and 'long' defect are shown next to the corresponding magnitude. (b) Side-by-side (left-to-right) comparison of a scratch defect in a microscope-imaged optic (left) and a cosmetic standard defect (right), both displayed at a magnification of 30X.

(a)   (b)   (c)

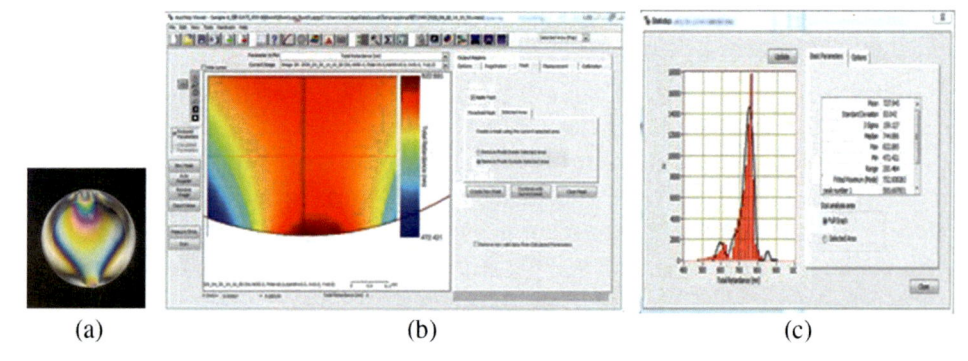

Figure 7.7 Example of modern birefringence assessment in polymer optics. (a) Image of a molded polymer lens analyzed by Axometrics AxoScan Mueller Matrix Polarimeter, with the gate oriented at the top. (b) Quantified stress (retardation) field in the region opposite the gate. (c) Statistics corresponding to the selected region of interest in the imaged part.

Such metrology provides the molding technician the feedback needed to optimize the molding process to diminish part stress.

One key trend in the metrology of polymer optics is the integration of metrology stations into the manufacturing cell. It is now possible for parts to be molded, degated, tested, and packaged without human intervention using robotics, machine vision systems, and programmable logic controllers (PLCs). This approach is of great interest in the healthcare and photonics industries, in which minimization of contaminants and unnecessary handling steps is important to achieve the performance and price targets of the program.

7.7 Ancillary Services

Polymer elements require an analogous set of ancillary services to create a completed system as other types of optics. Some of the techniques used to accomplish these tasks, however, are modified to suit the characteristics and limitations of the polymers used.

7.7.1 Gate vestige removal

For polymer optics mass produced using injection molding, there is a need to remove the gate vestige from the element. This is typically performed using either hand clippers or a heated clipper (for more substantial size and/or hardness). More recently, the option of using laser degating has gained in popularity. The benefit to this emerging option is that the degating process can be performed at the molding press. This can eliminate the unit-to-unit variability and potential impact on cleanliness of a human-performed task. As cleanroom molding expands in use, this means a fully degated part can be produced and packaged without leaving this environment. Laser degating can result in an exceptionally smooth surface around the gate vestige, which may

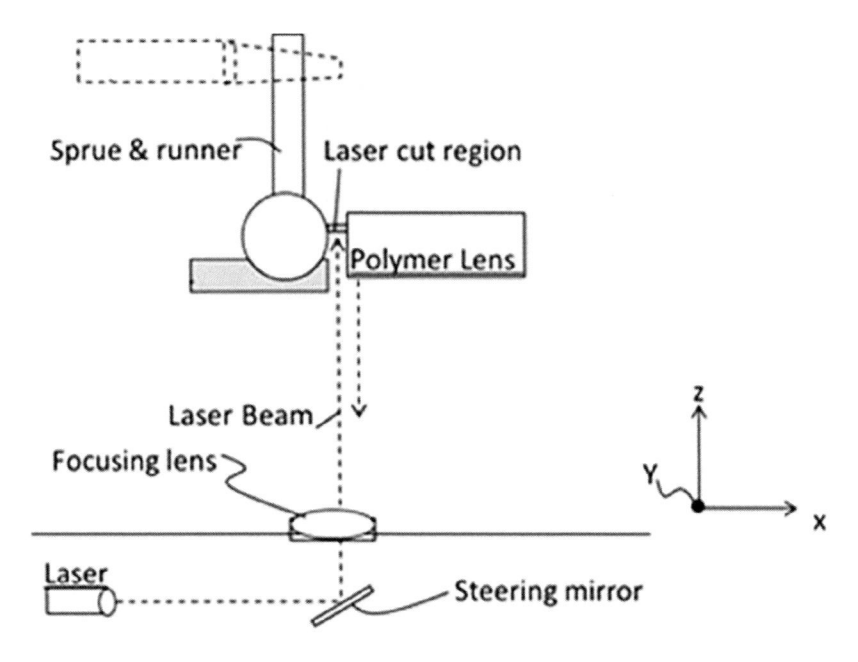

Figure 7.8 Schematic of laser degating of a polymer optic.

be an important criterion for selecting this method. The technology works best for polymers that are low in transmission at the laser wavelength, so that the lasing energy is contained at the cut location. This means changing a polymer when switching molds in a press may also require changing the type of laser. An example schematic of laser degating is shown in Fig. 7.8.

7.7.2 Optical thin film coating

Optical thin film coating is a common part of the polymer optic fabrication process. Whether for a high-reflectance surface, an anti-reflective surface, a beam-splitter, or other purpose, processes such as physical vapor deposition, chemical vapor deposition, and dip coating are directly applicable to polymer systems [105,106,107,108,109,110,111]. The key to success is the tailoring of the processes to the relatively low glass transition temperatures of most polymers. Polymers may be ultrasonically cleaned [112,113] to remove surface imperfections. In the physical vapor deposition process, the optical component may be pre-treated with an ionization process to increase adhesion of the coating. The coating process itself involves lower temperatures than typically employed for glass elements, meaning the resultant coating is generally less dense/compacted. The designer must accept looser performance tolerances than for glass elements, which can be an adjustment for some designers; the ability to achieve very narrowband performance is limited, since such performance often requires in excess of 20 distinct coating layers – some being

Table 7.6 Representative specifications for wideband coatings on polymer optics.

LightGray Parameter	Common Specifications
Radiometric Performance	**Wideband Visible Antireflection Coating:** Unpolarized reflectance ≤0.50% averaged across the spectral range 450–650 nm for incidence angles 0–25°. **Wideband Visible High-Reflection Coating:** Unpolarized reflectance ≥94% averaged across the spectral range 450–650 nm for incidence angles 0–25°, in front-surface geometry.
Durability & Environmental Performance	Per MIL-C-48497A, optical components cannot show signs of deterioration or abrasion after each of the following tests: • Adhesion: A 1/2" wide strip of cellophane tape is pressed against the coated surface and quickly removed • Humidity: Components are exposed to an atmosphere of 120°F 4°F and 95–100% relative humidity • Moderate Abrasion: Coated optical components are rubbed with a cheese cloth pad for 25 cycles (50 strokes) • Temperature: Components are exposed to temperatures of −80°F and +160°F for 2 hours at each temperature • Solubility and Cleanability: Coated components are immersed in ethyl alcohol and wiped with cheesecloth Pass/Fail Criteria: 60/40 scratch/dig specification
Clear Aperture	Rail fixturing will leave 0.5-mm strip of uncoated surface on opposite sides of part.

extremely thin (5–10 nm thick) [114]. Table 7.6 and Fig. 7.9 show common performance characteristics for the visible spectrum.

7.7.3 Alignment and joining

The alignment and joining processes for polymer optics leverage many of the same techniques as are used for non-polymer optical systems. Simple 'drop in' (e.g., 'snap-fit') approaches that make use of mechanical registration are applicable to lower-fidelity optics. Systems requiring a higher level of quality performance make use of active alignment. This involves the use of a laser source and spinning chuck to 'adjust the orientation of an optical element such that tilt and decentration are removed and the element's optical axis is aligned vertically. There are multiple instrument manufacturers that offer active alignment systems. Two well-known providers are TRIOPTICS [115,116,117] and Opto-Alignment [118,119,120]. Typical alignment tolerances for polymer optics are shown in Table 7.7. To avoid deforming the optics and introducing unwanted birefringence, discrete-point pressure approaches – such as threaded pins pressing directly onto the polymer component – should not be used. While a fuller exposition of active alignment techniques and state-of-the-art is beyond the scope of this work, the interested reader will find many suitable references on this topic [121,122,123,124].

Once properly aligned, the position of the element in the larger assembly must be fixed. The most common method is to use a mechanical bonding agent that is activated (cured) using ultraviolet radiation or heat. The

Polymer Optics 243

(a)

(b)

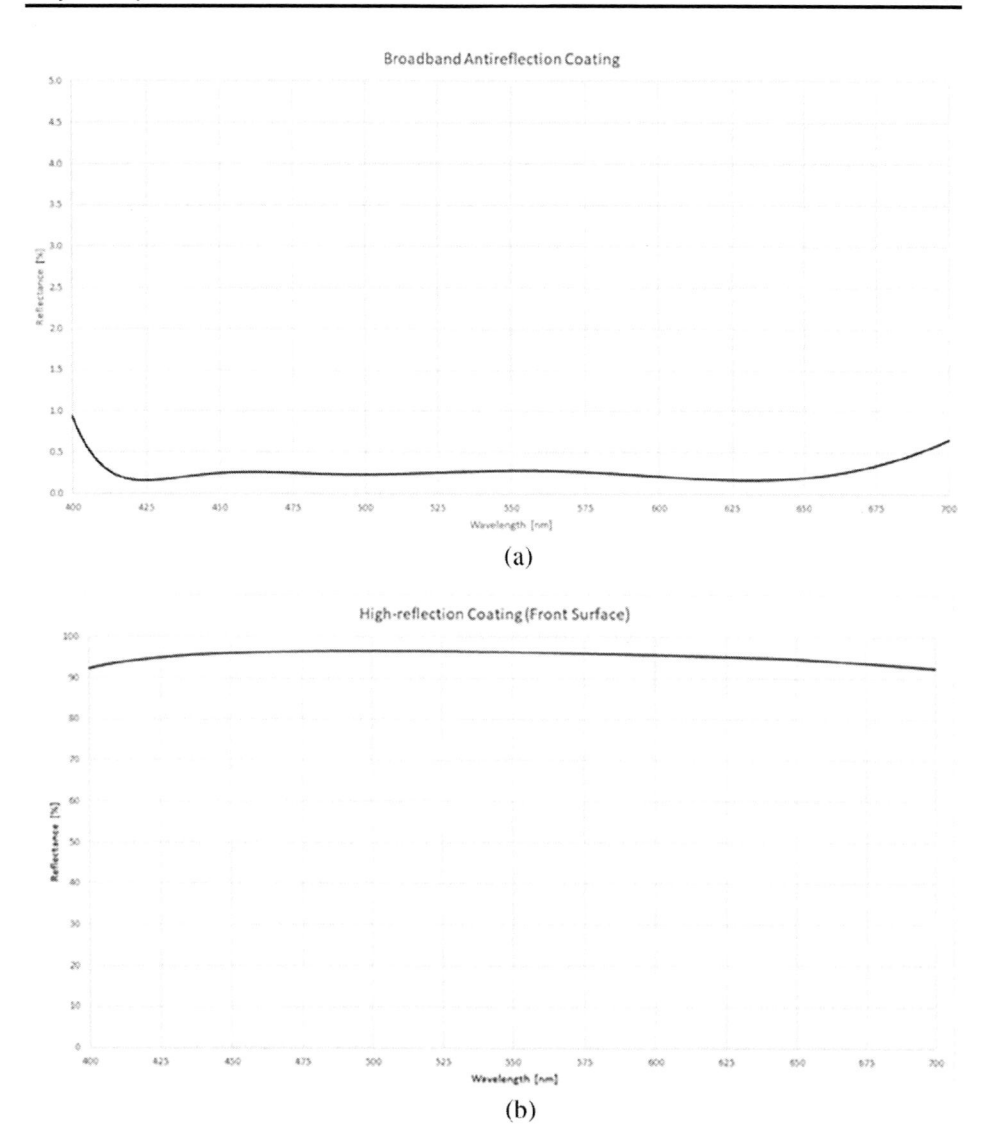

Figure 7.9 Representative spectral reflectance spectra for polymer substrates. (a) A broadband antireflection coating reflectance spectrum. (b) A front-surface mirror reflectance spectrum.

Table 7.7 Representative assembly tolerances for actively-aligned polymer optics.

Specification	Lower-Cost Tolerances	State-of-the-Art Tolerances
Axial Position	±10 μm	±1 μm
Tilt	±6 arcsec	±1 arcsec
Decentration	±10 μm	±0.5 μm

refractive indices of polymers are such that a variety of bonding agents exist that provide good index matching [125,126,127,128] to minimize unwanted surface reflectances.

In cases where a higher bond strength is desired – for example, in applications that will experience strong acceleration forces – a chemical bond [129,130,131] is preferable to a mechanical bond. A chemical bond can be achieved in multiple ways: one is by formulating the adhesive with monomers or oligomers that are in the same chemical family as the polymers in the optical element; another is by increasing the surface energy of the elements so that an ionic bond can supplement mechanical bond strength.

Polymer components are candidates for a wider array of bonding and joining techniques than glass, metal, or crystalline components. In systems in which two polymer components are being bonded together, sonic welding [132,133,134] and laser welding [135,136,137,138,139] may be viable options. Example instrumentation from one supplier of these technologies is shown in Fig. 7.10. These methods fuse the polymers of the elements to one another. This is particularly relevant for systems in which polymer optical elements are aligned and bonded within polymer optomechanical fixtures. In choosing the best bonding method, it is important to remember that polymer components have a relatively high CTE as compared to glasses and crystals.

7.7.4 Stress reduction

When assembling a hybrid system composed of multiple material types, it may be necessary to first de-stress the polymer components using an appropriate annealing process. An illustrative example of a basic annealing process for polymer materials is shown in Fig. 7.11. The relaxation ramp is more gradual

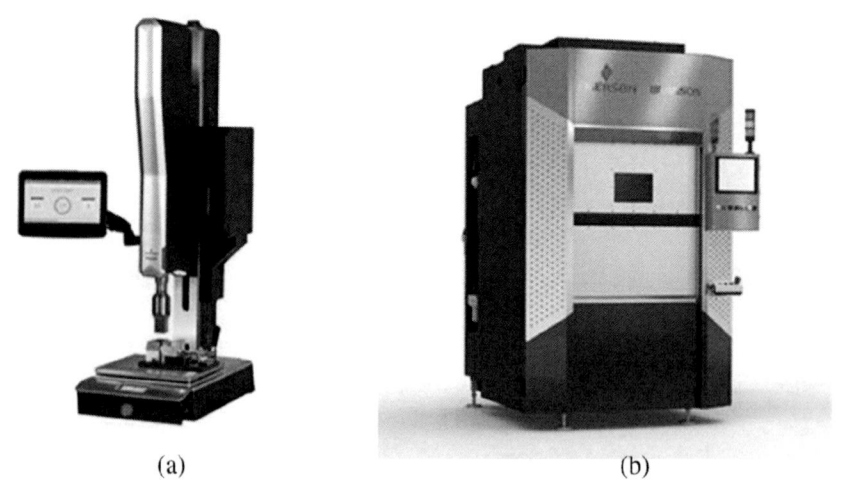

(a) (b)

Figure 7.10 Joining technology examples: (a) Emerson Branson GSX-E1 Ultrasonic Welder, and (b) Emerson Branson GLX-1.5 laser welding station.

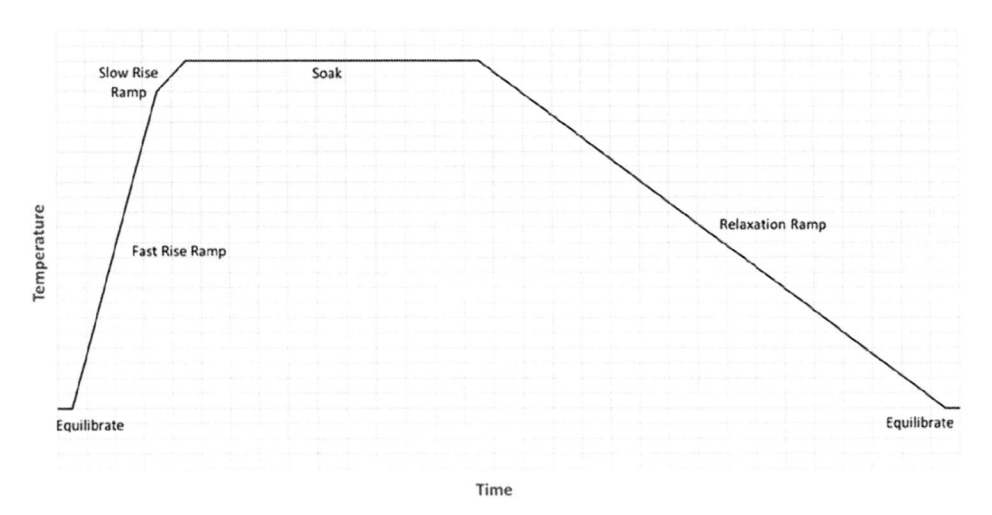

Figure 7.11 Schematic representation of a typical annealing cycle for optical-grade polymers.

than the rise ramp phase, so that internal stress has the time needed to be eliminated [140,141,142,143,144,145].

7.8 Emerging Techniques and Applications

In the preceding sections of this chapter, topic-specific areas of advancement have been mentioned. In this final section, a broader set of frontiers in polymer optics is discussed. These frontiers center on different application areas gaining prominence in the optics world.

7.8.1 Micro-optics

Polymers offer advantages in the diversity of component geometries that can be manufactured. Individual and bonded components with clear apertures less than 2 mm are now within the capability of many manufacturers. The mobile-phone industry is a particularly good example of the cutting edge in this trend: the associated sizes and tolerances at extremely-high volumes are pushing the metrology and manufacturing of polymer and glass optics to new limits. Industry leaders in manufacturing and metrology equipment are selling 80–90% of their latest-generation equipment to the Asia region where these products are currently being produced.

Polymer components also are finding their way into the designs of advanced medical systems that must operate effectively within the human body [10,11,12,13,14,15,16], sometimes being single-use ('disposable') modules that alleviate the cost and health risk of repeated sterilization. In the commercial market, polymer solutions are now in use in augmented reality / virtual reality (AR/VR) systems that power state-of-the-art entertainment and communication systems [146,147,148]. Many polymer optics applications are imaging in nature, requiring high levels of surface accuracy and alignment to achieve their promise. This implies that the infrastructure for the metrology and joining of such micro-optics must similarly improve. The challenge of manually assembling such small optics is nontrivial; achieving mass production can require the development of custom equipment and protocols [51,149,150].

7.8.2 Photonics

The trend in miniaturization is now continuing down to the wafer level. Photonics systems making use of optical components printed on a silicon component will power new solutions for optical computing and biomedical testing [151,152]. While the thermal properties of polymers make them an unlikely near-term candidate to be the materials used to manufacture these chips, their use as a means to couple radiation into and out of photonics components is an area of active research and experimentation [151,153]. Approaches range from more traditional lenslet arrays to more novel waveguides. An example of the latter is shown in Fig. 7.12.

7.8.3 Asymmetric form factors

The enablement of progressively more complex surface geometries is another area receiving heightened attention [147,154,155,156,157,158]. Design is a challenging aspect of the manufacturing process for freeform optics. Advancement in the capabilities of optical design tools is critical to the success of freeform production [63,64,65]. Likewise, techniques to measure form errors in freeforms are advancing, as is the precision of the instrumentation that will be used for such certification. Heads-up display

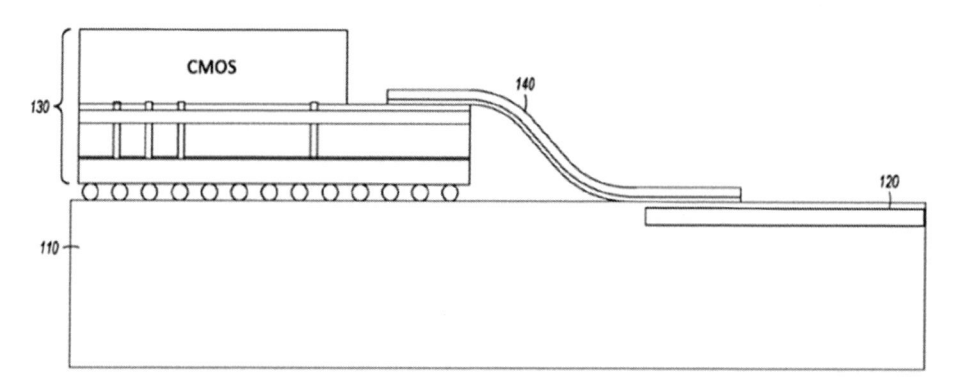

Figure 7.12 Example of an experimental polymer-based solutions for coupling photonics chips and components to bulk optics. Schematic from a waveguide patent application developed at the University of Arizona [153].

Polymer Optics 247

Figure 7.13 Examples of modern Heads-Up Display (HUD) and Head-Mounted Display (HMD) systems: (a) BAE Systems STRIKER® II helmet. (b) Elbit night-vision system. Future systems are likely to incorporate freeform components.

applications that seek to minimize form factor size and weight make use of such freeform components.

7.8.4 Novel materials

Another intriguing frontier in polymer optics is the area of novel materials and fabrication techniques, a few of examples of which are shown in Fig. 7.14.

As mentioned elsewhere in this text, the development of reliable, consistent methods for producing gradient-index materials is very active [35,36,37,38,39,41,42]. For these materials to be used in mainstream applications will require both the creation of optical design packages and scalable manufacturing processes for these layered composites.

The evolving ability to enprint nanostructures on polymer substrates offers interesting prospects for tailoring the bulk optical properties of a surface or element – including the elimination of unwanted surface reflections without the need for additional optical thin film coating [159,161,162,163,164,165]. These features may be fabricated through

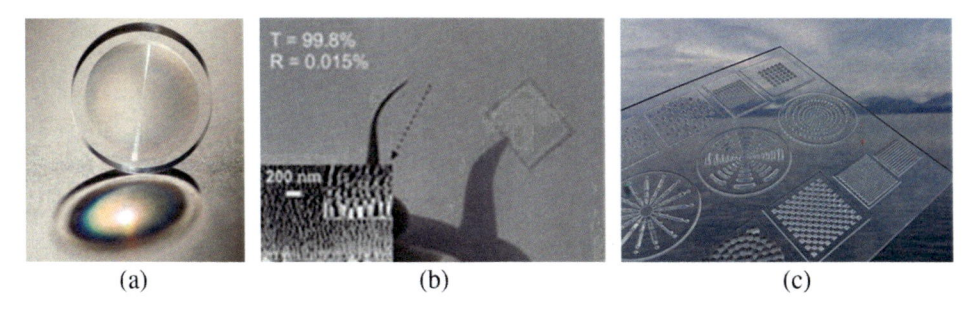

Figure 7.14 Examples of important frontiers in polymer optics: (a) A refractive optic produced from a gradient-index polymer [35]. (b) A nanostructure 'invisibility cloak' application demonstrating >99.8% transmittance across the visible and near-infrared spectral range [159]. (c) 3-D printed illumination optics [160].

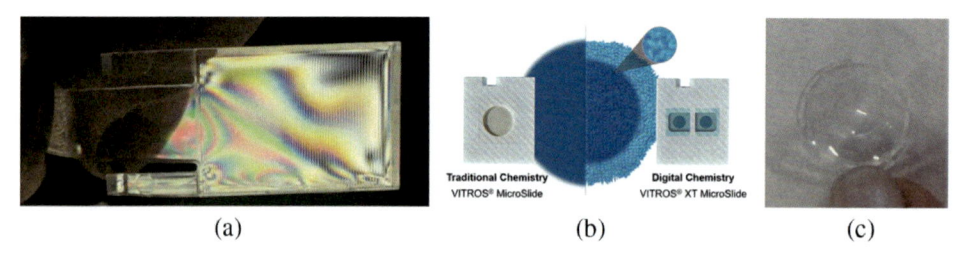

Figure 7.15 Examples of advancements in polymer optics for the healthcare market: (a) A disposable polymer surgical lightguide. (b) Polymer slide substrates used in microfluids testing. (c) An intraocular implant.

embossing operations, lithographic techniques, or newer options such as 3-D printing.

3-D printing also holds promise as a disruption technology for small-scale manufacturing of optical elements – particularly those having delicate features. Achieving this promise will require further advances in the homogeneity of the printed material; at the time this work is being published, use of 3-D printed optics is largely limited to low-fidelity applications such as basic illumination; the current level of wavefront/phase disruption makes such optics unsuitable for high-fidelity imaging systems [160,166,167,168].

7.8.5 Healthcare

Given recent concerns about global health issues such as COVID-19, proliferation of polymer optics in the medical field is to be expected. Properly sterilized single-use optics provide an intrinsically lower level of risk of infection during medical procedures than components that are reused. Low-cost polymer optical components are already in common use as surgical light guides and in microfluidics testing systems. As miniaturization trends continue, it should be expected that these components will find their way into higher-fidelity surgical imaging systems and human implants.

References

[1] M. Schaub, J. Swiegerling, E. Fest, A. Symmons, and R. Shepard, *Molded Optics: Design and Manufacture*, CNC Press, Boca Raton, first ed. (2011).

[2] M. Schaub, *The Design of Plastic Optical Systems*, SPIE Press, Bellingham, first ed. (2009).

[3] S. Baumer, *Handbook of Plastic Optics*, WILEY-VCH Verlag GmbH & Co. KGaA, Weinheim, second ed. (2010).

[4] Powell and Fisher, "Plastic optics: Polymer optics gain increased precision," (2007). LaserFocusWorld. https://www.laserfocusworld.com/optics/article/16552767/plastic-optics-polymer-optics-gain-increased-precision (accessed on 01 Oct 2020).

[5] Cavagnaro, "Polymer optics: Progress in plastic optics follows advances in materials and manufacturing," (2011). LaserFocusWorld. https://www.laserfocusworld.com/optics/article/16547849/polymer-optics-progress-in-plastic-optics-follows-advances-in-materials-and-manufacturing (accessed on 01 Oct 2020).

[6] V. Doushkina, "Advantages of polymer and hybrid glass-polymer optics," *Photonics Spectra* **44**(6), 54–58 (2010).

[7] P. Tolley, "Polymer optics gain respect," *Photonics Spectra* **37**(10), 76–79 (2003).

[8] W. Beich, "Specifying injection-molded plastic optics," *Photonics Spectra* **36**(3), 127–128 (2002).

[9] S. Fantone, "A user's guide to plastic optics," *Proc SPIE 0406: Optical Specifications: Components and Systems* (1983). https://doi.org/10.1117/12.935672.

[10] N. Sultanova, S. Kasarova, and I. Nikolov, "Optical polymers for laser medical applications," *Proc SPIE 1022: 19th International Conference and School on Quantum Electronics: Laser Physics and Applications*, 10226J (2017). https://doi.org/10.1117/12.2249843.

[11] N. Sultanova, S. Kasarova, and I. Nikolov, "Optical properties of plastic materials for medical vision applications," *J Phys Conf Ser* 398 (2012). https://iopscience.iop.org/article/10.1088/1742-6596/398/1/012030.

[12] V. Doushkina, "Polymer optics for thermally stable imaging," *BioPhotonics* **18**(1), 32–35 (2011).

[13] A. Méndez, "Optics in medicine," in *Optics in Our Time*, M. Al-Amri, M. El-Gomati, and M. Zubairy, Eds., Springer, Cham (2016). https://doi.org/10.1007/978-3-319-31903-2_13.

[14] N. Yu, F. Fang, B. Wu, L. Zeng, and Y. Cheng, "State of the art of intraocular lens manufacturing," *Int J Adv Manuf Technol* **98**, 1103–1130 (2018). https://doi.org/10.1007/s00170-018-2274-5.

[15] M. Maitz, "Applications of synthetic polymers in clinical medicine," *Biosurface and Biotribology* **1**(3), 161–176 (2015). https://doi.org/10.1016/j.bsbt.2015.08.002.

[16] C. Zeiss, *Handbook of Ophthalmic Optics*, Zeiss, Oberkochen, third ed. (2000).

[17] "Polymer optics for defense." GS Plastic Optics. https://www.gsoptics.com/defense/ (accessed on 01 Oct 2020).

[18] "Defense optics." Syntec Optics. https://syntecoptics.com/technologies/defense-military/ (accessed on 01 Oct 2020).

[19] "Defense optics." Edmund Optics. https://www.edmundoptics.com/knowledge-center/industry-expertise/defense-optics/ (accessed on 01 Oct 2020).

[20] "Defense optronics." Excelitas Technologies. https://www.excelitas.com/product-category/defense-optronics (accessed on 01 Oct 2020).

[21] "Common thermoplastic materials in aerospace & defense," (2017). AIP Precision Machining. https://aipprecision.com/common-thermoplastic-materials-in-aerospace-defense/ (accessed on 01 Oct 2020).

[22] "NVD BNVD night vision binocular – no gain." Night Vision Devices. https://www.nvdevices.com/products/night-vision/binoculars/bnvd-night-vision-binocular/ (accessed on 08 Dec 2020).

[23] "Solos smart glasses." Solos. https://www.solos-wearables.com/ (accessed on 01 Oct 2020).

[24] "Aer-o-scope disposable scanner (colonoscope component)." GI-View. http:/www.giview.com/crc-screening/aeroscope-colonoscope-overview.html (accessed on 01 Oct 2020).

[25] J. Lytle, "Polymeric optics," in *Handbook of Optics*, M. Bass, C. DeCusatis, J. Enoch, V. Lakshminarayanan, G. Li, C. MacDonald, V. Mahajan, and E. Stryland, Eds., 3.1–3.17, McGraw Hill (2010).

[26] H. Brinson and L. Brinson, "Characteristics, applications and properties of polymers," in *Polymer Engineering Science and Viscoelasticity*, H. Brinson and L. Brinson, Eds., 55–97, Springer, Boston (2008). https://doi.org/10.1007/978-0-387-73861-1_3.

[27] J. Brandup, E. Immergut, and E. Grulke, *Polymer Handbook*, John Wiley & Sons, New York, *fourth ed.* (2003).

[28] N. Sultanova, S. Kasarova, and I. Nikolov, "Dispersion properties of optical polymers," *Acta Physica Polonica A* **116**(4), 585–587 (2009).

[29] M. Mehr, W. Van Driel, F. De Buyl, and K. Zhang, "Study on the degradation of optical silicone exposed to harsh environments," *Materials* **11**(8), 1305 (2018). https://doi.org/10.3390/ma11081305.

[30] Zhang, "Polymer optics," (2007). College of Optical Sciences, University of Arizona. https://wp.optics.arizona.edu/optomech/wp-content/uploads/sites/53/2016/10/synopsis-for-polymer-optics-Rui-Zhang.pdf (accessed on 01 Oct 2020).

[31] "UV transmitting acrylic." EMCO Industrial Plastics. ~https://www.emcoplastics.com/uv-transmitting-sheet/#:~:text=While%20UV%20filtering%20acrylic%20is,short%20wavelength%20high%20frequency)%20light
(accessed on 01 Oct 2020).

[32] "Plexiglas acrylic resin - optical and weathering properties." Altuglas International – Arkema Inc. https://www.plexiglas.com/en/acrylic-resins/optical-and-weathering-properties/ (accessed on 01 Oct 2020).

[33] L. Anderson, T. Kleine, Y. Zhang, D. Phan, S. Namnabat, E. LaVilla, K. Konopka, L. Diaz, M. Manchester, J. Schwiegerling, R. Glass, M. Mackay, K. Char, R. Norwood, and J. Pyun, "Chalcogenide hybrid inorganic/organic polymers: Ultrahigh refractive index polymers for

infrared imaging," *ACS Macro Lett* **6**(5), 500–504 (2017). https://doi.org/10.1021/acsmacrolett.7b00225.

[34] J. Griebel, S. Namnabat, E. Kim, R. Himmelhuber, D. Moronta, W. Chung, A. Simmonds, K. Kim, J. Van der Laan, N. Nguyen, E. Dereniak, M. Mackay, K. Char, R. Glass, R. Norwood, and J. Pyun, "New infrared transmitting material via inverse vulcanization of elemental sulfur to prepare high refractive index polymers," *Adv Mater* **26**(19) (2014). https://doi.org/10.1002/adma.201305607.

[35] PolymerPlus. https://www.polymerplus.net/16-about (accessed on 01 Oct 2020).

[36] N. Shatz, "Gradient-index optics: Final report," in *Contract W911NF-09-C-0077*, Science Applications International Corp., US ARMY (Grant W911NF-09-C-0077), (Research Triangle Park, NC) (2011).

[37] J. Teichman, J. Holzer, B. Balko, B. Fisher, and L. Buckley, "Gradient index optics at DARPA," in *IDA Document D-5027*, Institute for Defense Analyses, DARPA, (Alexandria, VA) (2013).

[38] T. Yang, N. Takaki, J. Bentley, G. Schmidt, and D. Moore, "Efficient representation of freeform gradient-index profiles for non-rotationally symmetric optical design," *Opt Express* **28**(10), 14788–14806 (2020). https://doi.org/10.1364/OE.391996.

[39] A. Visconti, K. Fang, J. Corsetti, P. McCarthy, G. Schmidt, and D. Moore, "Design and fabrication of a polymer gradient-index optical element for a high-performance eyepiece," *Opt Eng* **52**, 112107 (2013). https://doi.org/10.1117/1.OE.52.11.112107.

[40] A. Visconti, K. Fang, G. Schmidt, and D. Moore, "All-plastic high-performance eyepiece design utilizing a spherical gradient-index lens," Classical Optics 2014, OSA Technical Digest (online) paper IW2A.6 (2014). https://www.osapublishing.org/abstract.cfm?URI=IODC-2014-IW2A.6.

[41] S. Ji, K. Yin, M. Mackey, A. Brister, M. Ponting, and E. Baer, "Polymeric nanolayered gradient refractive index lenses: Technology review and introduction of spherical gradient refractive index ball lenses," *Opt Eng* **52**, 112105 (2013). https://doi.org/10.1117/1.OE.52.11.112105.

[42] A. Boyd, M. Pointing, and H. Fein, "Layered polymer GRIN lenses and their benefits to optical designs," *Adv Opt Technol* **4**(5-6), 429–443 (2015).

[43] P. Tapaswi and C. Ha, "Recent trends on transparent colorless polyimides with balanced thermal and optical properties: Design and synthesis," *Macromolecular Chemistry and Physics* **220**(3), 1800313 (2019). https://doi.org/10.1002/macp.201800313.

[44] M. Hasegawa, Y. Hoshino, N. Katsura, and J. Ishii, "Superheat-resistant polymers with low coefficients of thermal expansion," *Polymer* **111**, 91–102 (2017). https://doi.org/10.1016/j.polymer.2017.01.028.

[45] M. Hasegawa, T. Ishigami, and J. Ishii, "Optically transparent aromatic poly(ester imide)s with low coefficients of thermal expansion (1). Self-orientation behavior during solution casting process and substituent effect," *Polymer* **74**, 1–15 (2015). https://doi.org/10.1016/j.polymer.2015.07.026.

[46] N. Mushtaq, Q. Wang, G. Chen, B. Bashir, H. Lao, Y. Zhang, L. Sidra, and X. Fang, "Synthesis of polyamide-imides with different monomer sequence and effect on transparency and thermal properties," *Polymer* **190** (2020). https://doi.org/10.1016/j.polymer.2020.122218.

[47] W. Smith, *Modern Lens Design*, McGraw-Hill, New York, second ed. (2005).

[48] R. Kingslake and R. Johnson, *Lens Design Fundamentals*, Academic Press, Burlington, *second ed.* (2010). https://doi.org/10.1016/C2009-0-22069-1.

[49] R. Shannon, *The Art and Science of Optical Design*, Cambridge University Press, New York, *first ed.* (1997).

[50] R. Fischer, B. Tadic-Galeb, and P. Yoder, *Optical System Design*, McGraw-Hill, New York, second ed. (2008).

[51] V. Doushkina and E. Fleming, "Optical and mechanical design advantages using polymer optics," *Proc SPIE* **7424**: Advances in Optomechanics, 74240Q (2009). https://doi.org/10.1117/12.832319.

[52] A. Palmer, "Practical design considerations for polymer optical systems," *Proc SPIE* **0306**: Contemporary Methods of Optical Fabrication (1982). https://doi.org/10.1117/12.932712.

[53] A. Voznesenskaya and A. Ekimenkova, "Modeling of hybrid polymer optical systems," *Proc SPIE* **10690**: Optical Design and Engineering VII, 1069014 (2018). https://doi.org/10.1117/12.2312706.

[54] G. Behrmann, J. Bowen, and J. Mait, "Thermal properties of diffractive optical elements and design of hybrid athermalized lenses," *Proc SPIE* **10271**: Diffractive and Miniaturized Optics: A Critical Review, 102710C (1993). https://doi.org/10.1117/12.170185.

[55] "OpticStudio." Zemax. https://www.zemax.com/products/opticstudio (accessed on 01 Oct 2020).

[56] "Code V optical design software." Synopsys. https://www.synopsys.com/optical-solutions/codev.html (accessed on 01 Oct 2020).

[57] "Light tools illumination design software." Synopsys. https://www.synopsys.com/optical-solutions/lighttools.html (accessed on 01 Oct 2020).

[58] "OSLO (optics software for layout and optimization)." Lambda Research Corporation. https://www.lambdares.com/oslo/ (accessed on 01 Oct 2020).

[59] TracePro. Lambda Research Corporation. https://www.lambdares.com/tracepro/ (accessed on 01 Oct 2020).

[60] "FRED optical engineering software." Photon Engineering. https://photonengr.com/fred-software/ (accessed on 01 Oct 2020).

[61] Diaz, "Optical systems benefit from aspherical lenses," (2006). optics.org. https://optics.org/article/26635 (accessed on 01 Oct 2020).

[62] "All about aspheric lenses." Edmund Optics. https://www.edmundoptics.com/knowledge-center/application-notes/optics/all-about-aspheric-lenses/ (accessed on 01 Oct 2020).

[63] K. Thompson and J. Rolland, "Freeform optical surfaces: A revolution in imaging optical design," *Opt Photonics News* **23**(6), 30–35 (2012). https://doi.org/10.1364/OPN.23.6.000030.

[64] T. Yang, J. Zhu, X. Wu, and G. Jin, "Direct design of freeform surfaces and freeform imaging systems with a point-by-point three-dimensional construction-iteration method," *Opt Express* **23**(8), 10233–10246 (2015). https://doi.org/10.1364/OE.23.010233.

[65] C. Menke and G. Forbes, "Optical design with orthogonal representations of rotationally symmetric and freeform aspheres," *Adv Opt Technol* **2**(1), 97–109 (2013). https://doi.org/10.1515/aot-2012-0072.

[66] Moore Nanotechnology Systems. http://www.nanotechsys.com (accessed on 01 Oct 2020).

[67] AMETEK Precitech Inc. https://www.precitech.comhttps://www.precitech.com (accessed on 01 Oct 2020).

[68] "NanoCAM4 5MT." Moore Nanotechnology Systems. http://nanotechsys.com/nanocam (accessed on 01 Oct 2020).

[69] K. Jagtap and R. Pawade, "Experimental investigation on the influence of cutting parameters on surface quality obtained in SPDT of PMMA," *Int Jour Adv Design and Manuf Technol* **7**(2), 53–58 (2014).

[70] T. Kwok, C. Cheung, L. Kong, S. To, and W. Lee, "Analysis of surface generation in ultra-precision machining with a fast tool servo," Proc Inst Mech Eng – Part B: Jour Eng Manuf **224**(9), 1351–1367 (2010). https://doi.org/10.1243/09544054JEM1843.

[71] H. Ding, S. Luo, X. Chang, and D. Xie, "Optimization algorithm of tool radius compensation in fast tool servo machining of microlens arrays," *Adv Mater Res* **1039**, 383–389 (2014). https://doi.org/10.4028/www.scientific.net/AMR.1039.383.

[72] W. Zong, Z. Li, T. Sun, K. Cheng, D. Li, and S. Dong, "The basic issues in design and fabrication of diamond-cutting tools for ultra-precision and nanometric machining," *Int J Mach Tools Manuf* **50**(4), 411–419 (2010). https://doi.org/10.1016/j.ijmachtools.2009.10.015.

[73] G. Gubbels, *Diamond Turning of Glassy Polymers*, PhD dissertation, Technische Universiteit Eindhoven (2006). https://doi.org/10.6100/IR613637.

[74] F. Xu, J. Wang, F. Fang, and X. Zhang, "A study on the tool edge geometry effect on nano-cutting," *Int J Adv Manuf Technol* **91**, 2787–2797 (2017). https://doi.org/10.1007/s00170-016-9922-4.

[75] H. Cai and G. Shi, "Tool path generation for multi-degree-of-freedom fast tool servo diamond turning of optical freeform surfaces," *Exp Tech* **43**, 561–569 (2019). https://doi.org/10.1007/s40799-019-00307-1.

[76] Q. Liu, X. Zhou, and P. Xu, "A new tool path for optical freeform surface fast tool servo diamond turning," *P I Mech Eng B-J Eng* **228**(12), 1721–1726 (2014). https://doi.org/10.1177/0954405414523595.

[77] M. Cheng, C. Cheung, W. Lee, and S. To, "A study of factors affecting surface quality in ultra-precision raster milling," *Key Eng Mater* **339**, 400–406 (2007). https://doi.org/10.4028/www.scientific.net/KEM.339.400.

[78] H. Sun, J. Qi, Z. Lin, S. Wang, and Z. Yan, "Factors affecting the machining precision of a micro-lens array on a spherical surface in slow tool servo machining," *Opt and Prec Eng* **26**(11), 2516–2526 (2018). https://dx.doi.org/10.3788/OPE.20182610.2516.

[79] W. Zong, Z. Cao, C. He, and T. Sun, "Theoretical modeling and FE simulation on the oblique diamond turning of ZnS crystal," *Int J Mach Tools Manuf* **100**, 55–71 (2016). https://doi.org/10.1016/j.ijmachtools.2015.10.002.

[80] D. Morrison, "Design and manufacturing considerations for the integration of mounting and alignment surfaces with diamond turned optics," *Proc SPIE* **0966**: Advances in Fabrication and Metrology for Optics and Large Optics (1989). https://doi.org/10.1117/12.948068.

[81] A. Sohn, "Fixturing and alignment of free-form optics for diamond turning," in *Winter Topical Meeting on Free-Form Optics: Design, Fabrication, Metrology, Assembly, Proceedings of the American Society for Precision Engineering* (2004).

[82] A. Hedges and R. Parker, "Low stress, vacuum - chuck mounting techniques for the diamond machining of thin substrates," *Proc SPIE* **0966**: Advances in Fabrication and Metrology for Optics and Large Optics (1989). https://doi.org/10.1117/12.948045.

[83] L. Chaloux, "Part fixturing for diamond machining," *Proc SPIE* **0508**: Production Aspects of Single Point Machined Optics (1984). https://doi.org/10.1117/12.944968.

[84] P. Tolley and R. Arndt, "Lens element from diamond-turned thermoplastic resin," US Patent 7413689, Syntec Technologies, Inc. (Pavillion, NY, US) (2008). https://www.freepatentsonline.com/7413689.html.

[85] X. Sun, J. Liu, W. Gao, G. Huang, and H. Huang, "Development of cnc software for single point diamond precision lathe based on umac," *Int J Nanomanuf* **14**(4) (2018). https://doi.org/10.1504/IJNM.2018.095322.

[86] M. Liman and K. Abou-El-Hossein, "Modeling and multiresponse optimization of cutting parameters in SPDT of a rigid contact lens polymer using RSM and desirability function," *Int J Adv Manuf Technol* **102**, 1443–1465 (2019). https://doi.org/10.1007/s00170-018-3169-1.

[87] D. Yu, S. Gan, Y. Wong, G. Hong, M. Rahman, and J. Yao, "Optimized tool path generation for fast tool servo diamond turning of micro-structured surfaces," *Int J Adv Manuf Technol* **63**, 1137–1152 (2012). https://doi.org/10.1007/s00170-012-3964-z.

[88] C. Cheung and W. Lee, "Study of factors affecting the surface quality in ultra-precision diamond turning," *Mater Manuf Process* **15**, 481–502 (2000). https://doi.org/10.1080/10426910008913001.

[89] "Polyester (okp optical plastic): For injection molding, raw material of the film." Osaka Gas Chemicals. http://www.ogc.co.jp/e/products/fluorene/okp.html (accessed on 01 Oct 2020).

[90] "Iupizeta EP, special polycarbonate resin." Mitsubishi Gas Chemical. https://www.mgc.co.jp/eng/products/kc/iupizeta_ep.html (accessed on 01 Oct 2020).

[91] T. Higashihara and M. Ueda, "Recent progress in high refractive index polymers," *Macromolecule* **48**(7), 1915–1929 (2015). https://doi.org/10.1021/ma502569r.

[92] J. Liu and M. Ueda, "High refractive index polymers: Fundamental research and practical applications," *J Mater Chem* **19**(47), 8907–8919 (2009). https://doi.org/10.1039/B909690F.

[93] E. Macdonald and M. Shaver, "Intrinsic high refractive index polymers," *Polymer International* **64**(1) (2014). https://doi.org/10.1002/pi.4821.

[94] W. Beich, "Injection molded polymer optics in the 21st century," *Proc SPIE* **5865**: Tribute to Warren Smith: A Legacy in Lens Design and Optical Engineering, 58650J (2005). https://doi.org/10.1117/12.626616.

[95] R. Mayer, "Precision injection molding: How to make polymer optics for high volume and high precision applications," *Optik & Photonik* **2**(4) (2011). https://doi.org/10.1002/opph.201190286.

[96] R. Boerret, J. Raab, and M. Speich, "Mold production for polymer optics," *Proc SPIE* **9192**: Current Developments in Lens Design and Optical Engineering XV, 91921L (2014). https://doi.org/10.1117/12.2060670.

[97] "Moldex3D." CoreTech System. https://www.moldex3d.com/ (accessed on 01 Oct 2020).

[98] S. Kulkarni, *Robust Process Development and Scientific Molding: Theory and Practice*, Carl Hanser Verlag GmbH & Co. KG, Munich, second ed. (2017).

[99] D. Yao, "Micromolding of polymers," in *Advances in Polymer Processing: From Macro- to Nano- Scales*, S. Thomas and Y. Weimin, Eds., 552–578, Woodhead Publishing (2009). https://doi.org/10.1533/9781845696429.4.552.

[100] M. Heckele and W. Schomburg, "Review on micromolding of thermoplastic polymers," *J Micromechan and Microeng* **14**(3), 1–14 (2004). https://iopscience.iop.org/article/10.1088/0960-1317/14/3/R01.

[101] J. Giboz, T. Copponnex, and P. Mélé, "Microinjection molding of thermoplastic polymers: A review," *J Micromechan and Microeng* **17**(6), 96–109 (2007). https://doi.org/10.1088%2F0960-1317%2F17%2F6%2Fr02.

[102] M. Martyn, B. Whiteside, P. Coates, P. Allen, G. Greenway, and P. Hornsby, "Aspects of micromolding polymers for medical applications," *Annual Technical Conference* (ANTEC) of the Society of Plastics Engineers (SPE) (2004).

[103] V. Piotter, K. Mueller, K. Plewa, R. Ruprecht, and J. Hausselt, "Performance and simulation of thermoplastic micro injection molding," *Microsys Technol* **8**, 387–390 (2002). https://doi.org/10.1007/s00542-002-0178-6.

[104] M. Pazos, J. Baselga, and J. Bravo, "Limiting thickness estimation in polycarbonate lenses injection using CAE tools," *J Mater Process Technol* **143-144**, 438–441 (2003). https://doi.org/10.1016/S0924-0136(03)00425-4.

[105] U. Schulz, "Coating on plastics," in *Handbook of Plastic Optics*, S. Baumer, Ed., 161–195, WILEY-VCH Verlag GmbH & Co. KGaA, Weinheim, second ed. (2010).

[106] T. Bauer, "Optical coatings on polymers," *Proc SPIE* **5872**: Adv in Polymer Opt Design, Fabrication, and Mater (2005). https://doi.org/10.1117/12.614089.

[107] L. Martinu and J. Klemberg-Sapieha, "Optical coatings on plastics," in *Optical Interference Coatings*, N. Kaiser and H. Pulker, Eds., Springer, Berlin (2003). https://doi.org/10.1007/978-3-540-36386-6_15.

[108] H. Noh, "Changes of thin film coating on polymer lenses with varying temperature," *J Korean Ophthalmic Optics Society* **19**(1), 1–8 (2014). http://dx.doi.org/10.14479/jkoos.2014.19.1.1.

[109] U. Schulz, "Review of modern techniques to generate antireflective properties on thermoplastic polymers," *Appl Opt* **45**(7), 1608–1618 (2006). https://doi.org/10.1364/AO.45.001608.

[110] T. Schmauder, K. Nauenburg, K. Kruse, and G. Ickes, "Hard coatings by plasma CVD on polycarbonate for automotive and optical

applications," *Thin Solid Films* **502**(1-2), 270–274 (2006). https://doi.org/10.1016/j.tsf.2005.07.296.

[111] Overton, "Photonics applied: Thin-film applications: Coatings for plastic optics step up to meet higher-end applications," (2010). LaserFocusWorld. https://www.laserfocusworld.com/optics/article/16567794/photonics-applied-thinfilm-applications-coatings-for-plastic-optics-step-up-to-meet-higherend-applications (accessed on 01 Oct 2020).

[112] W. Gang, M. Ping, Q. Fumin, H. Jiangchuan, and Y. Dingyao, "Optimization of ultrasonic cleaning of optics," *High Power Laser and Particle Beams* **24**(7), 1761–1764 (2012).

[113] K. Guenther and H. Enssle, "Ultrasonic precision cleaning of optical components prior to and after vacuum coating," *Proc SPIE* **0652**: Thin Film Technologies II (1986). https://doi.org/10.1117/12.938355.

[114] D. Cushing, *Enhanced Optical Filter Design*, SPIE Press, Bellingham, *first ed.* (2011).

[115] TRIOPTICS. https://trioptics.com/ (accessed on 01 Oct 2020).

[116] C. Wilde, F. Hahne, P. Langehanenberg, and J. Heinisch, "Reducing the cycle time of cementing processes for high quality doublets," Proc SPIE **9628**: Optical Systems Design 2015: Optical Fabrication, Testing, and Metrology V, 96280X (2015). https://doi.org/10.1117/12.2196876.

[117] D. Winters and P. Erichsen, "Image quality testing of assembled IR camera modules," *Proc SPIE* **8896**: Electro-Optical and Infrared Systems: Technology and Applications X, 88960K (2013). https://doi.org/10.1117/12.2029541.

[118] Opto Alignment. https://optoalignment.com/ (accessed on 01 Oct 2020).

[119] C. Wenzel, R. Winkelmann, R. Klar, P. Philippen, R. Garden, S. Pearlman, and G. Pearlman, "Advanced centering of mounted optics," *Proc SPIE* **9730**: Components and Packaging for Laser Systems II, 973012 (2016). https://doi.org/10.1117/12.2213125.

[120] M. Green and R. Garden, "Novel active alignment technique for measuring tilt errors in aspheric surfaces during optical assembly using lens alignment station (LAS)," *Proc SPIE* **11103**: Optical Modeling and System Alignment, 1110305 (2019). https://doi.org/10.1117/12.2528822.

[121] J. Bala, "Automated optical assembly," Proc SPIE **2622**: Optical Engineering Midwest '95 (1995). https://doi.org/10.1117/12.216803.

[122] P. Yoder, *Mounting Optics in Optical Instruments*, SPIE Press, Bellingham, *second ed.* (2008).

[123] P. Yoder, *Opto-Mechanical Systems Design*, Taylor & Francis Group LLC, Boca Raton, *third ed.* (2005).

[124] D. Vukobratovich and P. Yoder, *Fundamentals of Optomechanics (Optical Sciences and Applications of Light)*, Taylor & Francis Group LLC, Boca Raton, *first ed.* (2018).

[125] "Dymax light-cure plastic bonding adhesives and sealants." Dymax. https://dymax.com/products/formulations/light-curable-materials/bonding/plastic-bonding (accessed on 01 Oct 2020).

[126] "Norland adhesive selector guide." Norland Products. https://www.norlandprod.com/adhchart.html (accessed on 01 Oct 2020).

[127] "Optical adhesive solutions – strong bonds to glass, metal, ceramics and plastics," (2015). AMS Technologies. http://www.amstechnologies.com/fileadmin/amsmedia/AMS_Brochures_2016/09-16-03_Optical_Adhesive___Dispensing_Solutions_web.pdf (accessed on 01 Oct 2020).

[128] Elgin, "Selection of index matching materials," (2015). Fiber Optic Center. https://focenter.com/selection-of-index-matching-materials/ (accessed on 01 Oct 2020).

[129] Fisher, "Adhesives for fiber optics assembly: Making the right choice." photonics.com. https://www.photonics.com/Articles/Adhesives_for_Fiber_Optics_Assembly_Making_the/a25147 (accessed on 01 Oct 2020).

[130] Talbot, "Tutorial on adhesives and how to use them for mounting," (2016). College of Optical Sciences, University of Arizona. https://wp.optics.arizona.edu/optomech/wp-content/uploads/sites/53/2016/12/Tutorial_Talbot_Jared.pdf (accessed on 01 Oct 2020).

[131] Clements, "Selection of optical adhesives," (2006). College of Optical Sciences, University of Arizona. https://wp.optics.arizona.edu/optomech/wp-content/uploads/sites/53/2016/10/ClementsTutorial1.doc (accessed on 01 Oct 2020).

[132] M. Troughton, *Handbook of Plastic Joining: A Practical Guide*, William Andrew Inc., Norwich, *second ed.* (2008).

[133] S. Volkov, G. Bigus, and A. Remizov, "Ultrasonic welding of dissimilar plastics," *Russ Engin Res* **38**, 281–284 (2018). https://doi.org/10.3103/S1068798X18040238.

[134] J. Sackmann, K. Burlage, C. Gerhardy, B. Memering, S. Liao, and W. Schomburg, "Review on ultrasonic fabrication of polymer micro devices," *Ultrasonics* **56**, 189–200 (2015). https://doi.org/10.1016/j.ultras.2014.08.007.

[135] P. Tres, *Designing Plastic Parts for Assembly*, Hanser, Munich, *eighth ed.* (2017). https://doi.org/10.3139/9781569906699.005.

[136] P. Hilton, I. Jones, and Y. Kennish, "Transmission laser welding of plastics," *Proc SPIE* **4831**: First International Symposium on High-Power Laser Macroprocessing (2003). https://doi.org/10.1117/12.486499.

[137] M. Troughton, *Handbook of Plastic Joining: A Practical Guide*, William Andrew Inc., Norwich, *second ed.* (2008).

[138] A. Benatar, "Plastics joining," in *Applied Plastics Engineering Handbook*, M. Kutz, Ed., 575–591, Plastics Design Library (2017). https://doi.org/10.1016/B978-0-323-39040-8.00027-4.

[139] J. Martan, J. Tesa, M. Kučera, P. Honnerová, M. Benešová, and M. Honner, "Analysis of short wavelength infrared radiation during laser welding of plastics," *Appl Opt* **57**, D145–D154 (2018). https://www.osapublishing.org/ao/abstract.cfm?URI=ao-57-18-D145.

[140] J. Rault, M. Sotton, C. Rabourdin, and E. Robelin, "Crystallization of polymers. Part I: Polydispersed polymers quenched from the liquid state," *Journal de Physique* **41**(12), 1459–1467 (1980). https://doi.org/10.1051/jphys:0198000410120145900.

[141] Y. Shangguan, F. Chen, E. Jia, Y. Lin, J. Hu, and Q. Zheng, "New insight into time-temperature correlation for polymer relaxations ranging from secondary relaxation to terminal flow: Application of a universal and developed wlf equation," *Polymers* **9**(11), 567 (2017). https://doi.org/10.3390/polym9110567.

[142] W. Vogel, *Glass Chemistry*, Springer-Verlag, Berlin, *second ed.* (1994).

[143] W. Wang, G. Zhao, X. Wu, and Z. Zhai, "The effect of high temperature annealing process on crystallization process of polypropylene, mechanical properties, and surface quality of plastic parts," *Journal of Applied Polymer Science* **132**(46) (2015). https://doi.org/10.1002/app.42773.

[144] O. Hasan and M. Boyce, "Energy storage during inelastic deformation of glassy polymers," *Polymer* **34**(24), 5085–5092 (1993). https://doi.org/10.1016/0032-3861(93)90252-6.

[145] J. Butt and R. Bhaskar, "Investigating the effects of annealing on the mechanical properties of FFF-printed thermoplastics," *J Manuf Mater Process* **4**(2), 38 (2020). https://doi.org/10.3390/jmmp4020038.

[146] M. Kyrish, J. Miller, M. Fraelich, O. Lechuga, R. Claytor, and N. Claytor, "Metrology of injection molded polymer optics for a commercial VR system," *Proc SPIE* **10742**: Optical Manufacturing and Testing XII, 1074213 (2018). https://doi.org/10.1117/12.2325018.

[147] R. Huxford, "Wide-FOV head-mounted display using hybrid optics," *Proc SPIE* **5249**: Optical Design and Engineering (2004). https://doi.org/10.1117/12.516541.

[148] "The 30 best VR companies in the world," (2020). GameDesigning. https://www.gamedesigning.org/gaming/virtual-reality-companies/ (accessed on 01 Oct 2020).

[149] F. Lamontagne, N. Desnoyers, M. Doucet, P. Côté, J. Gauvin, and G. Anctil, "Disruptive advancement in precision lens mounting," *Proc SPIE* **9582**: Optical System Alignment, Tolerancing, and Verification IX, 95820D (2015). https://doi.org/10.1117/12.2196441.

[150] Tri-Power Design. https://www.tripowerdesign.com/ (accessed on 01 Oct 2020).

[151] R. Bailey, A. Washburn, A. Qavi, M. Iqbal, M. Gleeson, F. Tybor, and L. Gunn, "A robust silicon photonic platform for multiparameter

biological analysis," *Proc SPIE* **7220**: Silicon Photonics IV, 72200N (2009). https://doi.org/10.1117/12.809819.

[152] R. Vázquez, G. Trotta, A. Volpe, M. Paturzo, F. M. anbd, V. Bianco, S. Coppola, A. Ancona, P. Ferraro, I. Fassi, and R. Osellame, "Plastic lab-on-chip for the optical manipulation of single cells," in *Factories of the Future: The Italian Flagship Initiative*, T. Tolio, G. Copani, and W. Terkaj, Eds., 339–363, Springer Open (2018). https://doi.org/10.1007/978-3-319-94358-9.

[153] T. Koch, R. Norwood, S. Pau, and N. Peyghambarian, "Optical printed circuit board with polymer array stitch," U.S. Patent Application 20190098751, Arizona Board of Regents on Behalf of the University of Arizona (Tucson, AZ, US) (2019). https://www.freepatentsonline.com/y2019/0098751.html.

[154] A. Yee, W. Song, N. Takaki, T. Yang, Y. Zhao, Y. Ni, S. Bodell, J. Rolland, J. Bentley, and D. Moore, "Design of a freeform gradient-index prism for mixed reality head mounted display," *Proc SPIE* **10676**: Digital Optics for Immersive Displays, 106760S (2018). https://doi.org/10.1117/12.2307733.

[155] S. Wei, Z. Fan, Z. Zhu, and D. Ma, "Design of a head-up display based on freeform reflective systems for automotive applications," *Appl Opt* **58**(7), 1675–1681 (2019). https://doi.org/10.1364/AO.58.001675.

[156] "Night vision." Elbit Systems of America. http://www.elbitsystems-us.com/night-vision (accessed on 01 Oct 2020).

[157] F. Fang, X. Zhang, A. Weckenmann, G. Zhang, and C. Evans, "Manufacturing and measurement of freeform optics," *CIRP Annals* **62**(2), 823–846 (2013). https://doi.org/10.1016/j.cirp.2013.05.003.

[158] "U.S. Army Manufacturing Technology: Fiscal year 2018," 26, US Army Manufacturing Technology (ManTech) Program, US ARMY, (Aberdeen Proving Ground, MD) (2017).

[159] Z. Diao, M. Kraus, R. Brunner, J. Dirks, and J. Spatz, "Nanostructured stealth surfaces for visible and near-infrared light," *Nano Lett* **16**(10), 6610–6616 (2016). https://pubs.acs.org/doi/10.1021/acs.nanolett.6b03308. Printed with permission. Further permissions related to the material in this source should be directed to the ACS.

[160] "3D printed optics: The digitization of illumination optics fabrication." 3DPrinting. Lighting. http://www.3dprinting.lighting/category/3d-printed-optics/ (accessed on 01 Oct 2020).

[161] M. Gupta, C. Ungaroa, J. Foley IV, and S. Gray, "Optical nanostructures design, fabrication, and applications for solar/thermal energy conversion," *Solar Energy* **165**, 100–114 (2018). https://doi.org/10.1016/j.solener.2018.01.010.

[162] F. Flory, L. Escoubas, and G. Berginc, "Optical properties of nanostructured materials: A review," *Journal of Nanophotonics* **5**(1), 052502 (2011). https://doi.org/10.1117/1.3609266.

[163] L. Lee and R. Szema, "Inspirations from biological optics for advanced photonic systems," *Science* **310**(5751), 1148–1150 (2005). https://doi.org/10.1126/science.1115248.

[164] R. Kurahatti, A. Surendranathan, S. Kori, N. Singh, A. Kumar, and S. Srivastava, "Defence applications of polymer nanocomposites," *Defence Science Journal* **60**(5), 551–563 (2010). https://doi.org/10.14429/dsj.60.578.

[165] M. Foley, "Technical advances in microstructured plastic optics for display applications," *SID Symposium Digest of Technical Papers* 30 (1999). https://doi.org/10.1889/1.1833962.

[166] A. Heinrich and M. Rank, 3D Printing of Optics, SPIE Press, Bellingham, first ed. (2018). https://doi.org/10.1117/3.2324763.

[167] Schott, "New technique creates smoother 3D printed optical components," (2018). 3dPrint.com. https://3dprint.com/221118/3d-printed-optical-components/ (accessed on 01 Oct 2020).

[168] N. Vaidya and O. Solgaard, "3D printed optics with nanometer scale surface roughness," *Microsyst Nanoeng* **4**(18) (2018). https://doi.org/10.1038/s41378-018-0015-4.

Robert Parada is an Advisory Board Member at Syntec Optics. He joined Syntec in 2014 as VP of Operations. From 2015–2017, he led Syntec's customer product development division which encompasses manufacturing engineering, program management, tooling design, optical and opto-mechanical design, and optical thin film coating. From 2018–2020, he served as General Manager of the corporation. Before joining Syntec, Rob held numerous R&D leadership roles at Eastman Kodak Company, and he was VP of Engineering at a high-technology panoramic imaging company in New England. Rob received his B.S. degree in Imaging Science from Rochester Institute of Technology, his M.S. and Ph.D. degrees in Optical Sciences from the University of Arizona, and a second doctoral degree in optics from the Université du Littoral Côte d'Opale in France.

Douglas Axtell is Chief Technology Officer at Syntec Optics. He joined Syntec in 2016 as Program Manager in Syntec's polymer optics division, which encompasses process engineering, quality engineering, optical and opto-mechanical assembly. Before joining Syntec, Doug held various project, chemical and coating engineering positions in Kodak's Research Laboratories and later at Reflexite Display Optics as Sr. Material Scientist in Product Development where he contributed IP in UV cure formulation and coating performing toll R&D work. Doug achieved his Six Sigma Black Belt certification while at Kodak and during his tenure at Kodak and Reflexite. He has been awarded ten patents. He received his B.S. in Imaging Science and M.S. in Product Development from Rochester Institute of Technology.

Dan Morgan is Optical Quality Director at Syntec Optics. He joined Syntec in 2003 to manage the corporation's Optical Metrology Lab and has been integral in achieving ISO 9001 and ISO 13485 certifications. Before joining Syntec, Dan spent 15 years in precision optical molding, assembly, and metrology. Dan received a degree in Applied Optics from Monroe Community College and has completed additional studies at the Rochester Institute of Technology Center for Quality and Applied Sciences.

Chapter 8
Optical Fibers and Optical Fiber Assemblies

Devinder Saini, Kevin Farley, and Brian Westlund
Fiberguide Industries, 3409 East Linden St., Caldwell, ID, 83605, USA

8.1 Optical Fibers

The guiding of light in various materials (glass and water) using the principle of refraction has been demonstrated throughout history, in fact this phenomenon was known in ancient Egypt and Mesopotamia where colored glass was used for decoration. In the late 19th and 20th century, various pioneers demonstrated the use of bent glass rods to illuminate cavities in the body and for early form of television. The bare glass fibers/rods that were used in these experiments were, however, very lossy which meant that they did not transmit a lot of light. The loss of light in these fibers was due to light escaping when these fibers touched each other or when they had scratches on the surface. These issues were solved during the 1950s when the idea of cladding a fiber was demonstrated by Bram van Heel [1], and in the same year N. S. Kapany and H. Hopkins demonstrated image transmission in a bundle of 10,000 fibers 75 cm long [2]. The theory of light propagation into fibers was described by N.S. Kapany and improved later by E. Snitzer [3]. The theory of the use of optical fiber for communication was developed and promoted by C. K. Kao and G. A. Hockham in 1965 [4,5].

Since 1970, low-loss optical fiber (invented by Corning) has steadily replaced the use of copper wire for long-range telecommunication and data transmission [6]. Although most of the optical fiber produced (in terms of length) is used for telecommunication and data transmission, optical fibers are finding uses in many other applications, such as astronomy, biomedical instrumentation, communication, defense, digital projection, industrial laser, medical, semiconductor manufacturing, spectroscopy, and many others. For these applications, specialty optical fibers are required. A specialty optical fiber is generally defined as any fiber that is not the standard fiber used in telecommunications. Due to the needs of various applications special fibers have developed that have large cores (ranging from a few microns up to 2500 microns), large numerical apertures, special coatings (polyimide, tefzel, nylon, acrylate, hard clad), metal coatings (aluminum, gold and other metals), radiation hardened fibers, biocompatible fiber, multicore fibers, photonic crystal fibers and many others.

8.1.1 Preform manufacturing

Most optical fibers produced are made from glass, more specifically silica-based glasses. This material is rather abundant and has been used in the manufacturing of glass for thousands of years. To produce low loss, high purity optical preforms, only the purist precursors (99.9999%) are used to make single-mode (SM), multimode (MM) and other specialty glass preforms. This has led to the creation of multiple glass making techniques that specialize in specific areas. They have evolved as the need for different fibers has evolved over the time: In the 1980s it was SM fibers for telecommunications, and in the 1990s, the need for fibers for erbium-doped fiber amplifiers (EDFAs) was a huge motivation. Once the fiber bubble burst in 2001, the types of optical fibers and preforms shifted yet again to the specialty area focusing on rare-earth-doped preforms for fiber lasers & amplifiers, beam delivery, photosensitive, polarization maintaining (PM) fibers and more. This section will review some of the glass-making techniques used to produce these fibers.

8.1.1.1 Multimode preforms

Multimode fibers can transmit hundreds to thousands of modes within a fiber based on its core diameter and numerical aperture (NA). These preforms are typically a pure silica core which is then surrounded by a layer of fluorine doped silica as the optical cladding. The fluorine-doped silica has an index of refraction that is lower than silica, producing total internal reflection. Standard NAs are 0.22, but as applications have matured, so have the requirements. Presently, these fibers are used within many different fields such as the medical industry, sensing, oil and gas, and microscopy. There is not one fiber size or NA that will suit the requirements of all these potential applications. This has created a demand for tighter tolerances within the industry. On the preform side, the main areas are purity of the glass, affecting transmission, and the NA control.

High-purity silica is the #1 component to make an optical fiber, no matter the application. With MM preforms, the starting point is to produce the pure silica core and then to produce the optical cladding after. This type of process permits for different sizes of preforms to be manufactured and is extremely scalable. The core material can be made from a few different methods but mostly through flame hydrolysis as previously shown in Figure

2.31(a) of Chapter 2. It consists of a target or bait rod that is attached to a lathe to rotate it. A high-purity silica precursor, such as $SiCl_4$, is used to produce a soot layer across the length of the bait rod through a H_2/O_2 flame. Based on the size of the fused silica that is being produced, this can take hours to days to complete. After the soot process is completed, it can be dehydrated by a chlorine-based step. This is done to a certain degree based on the type of glass, as both low- and high-OH silica is commonplace for MM glasses. The chlorine will bond with the OH ions, creating a vapor that can be removed. The subsequent process is the vitrification of the soot into a high purity fused silica at very high temperatures, >2000°C. This method is scalable so it can produce completed fused silica rods from 15 mm to >200 mm in diameter.

This technique has produced high quality silica for many years, but the quality of the glass has continued to improve. Not completely through the raw materials, but from other factors as well. The process described previously would be performed on an outside vapor deposition (OVD) lathe. This could also be done in a vertical manner through vapor axial deposition (VAD). In either way, the rotation, chemical flows, flame control and traverse are all recipe controlled, permitting for very accurate and precise deposition layers. The improvements of mass flow controllers from analog to digital control over 20 years ago was a huge advancement in the industry. As time has elapsed, the mass flow controllers (MFCs) continue to get better in performance, creating the ability to deposit even thinner layers of silica, which can be correlated with reduced losses in the fiber. Programmable logic controllers (PLCs) and the machinery itself have all improved as well, which has also improved the product lines as well.

The quality of the core improvement is extremely vital to MM optical fibers, but so is the silica-doped cladding material. The purity of the glass has been discussed, but another area that is key in the MM field is the control of the NA. MM fibers have NAs that range from 0.10 up to 0.29. The higher the NA, the more fluorine is required to depress the index of refraction. As the need for increased NAs continues to grow, the preform manufacturers also struggle to increase the fluorine efficiency of deposition without inducing etching of the glass due to the presence of fluorine. The standard fiber will have an NA of 0.22 ± 0.02 as a specification. For most applications, this is more than suitable, but as MM fibers have increased in use in lasers and amplifiers, tighter controls are required. Multimode fibers are used to carry light from diodes into combiners, which lend themselves to tighter NA tolerances. It is extremely important to control the NA so that the brightness is conserved, shown in Eq. (8.1). The input fibers need to have an NA with a ±0.01 or even ±0.005 tolerance, while the output fibers require the same tolerance but on the opposite end of the total NA specification. This requires different preforms with extremely tight NA tolerances. This is one example, but preform manufacturers have to improve their processes to deliver on these tolerances.

$$\eta_{br} = N\eta_P \left(\frac{d_{in} \text{NA}_{in}}{d_{out} \text{NA}_{out}} \right)^2 \quad (8.1)$$

where η_{br} is brightness loss in the combiner, N is the number of multimode pump fiber inputs, η_P is the power throughput in the combiner, d_{in} and d_{out} are the core diameter of input and output fibers, respectively, and NA_{in} and NA_{out} are the numerical aperture of input and output fibers, respectively.

The efficiency and repeatability of the plasma outside deposition (POD) has permitted manufacturers to produce fluorine-doped silica preforms with improved tolerances. The POD process is shown in Fig. 8.1. The substrate material can be a silica rod or tube that is rotated, and then the plasma torch has oxygen, a silicon compound and a fluorine compound that is deposited on the outside of the substrate material. This is the preferred method to make MM preforms as it can produce large diameters. Much like the fabrication of high-purity silica rods, the control mechanisms have improved to the point where glass manufacturers can provide tolerances of ±0.01 or ±0.005 for the right price.

The POD process permits for even more unique preforms to be developed as the substrate is not required to be circular. Over the past 15 to 20 years, there has been more and more interest in shaped-core optical fibers. The dominant shape has been a square core but is not limited to just that. Rectangular, octagonal and other polygon shapes can be utilized, and the POD process can deposit a fluorine-doped layer as the cladding. The NA

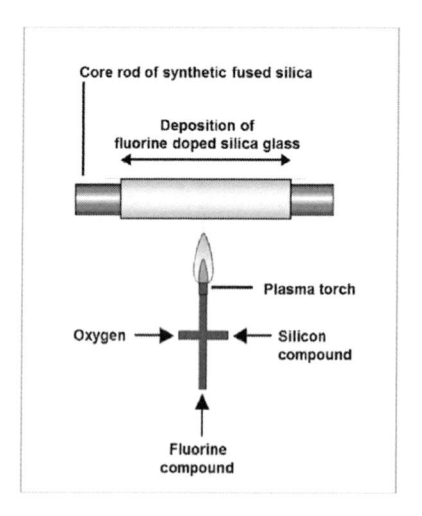

Figure 8.1 Plasma outside deposition (POD) of fluorine-doped silica for high-power laser applications [7].

control is not as repeatable due to the fact that the efficiency of the fluorine doping is changing as the different shapes are rotating, but these shaped-core MM preforms are another area that has produced a large interest in the specialty fiber market and continues to grow.

8.1.1.2 Rare-earth-doped preforms

Another area that optical fibers have penetrated over the past 20 years is in doped fibers and amplifiers. They were used prior to the 21st century, but in the 2000s, the improvements have been astronomical in making these fibers repeatable enough to be used in industrial laser applications. The work performed in making EDFAs was the starting point to doping rare earth elements into a silica matrix. Erbium itself cannot be added to a silica host without causing devitrification, but through the use of co-doping with homogenizers, such as aluminum, it easily can be incorporated. Research with other rare earth elements led to more possibilities in the defense and industrial applications. The advantage of active ions such as rare earths is to absorb light at one wavelength and have emission at a longer wavelength. In the case of ytterbium it absorbs light at 915 and 976 nm, where diodes are readily available, and can have emission from 1030 nm to 1100 nm, which are wavelengths of interest in many industrial applications. Table 8.1 displays the absorption and emission wavelengths for some common rare-earth ions in silica glass.

These types of preforms can be fabricated using multiple machines such as an OVD or modified chemical vapor deposition (MCVD) systems, shown in Fig. 8.2. Modified Chemical Vapor Deposition (MCVD) is a preferred method to fabricate high-purity glass core rods that are the precursor to fiber optics. This method produces high-purity, low-loss SiO_2, GeO_2 and P_2O_5 doped cladding and core materials with low water content across the wavelengths of interest in the telecommunications industry. With the advent of fiber lasers and amplifiers, different core materials with active rare-earth-doped ions (Yb, Er, Tm, Nd, etc.) needed to be incorporated. Liquid chloride precursors such as $SiCl_4$ and $GeCl_4$ with vapor pressures at low temperatures are not available, so other methods needed to be created. Solution doping,

Table 8.1 List of rare earth ions in silica glasses with pump and emission wavelengths.

Rare Earth Ion	Pump Wavelengths	Emission Wavelengths
Neodymium	808, 869 nm	1030 to 1100 nm, 900 to 950 nm
Ytterbium	915, 920, 976 nm	1030 nm to 1100 nm
Erbium	980 nm	1530 to 1580 nm
Erbium/Ytterbium	976 nm	1530 to 1580 nm
Thulium	793 nm	1700 to 2100 nm
Holmium	1610, 1950 nm	2050 to 2150 nm

Figure 8.2 Image of a Nextrom OFC 12 MCVD system (www.rosendahlnextrom.com).

vapor deposition and chelate delivery are methods to incorporate rare-earth-doped ions into the glass.

8.1.1.3 Solution doping

One of the least evasive ways to perform rare earth doping is via solution doping. Standard MCVD glass is deposited in vapor form with chloride precursors ($SiCl_4$, $GeCl_4$ & $POCl_3$) along with oxygen and helium to make a single layer of glass deposition. The main burner traverses the length of the deposition tube and consolidates the layer into glass at various temperatures based on the chemical composition, with temperatures generally in the range of 1600 to 2000°C. Solution doping requires a different methodology which begins with the deposition of a porous soot core.

In principle, the soot core deposition process is a simple step. An un-sintered SiO_2 layer is deposited along the length of the tube. The degree of "sintering" is controlled both by the burner temperature and by chemical flows. The un-sintered layer manifests itself as a porous soot frit that is an interconnection of particles of various sizes. The soot is cooled to approximately room temperature (for consistency purposes) and then will

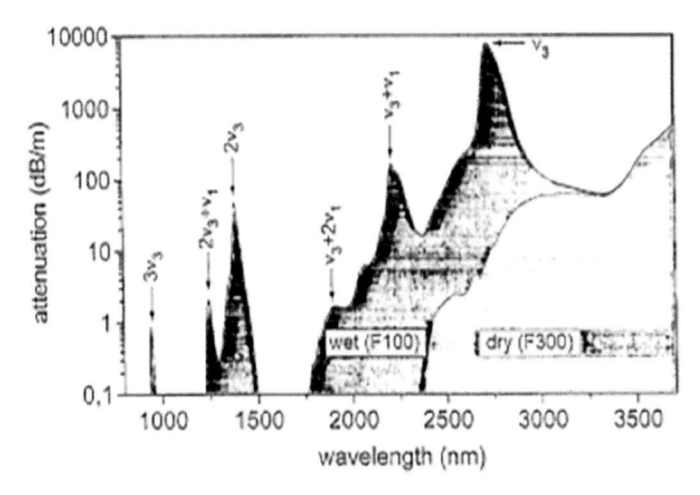

Figure 8.3 Wet vs. dry F-300 [8].

be subsequently soaked in a salt solution to achieve the appropriate chemical constituents. While very simple in practice all the parameters used during the deposition of this layer require stringent attention and have consequences regarding the quality of the soot core along the length of the deposition tube.

There are certain absorption bands due to the formation of hydroxyl (OH) in the glass. These bands grow in intensity as more and more OH formation occurs. Figure 8.3 displays a typical attenuation curve of SiO_2 glass with OH impurities. The fundamental absorption band exists at ~2.7 μm, and this is the strongest absorption peak. The OH has other overtone and combinational bands that exist in wavelengths of interests for specific rare-earth-doped fibers. While additional attenuation at 2.7 μm is of no consequence for operating at 1 μm for Yb fibers, the attenuation at 980 and 1400 nm is consequential. The OH loss at 1400 nm can increase the overall background loss of the fiber resulting in lower efficiencies. The loss at 980 nm can detrimentally affect the absorption of the Yb ions at 976 nm, thus resulting in less efficiency and poorer overall performance, so the reduction of these peaks is vital.

In standard MCVD, the deposition tube is not taken off the lathe, so there is no exposure to the atmospheric environment. These glasses will have low OH attenuations, typically 0.1 ppm of OH. Once the tube is exposed to the atmosphere when taken off the lathe, the magnitude of OH will increase significantly. This is the reason that all soot cores are chlorine dried: to remove the OH impurities. Based on the product type, the severity of the chlorine drying can be altered. For glasses that contain thulium or holmium and operate at 2 μm, the OH impurities are of even more concern since there is an OH absorption band at 2200 nm.

After the drying step, the soot undergoes an oxidation step to prepare the rare earth chlorides to form into an oxide layer. The consolidation of the layer is final phase in which the soot core, solutes and additional co-dopants are sintered into a glass layer. This is also the phase that one can determine if the temperature and mixture of the chemical constituents is correct. Based on the composition of the glass layer there will be different temperature and flow set points. Each soot layer contains a layer of glass after all the steps are completed. The advantage is that the process of incorporation is quite easy in terms of soaking the soot in an aqueous solution. The disadvantages are that it is a slower process since only one layer can be achieved at a time, possible exposure to the atmosphere and longer processing times based on the size of the final preform required. Another challenge is the axial variation along the length of the preform due to the soot process. It is also limited in terms of certain dopants and cannot achieve the same concentrations of Al_2O_3 as other techniques.

8.1.1.4 Vapor phase

Another methodology to incorporate rare earths is to heat them to a vapor phase and either incorporate them into a deposited soot or have them in the gas stream with the standard deposition. This technique has inherent advantages in that there are not as many OH concerns, higher dopant concentration, shorter process times and also larger doped cores. Some of the negatives are that manufacturers need to modify the setup and spend more on equipment as more MFCs, heaters and/or glass equipment based on the setup of choice. End users are able to perform this via either an internal, within the glass system setup, or external system. This section will review some of the known methods.

The key to inducing a vapor phase out of rare earth chlorides is to ensure that there is a sufficient vapor pressure. Researchers and investigations focused on both delivery using organometallic chelate compounds, $M(thd)_3$, or halide compounds. These are both very different in their behavior as chelates will sublime at 200°C but degrade rapidly with temperature, and the rare earth halide compounds will sublime between 900 and 1200°C based on the exact components. Some empirical calculations need to be performed to ensure the vapor pressure will be sufficient to incorporate within the gas stream to make a consolidated glass layer [9].

Figure 8.4 displays various ways to perform internal vapor phase of rare earth materials. For the rare earth source chamber, Fig. 8.4(a), the rare earth chloride is melted on the wall of an internal dopant chamber and then dehydrated in Cl_2 to remove any OH impurities. During the core layer deposition, the dopant chamber is heater to induce a high vapor pressure (typically 1000 to 1200°C) with a second torch. The vapor is then carried downstream to the reaction point and is oxidized at elevate temperatures (1700 to 2000°C) based on the composition to form a vitreous consolidated layer.

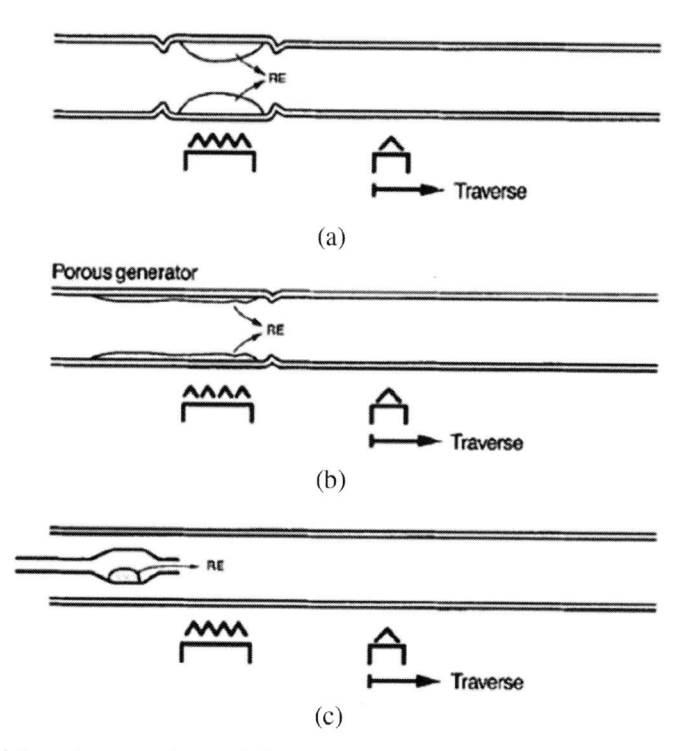

Figure 8.4 Internal vapor phase delivery systems: (a) rare earth source chamber, (b) porous soot generator, and (c) a heated source injector [12].

Figure 8.4(b) displays the case for a porous soot generator where the rare earth chloride (RECl$_3$) is impregnated in a porous silica frit before the deposition point. The RE generator is created by depositing a layer of 0.5-mm-thick low-density (0.5 g/cm^3) porous silica powder inside a section of the substrate tube at a sufficiently low temperature to inhibit complete sintering. This layer is impregnated by soaking the tube in an aqueous or alcoholic solution containing RE salt. The tube is then dried and baked in chlorine atmosphere to remove OH. To introduce RECl$_3$ vapor into the precursor mixture, the generator is then heated to 900°C by a stationary second burner producing a vapor pressure of a few Torr. The evolved vapor is then mixed with the other precursors and forms rare earth oxide (RE$_2$O$_3$) at the reaction point [10].

The final internal method is that of a heated source injector, depicted in Fig. 8.4(c). This method has the advantage having AlCl$_3$ vapor supplied through the injection tube to the rare earth chloride in the enclosed ampoule. The RECl$_3$ is placed in the ampoule and dehydrated by flowing Cl$_2$ gas. Once the process begins, the AlCl$_3$ vapor flows into the ampoule and mixes with the RECl$_3$ to form an Al-RECl$_3$ complex with higher vapor pressures compared to RECl$_3$. Also, flow of the AlCl$_3$ through the injection tube allows for better control of the rare earth concentration and results in improved homogeneity [11].

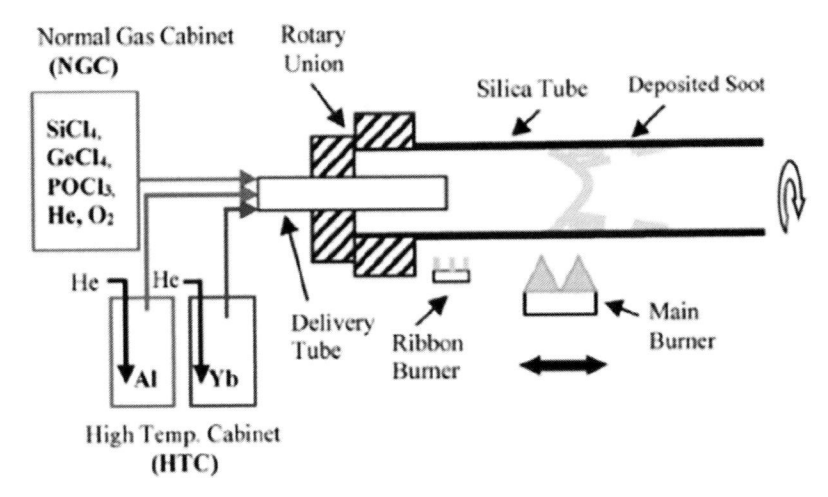

Figure 8.5 External vapor phase delivery system [14].

The final method is that of an external delivery system that is connected to the glass-making system thorough an injection tube and connected via a rotary union. The high-temperature cabinet as seen in Fig. 8.5, contains independent vessels of the AlCl$_3$ and rare earth that are heated to their sublimation temperatures and carried the gas stream by He or another inert gas. Here, a low-temperature ribbon burner can be used to heat the area between the injection tube and the glass deposition tube to avoid condensation of the vapors. This method has been used for many years, but there were always a lot of fundamental issues with blockages, the rare earths losing vapor pressure are other factors that limited the use of this technology and severely limited core sizes. Over the past 8 years, advancements have been made which makes this style repeatable, improved axial variations and the ability to dope higher rare earth concentrations and larger core diameters [13].

These advancements in the ability to dope rare earth ions led to a large boom within specialty fiber applications. Once the other challenges such as photodarkening of Yb-doped glasses were set to acceptable levels to achieve long operating times, these fibers were used in industrial fiber and laser systems. However, improvements at the glass preform level still need to be realized before optimization of the systems can be achieved.

In a double-clad fiber, there are two waveguides – the rare-earth-doped core that forms the signal waveguide, and the inner cladding waveguide for the pump light. The inner cladding of active fiber is often shaped to scramble the cladding modes and increase pump overlap with the doped core. The matching of active and passive fibers for improved signal integrity requires optimization of the core/clad concentricity and of the mode field diameter (MFD) through the core diameter and NA, which reduces splice loss. This is primarily done by tightening all the pertinent fiber specifications [15]. Both the

NAs of the active and passive preforms need to be extremely well controlled to reduce splice losses and to ensure that there is batch-to-batch consistency. This is also true of photosensitive fibers for gratings, multi-mode fibers for beam delivery from the diodes and from the lasers and amplifiers. All these factors at the preform level, NA, refractive index profile and control, and the composition of dopants are the key to successful systems within active fiber systems. This also lends itself to tighter control of the actual fiber dimensions leading to improvements of the fiber draw process.

8.1.2 Fiber draw

End users continually ask for tighter and tighter specifications on all properties of the fiber, specifically on geometrical tolerances which are dictated by the drawing process. This technology has evolved over the past 50 years and continues to make remarkable strides. Single-mode fibers drawn to 125 μm can now be held to ±0.5 μm with general ease, and to 0.3 μm in certain cases. This represents a major improvement in the technology, and it has been carried over into the specialty industry where many different fiber diameters are drawn in smaller volumes. The rise in active fibers has created many challenges on the fiber draw side as well, such as drawing octagonal and other shaped claddings to smaller and smaller tolerances. Improvements in the preform quality, feedback loops between the diameter gauges and draw equipment and draw furnaces themselves has created longer lengths, stable geometry across the draw and improved performance from the fiber itself.

Fiber optic draw towers come in many shapes and sizes and can be customized to meet almost any need for a manufacturer or research facility. Each tower will have different parts, but there are some standard parts for a draw process when using a preform. Figure 8.6 displays a simple schematic of a draw tower. It consists of the preform feed mechanism, a draw furnace, diameter gauges, coating station followed by a curing unit (either thermal or UV) and another diameter gauge. It then goes through a capstan and then will be collected on a drum winder. Maintaining the center line of the furnace to the capstan and then centering the coating stages and curing units will ensure concentric fiber to the coatings of choice.

Optical fibers coatings are another field unto itself nowadays. Not only can acrylate coatings be applied in line during the process but polyimide, silicone, low index polymers, metals and others that extend either the temperature range of the fiber, improved mechanical reliability and optical performance. Some specialty manufacturers produce fibers in line with extruded materials such as nylons, Teflon and Tefzel materials as a jacket material. This improves quality, yield and lead time as opposed to sending fibers to extrusion companies as a secondary process. The advancements in the coating industry have created more application space for specialty optical fibers.

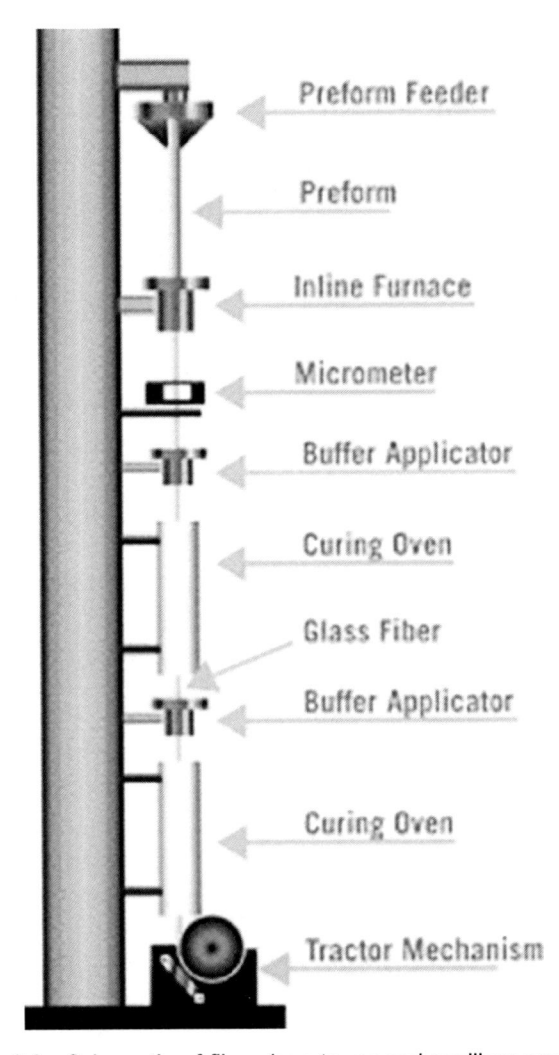

Figure 8.6 Schematic of fiber draw tower and ancillary equipment.

8.1.2.1 Draw furnaces

The draw furnace is the essential part of this process. Typically, the draw furnaces consist of high-purity graphite heating elements which create resistive heating or induction furnaces with zirconia parts that can heat up to 2300 to 2400°C. These devices can control to within ±0.2°C from the desired setpoint. In order to prevent any oxidation of the graphite elements, argon or other inert gases are sued within the furnace heat zones to combat that effect. The more recent and updated furnace models have also put additional features to limit the particulate size that can be produced. This is a key in producing high-strength fiber with long cut lengths after proof-testing. The size of these furnaces is also tailored based on the manufacturers' needs and capabilities.

Optical Fibers and Optical Fiber Assemblies 275

Figure 8.7 Nextrom 230-mm draw furnace (www.rosendahlnextrom.com).

Single-mode preforms can be made extremely large, over 200 mm in diameter, and require draw furnaces as seen in Fig 8.7. Larger furnaces such as that allow for large quantities of 125 μm fiber to be drawn >500 km at extremely fast speeds. For optical fiber manufacturers that only draw a few products with only one coating type, this is a popular furnace type.

However, for specialty manufacturers, smaller furnaces that have the capability to draw preforms ranging from 20 to 60 mm in diameter, shown in Fig. 8.8, are preferred. A large majority of these are resistive furnaces that use graphite components (elements, locking nuts and flow tubes) and can provide more flexibility with wider assortments of element sizes and flow tubers for subsequent fiber draws. Specialty preforms can be limited to smaller sizes based on the glass-making capabilities that are available so being able to draw a 20-mm preform and a 50-mm preform requires different elements and flow tubes to ensure that the gaps between the glass preform and flow tubes in the heating zone are consistent between various sizes. Otherwise, draw tower operators need to compensate with temperature and/or argon flows to maintain a consistent neck down region, which affects the fiber draw tension.

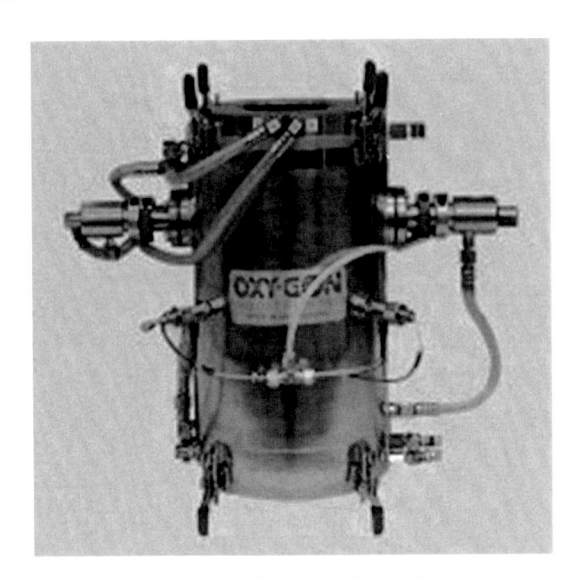

Figure 8.8 Resistive draw furnace (50 mm) (https://www.oxy-gon.com/fd1.htm).

This is extremely important in single-mode fibers as the cutoff wavelength and attenuation can be affected by changes in draw tension. In many towers, an inline draw-tension monitor can be used to ensure that the appropriate tension is applied. If it is out of range, either the draw speed or temperature can be modified to compensate. In multimode fibers, the draw tension is important but does not have as dramatic an impact as on single-mode or large-mode-area (LMA) fibers that inherently contain only a few modes. High tensions can lead to increased losses due to attenuation which affects the fiber performance and needs to be maintained.

The draw process, while fundamentally simple, has many complexities. Ensuring that a continuous feedback loop mechanism is fundamental to producing and manufacturing a quality optical fiber. Non-contact diameter gauges can take hundreds to thousands of scans per second and relay them back to the PLCs of the draw tower. There the speed of the capstan can be changed to constantly maintain the fiber diameter. Optionally, a caterpillar capstan can be installed in line with the drawing system, which is designed to pull glass rods, thin-walled capillaries, or sub-structured canes with minimal surface pressure. All these devices are aligned vertically along the drawing line in a mechanically stable tower that is usually higher than 4 m. Larger heights (e.g., >20 m) are necessary for high-speed fiber drawing [16] at drawing speeds of >1000 m/min in order to provide sufficient fiber cooling across the distance between the furnace outlet and coating applicators. Appropriate fiber cooling is necessary to prevent overheating the coating material from contact of the coating material with a high-temperature fiber, which can result in improper wetting [17].

8.1.2.2 Optical fiber coatings

Coatings for optical fibers vary by nature due to the application of the product. Traditionally, they have been in place for mechanical protection of the glass fiber from atmospheric impurities such as OH and from effects such as micro-bending and other harsh environments. The coatings can also be manufactured and tailored to ensure optical reliability by NA adjustment with low-index polymers, mode stripping and decrease in optical attenuations. In addition, the fiber coating might address other functional aspects, such as electrical or thermal conductivity or transmission of mechanical stress parameters from embedding media in the fiber (e.g., mechanical sensing) [18].

There are three main categories of coating types available: organic, inorganic and hybrid coatings. The organic category covers a wide range of coatings with UV curable and thermally cured resins. These include acrylate (and fluor-acrylate polymers), perfluoro polymers (e.g., Teflon), polyacrylates, polyimides, silicones, and their derivatives. Fluor-acrylate polymers have a refractive index lower than that of silica glass and have been used on active fibers for lasers and amplifier applications as they produce an NA >0.46 which is much larger than fluorine-doped silica glass can provide as a cladding material (~0.30). This improves the brightness factors that lasers and amplifiers can achieve.

Hybrid coating materials for optical fibers are based on organically modified ceramic precursors (ORMOCERs) and combine the properties of organic and inorganic components. Such hybrid materials are applied where properties of both polymers and inorganics are required (e.g., to achieve high temperature stability or lower hardness) [19].

Metal coatings are used when fiber applications are envisioned under harsh conditions (e.g., aggressive media, mechanical impacts), or when special properties are necessary (e.g., rapid fiber cooling during fiber operation) [20]. Gold-coated fibers, shown in Fig. 8.8, are of particular interest in high-temperature and also cryogenic-temperature sensing applications. This coating can operate over a temperature range of almost 1000°C, which is a rare ability with optical fiber coatings. Due to the nature of the process, the gold thickness is only ~15 μm for a 125-μm fiber. Aluminum is also a popular metal coating that is commercially available. It maintains very high strength and long life at extended stress levels in applications that require tight bends. Also, the strong chemical bond between the silica cladding and the aluminum enables direct termination without pistoning. This bond also makes aluminum coating the ideal choice to preserve deep-UV performance.

Carbon is an inorganic coating that can be applied to fibers that provides a hermitic protection against moisture or hydrogen. These are used primarily in the oil and gas industry and other down-hole applications where the optical fiber needs to be impervious to hydrogen impregnation. Carbon layers on

Figure 8.9 Image of gold coated optical fiber (www.fiberguide.com).

silica also influence surface properties such as wettability, which is of great importance for additional organic or metallic layers [21].

All coatings have different physical and chemical properties and are very specific to end applications. Figure 8.10 shows a table taken from K. Schuster et al. [17] with the optical, chemical and some physical properties of common optical fiber coatings. Not only does the viscosity of the materials vary, so does the coating diameters that are applied. Acrylate coatings are typically 50 to 60 μm thick, while polyimide coatings are only 6 to 7 μm per layer. While this does make them mechanically reliable and able to withstand >100-kpsi proof test levels, it does not permit for >200 or 300-kpsi levels such as standard acrylates. Both coatings in particular, polyimide and acrylate, utilize

	Curing	Viscosity (uncured)[a] [Pas]	n (cured)[b]	NA[c]	Max. coating thickness[d] (μm)
Polyamidimide	T	n.a.	1.81		6
Polyimide (Microquartz)	T	26.2	1.68		7
Polyimide (PMGI)	T	9.8	1.57		6
ORMORCER®	UV	3.0	1.51		100
Urethane-Acrylate DSM 3471-3-14	UV	9.2	1.505		100
Silica			1.4469		
Silicone LR7665 (Wacker)	T	14.2	1.415	0.30	100
Silicone RT601 (Wacker)	T	2.9	1.409	0.33	100
F-Acrylate Opticlad	UV	3.2	1.38	0.43	50
F-Acrylate (SSCP) PC 373	UV	3.5	1.376	0.45[e]	60
F-Acrylate (SSCP) PC 370	UV	5.7	1.372	0.46	60
F-Acrylate (SSCP) PC 363	UV	5.0	1.363	0.49	60
OF-133 (MyPolymers)	UV	n.a.	1.326	0.61[e]	50
Teflon AF (Du Pont)	T	n.a.	1.314	0.61	3

[a]25°C, [b]measured at 1300 nm (prism coupling device), [c]calculated from n, [d]single layer, depending on required layer properties (e.g., bubble-free) and technology (dip coating vs. pressure coating), [e]measured using far field method.

Figure 8.10 Table of chemical and optical properties of various coatings [17].

different curing techniques and coating types. Acrylates are applied either as a single- or dual-layer process where after each layer is applied, it then becomes cured after passing through UV curing systems. Polyimide coatings are more complex as it needs anywhere from 3 to >5 coating applications to occur to achieve the final diameter. After each layer is applied, it requires the fiber to pass through thermal oven to cure it, before the next coating application can begin. This requires an extremely tall draw tower or multiple stages to coat the fiber in succession before the fiber is collected. Not all facilities have the space or resources to perform this type of operation.

For the organic, hybrid and polymer materials, they are applied in line during the drawing process via a gravity cup or a pressurized system. Using a gravity cup with a shaped conical die is a simple method that is very convenient for the operators to utilize. The fiber goes through the gravity cup and die, and once the fiber is at the desired diameter, the material is poured very carefully into the cup, and the coating process begins. This method has drawbacks as bubbles can be trapped as well as contaminants can enter the cup and cause spot defects after the curing process. A pressurized coating system removes these issues as the coating is not exposed and high-quality coating can be applied. The material can stay in a bottle and be heated and/or rolled to maintain its consistency while also removing any air bubbles. Larger continuous holes can be drawn without the concern of refilling as in a gravity cup. This process can also have some limitations once the fiber diameter reaches sizes >1500 μm as it is possible to have difficulty to maintain concentricity of the coating. The other drawback is that the cleaning process of pressurized systems takes longer and is more detailed.

Independent of how the coating material is applied, the sheer number of different materials that can be applied has grown. Each of these materials can be used in different application spaces with different temperature requirements. Table 8.2 displays some of the more common industrial coatings with their application space and temperature limits.

Table 8.2 Listing of coating materials with their temperature ranges and applications.

Coating Material	Temperature Range	Application Space
Standard Dual Acrylate	−40°C to +150°C	Standard Telecom Coatings, Data Centers, Long-Distance Network Communications
Low-Index Acrylates	−40°C to +150°C	Fiber Lasers & Amplifiers, Sensing Systems, Medical Devices
Polyimide	−190°C to +350°C	Oil and Gas, Medical & Data Communications Apps
Silicone/Nylon	−40°C to +100°C	Fiber Bundles, Medical Applications
Tefzel	−40°C to +150°C	Disposable Medical Products
Aluminum	−269°C to +400°C	Solarization & Long-Term Stress Apps
Gold	−269°C to +700°C	Solarization & Down-Hole Sensing, High Vacuum and Pressure Apps

The combination of improved optical fiber coatings and advancements in fiber draw technology has allowed for more solid-state technologies to be replaced by optical fibers in many markets. Fiber lasers & amplifiers have grown in industrial applications for welding, marking and medical uses. All fiber systems have been developed for sensing with coatings that can withstand harsh environments. Disposable medical probes are made with optical fiber assemblies and coatings that are inert when entering the human body. These advancements continue to push the boundaries of optical fibers. Another development has also been the testing of fibers to go into new markets and applications. As the coatings have improved, so has the testing and quantification of the fibers as well as the coatings.

8.1.3 Fiber testing capabilities

For any manufactured product, there is always a specification with tolerances that is listed and measured. For any optical fiber there are basic parameters that are always measured: core/cladding/coating diameters, NA, core/clad offset and proof test level. For single-mode fibers, the cutoff wavelength and mode field diameter (MFD) are also recorded. There are industry-standard devices, such as the Photon Kinetics PK 2300 (shown in Fig. 8.11), that can perform geometrical and spectral attenuations on optical fibers as small as 40 μm in diameter to as large as 440 μm. This provides information of the diameter of every region, non-circularity and the offsets and concentricities. For the optical side, the cutoff, MFD and NA of low-index polymers can be measured as well. With additional modules, micro-structured fibers, multicore and polarization maintaining (PM) fibers can also be analyzed and properly

Figure 8.11 PK2300 Fiber Analysis System (http://www.pkinetics.com/products/product detail.aspx?model=2300).

quantified using one device. Benches such as these can perform the majority of testing required for optical fibers, but within the specialty fiber market more testing is required in which different benches need to be used or built from scratch to properly quantify fiber properties.

For fibers that have cross-sectional geometries over 440 μm, the method of measuring the diameter needs to be modified. There are a few ways to accomplish this: optical microscope with attached micrometers, more advanced microscopes with automated focusing systems and geometrical software or producing your own bench to work in similar fashions to that of commercially available units. Attenuations can be achieved via optical time-domain reflectometer (OTDR) at specific wavelengths of interest or with optical spectrum analyzers (OSAs) and multiple light sources to cover the ranges from 300 to 1700 nm. More specialty units can be used to go further into the UV and IR, if needed. For larger-diameter fibers, >1200 μm, a mechanical bend method of proof testing may be required as tension-based units may not have loads large enough to achieve the desired kpsi, typically 100. However, there are more specialty measurements that are currently required as the fibers have advanced in complexity.

High-temperature coatings such as polyimide and metal coatings have permitted optical fibers to be used in down-hole applications for the oil and gas industry along with other applications where the fibers need to survive harsh environments. Testing needs to be performed to confirm that the fibers can operate for long periods of time before moisture or hydrogen permeates into the fiber core and manifests as increased attenuation. For the oil and gas industry, hydrogen-loading chambers are utilized in which reels of fiber are placed into a chamber and loaded with hydrogen to a specific pressure and elevated temperatures to allow for H_2 diffusion (but lower than the maximum temperature of optical fiber coating). Optical sources are utilized to launch light into the fiber through feedthroughs for continual monitoring through spectrum analyzers or OTDRs. This provides timed data and how the fiber can resist the diffusion of hydrogen into the core. Figure 8.12 displays data from Coherent-Nufern from their NuSensor brochure that compares the induced attenuation of hermitic versus non-hermetic fibers. This shows that the addition of hermitic layers such as carbon play a major role in preventing hydrogen from getting into the glass. Since hydrogen atoms are so small and can easily diffuse out of glass at temperatures >100°C and increased pressures, the carbon hermetic layers are essential there to keep the hydrogen in the glass. They essentially provide the best method to significantly reduce the hydrogen diffusion. It is extremely vital for fiber manufactures to test the different fiber types with various coatings to see how the fibers perform in extreme environments. This testing shows the importance of hermetic coatings for these applications, and end users require this data before purchasing and testing new fibers.

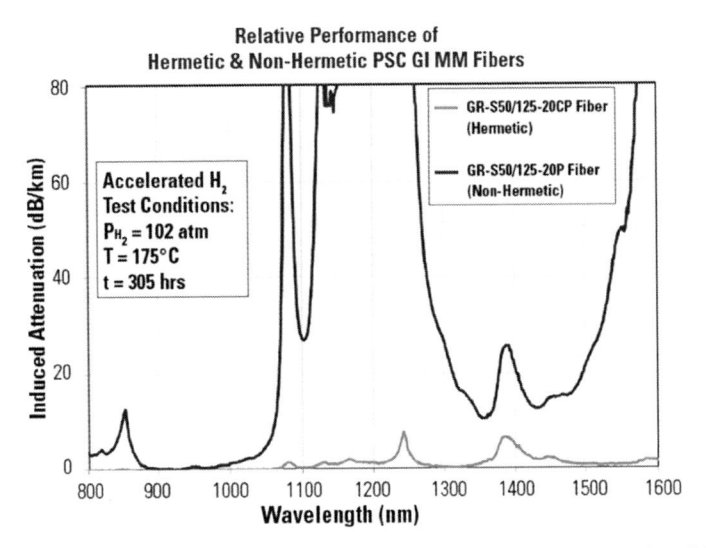

Figure 8.12 Induced attenuation of hermitic and non-hermitic fibers after 300 hours of Hydrogen loading at 175°C and H₂ pressure of 102 atm (www.nufern.com).

As active fibers have increased in popularity, the testing requirements have also evolved to provide more information on how the fiber will perform in a system. In addition to the standard geometries, offsets and attenuations, pump absorption needs to be measured, and in some cases end users may require slope efficiencies (SEs). These are not standard measurements and benches or low-power benchtop laser systems are made to measure the pump absorption and the SE of the fiber.

The slope efficiency is a good test to perform on active fibers. It obtains two vital pieces of information: the ability of the fiber to convert pump light into signal light and the pump absorption per meter (dB/m) which end users require to use the appropriate length in their lasers or amplifiers. The setup for this test can vary based on which type of fiber and how much power is needed, but in its simplest form a pump diode is used and spliced to the active fiber and then a power meter to record the signal power. Figure 8.13 shows an example of the SE result of a Yb doped fiber. Multiple different pump powers are used and plotted against the corresponding signal power. The slope of this plot is defined as the slope efficiency of the fiber, which is the ability of the fiber to convert pump light launched into the cladding to signal light from the core. The higher the value, the more efficient the fiber. Of course, the SE value can not be higher than the quantum defect for that fiber, which is the pump λ/signal λ. In this example, it would be 920 nm / 1060 nm = 86.8%. Also, from this data the total gain is recorded in dB, and the length of fiber is defined. From this, the absorption per meter can be calculated. These additional tests provide end users with the information they require to effectively select fibers to use in their systems.

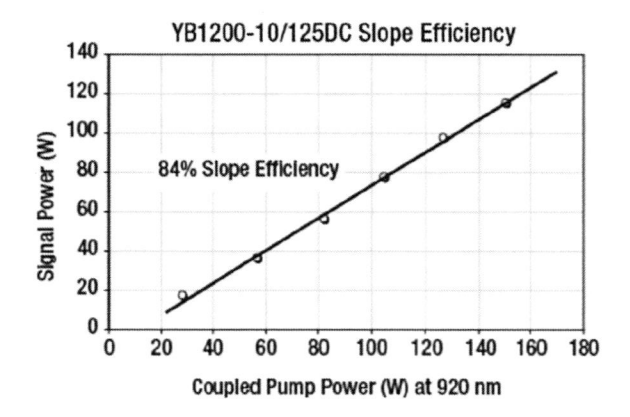

Figure 8.13 SE plot for a Yb-doped fiber pumped at 920 nm (www.thorlabs.com).

Many of these additional tests have been developed to allow fibers to penetrate markets that traditionally have been dominated from other means. Device users require data to confidently switch to an optical fiber solution, even if they are more efficient or provide longer lifetimes. As the fiber technologies continue to grow, so do the testing requirements and the ability to adapt to new markets and applications.

8.1.4 Advanced fibers

There has been an increasing need for more advanced fibers to deliver light, especially in the form of multicore fibers for applications in communication and sensing.

8.1.4.1 Multicore fibers

Over the past few decades, the demand for data has been increasing exponentially particularly due to video streaming and cloud computing. Typically, the data traffic in the backbone networks is transmitted along a single mode fiber using wavelength division multiplexing (WDM) to send multiple wavelengths of light along a single fiber. As the demand for data is increasing, the industry is now approaching the capacity limit. To overcome this limit, multicore fibers have been developed in which there are multiple cores (suitably separated to reduce crosstalk) with a common cladding [22].

The design freedom and flexibility in conventional single-core fibers is limited due to the core being placed at the center of the cladding. In these fibers, the core has some limited flexibility in design where the shape of the core (round, elliptical, square, hexagonal) can be altered to achieve the desired effect for the application the fiber is being used for. Multicore fibers have a greater degree of freedom in core design, number of cores, core layout, outer cladding thickness and cladding diameter can be optimized with regards to optical design required for the application at hand [23].

Typically, the distance between the cores is such that there is no crosstalk, but for certain applications the cores are placed closer together to allow for interaction between the cores. This interaction can be used to make temperature and strain sensors [24].

8.1.4.2 Photonic crystal fibers

Photonic crystal fibers (PCFs) are specialty optical fibers that have microstructures that are built into the length of the fiber that are essentially air holes. PCFs are different from the conventional optical fiber in that they use total internal reflection for light confinement in a hollow core to propagate the light along the fiber. Light propagation in these fibers is superior to that of light propagation in standard fibers [25].

The photonic crystal is the low-loss, periodic-structured dielectric medium created by the microstructures which have photonic band gaps that prevent light propagation in certain directions and within a certain range of wavelengths. There are numerous applications for these fibers such as

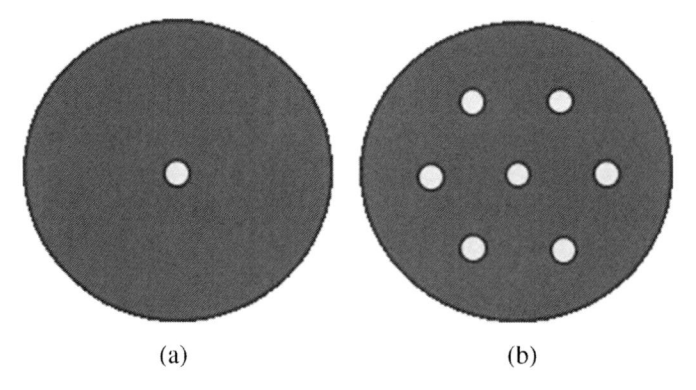

(a)   (b)

Figure 8.14 Cross-sections of fibers: (a) Single-core fiber; (b) Multicore fiber.

Figure 8.15 Photonic crystal fiber (public domain image: https://commons.wikimedia.org/w/index.php?curid=48680862).

spectroscopy, metrology, biomedicine, imaging, telecommunication, industrial machining and defense technologies.

8.2 Optical Fiber Applications and Assemblies

Optical fibers are used to collect and deliver light in numerous applications, and they are also used to make instruments more sensitive as well smaller and easier to use.

8.2.1 Applications

8.2.1.1 Medical

From the beginning, optical fibers were used for medical imaging (endoscopes) to look inside the body to determine treatments. The use of fibers in the medical field has increased considerably: they are now used for laser surgery, laser scalpels, optical coherence tomography, kidney stone ablation, tumor ablation, skin treatment and many other treatments.

8.2.1.2 Analytical/biomedical instrumentation

Fibers are finding uses in analytical instruments such as flow cytometry, DNA analysis, cell counting, confocal microscopy amongst others. Most of these instruments use fluorescent tags to detect abnormalities in cells. The intensity of fluorescent light from these is very small, in some cases in pW range. Fibers are used to collect the light and deliver it to photodetectors as well as deliver light from a laser to cell samples.

8.2.1.3 Fiber sensors

Optical fibers make excellent sensors as they are immune from electromagnetic radiation and can be used in environments where electronic sensors cannot be used. There are basically two classes of fiber sensors: (1) intrinsic, where the fiber is the sensing element, and (2) extrinsic, where the fiber is used to relay the signals from a sensing element outside of the fiber. Fibers can be used to sense almost everything and have been used to monitor stress, strain, temperature, pressure, current, voltage and used to measure chemicals. They are very useful in hazardous environments and have been used in diverse applications such as down-hole monitoring in oil and gas fields, structural monitoring of bridges, dams, buildings, and aircraft, to name a few.

8.2.1.4 Optical switching

Fibers form an integral part of optical switches where they are used as the conduit for information being switched. Optical switches are particularly important in data centers where the transfer of information needs to be performed as quickly as possible. Generally, information is transmitted using optical fibers and in conventional switches this information is switched from one channel to another by converting the optical signal to an electrical signal which is then switched to the required channel and then converted back to an optical signal which is then transmitted via another fiber. This process can be time consuming. To speed up the process of switching the information from one channel to another, a fully optical process is needed and used where the information is not converted from optical to electrical. This can be achieved via micro-electromechanical system (MEMS) technology where mirrors on a silicon chip can be used to switch the optical signal from one fiber to another in a 2D fiber array.

8.2.1.5 Industrial uses of optical fiber

Fibers are finding increasing uses in the industrial setting where fiber lasers are used for cutting and welding as well as sensors on the industrial production lines. Laser welding and cutting requires the delivery of high power and the cutting/welding head needs to move in various patterns (depending on the cut or the weld required) therefore the fibers tend to have a large core diameter.

8.2.2 Mode mixing and de-speckling

Due to the increasing use of lasers in medical and analytical instrumentation as well digital projection applications optical fibers are being used to deliver the laser light within these systems. One of the major issues for these applications is the laser speckle. This is seen as noise in the system as well as providing a grainy image in digital projection and needs to be reduced.

When observing light as a spot on a screen a random granular pattern is observed which is constantly changing. The pattern, also known as speckle, is due to fluctuations in the intensity of the light hitting the spot. The intensity fluctuations are the result of random differences in the phase of the light being reflected from a surface. Since most surfaces can be considered rough on the scale of optical wavelength, when light is scattered from a surface the phases will constructively and destructively interfere leading to contrasts in the intensity that we see. The changes in contrast have no visible pattern and appear completely random, leading to the static-like appearance of speckle. Speckle also appears when light is transmitted through materials since the surface structures of the materials will randomly change the phase and polarization of the light, leading again to changes in the intensity seen within the illuminated area [26].

Speckle is easy to observe with the naked eye, but it is important to know quantitatively the degree of speckle that a laser spot has. Speckle contrast is a standard way that laser speckle is measured. It is a measurement of how much the signal varies around the average output of the laser. It provides a measurement of the roughness of the output across the laser-beam cross-section. The mathematical expression is

$$C = \frac{\sigma}{\langle I \rangle} \qquad (8.2)$$

where C is the speckle contrast, σ is the standard deviation of the intensity over the radial distance, and <I> is the average intensity over the radial distance and is reported as a percentage. If there is a lot of variation in the signal, i.e., the signal looks very rough with a lot of speckle, the speckle contrast will approach 1 or 100%. If there is very little variation in the signal, i.e., the signal looks very smooth or has very little speckle, the speckle contrast will approach 0. Speckle contrast can be measured in the near field (top-hat profile) or far field (Gaussian profile). Measuring speckle contrast is usually done by either imaging a laser spot on a screen or imaging the beam with a laser-beam profiler. To be able to measure the speckle contrasts of either the near or far field, images can be broken up into small groups of pixels that contain the intensity data of the light stored by the pixels. If a square of pixels contains the beam, speckle contrast can be measured for that local area of the beam. This can be done over the entire beam area and averaged to get an overall speckle contrast for the entire beam profile.

The speckle can be reduced actively by using vibrating diffusers, but this causes an issue when used with fibers as light from a laser (particularly from pigtailed laser diodes). Pigtailed lasers output the light via optical fibers and for applications where speckle is an issue the light has to be de-speckled by taking light out of the fiber sending it through lenses and diffusers (where it is de-speckled) and then launched back into a fiber. This causes an increase in loss and an increase in cost due to the need for additional optics.

Speckle can also be reduced within a fiber by actively agitating it, as shown in Fig. 8.16. In this way there is zero loss, and the costs are reduced.

8.2.3 Anti-reflection technologies

In all applications there is a reflection from the end face of the fiber which is generally unwanted and can cause problems. Typically, for glass in air, 4% of the light is reflected. For many applications this reflection can be problematic whether using high power (reflections can cause damage) or low power (where the reflection is reducing the light transmitted). This reflection is normally reduced by using anti-reflection coatings.

This increased use of AR coatings has thrown up some limitations particularly in high-power delivery where the low damage threshold of conventional AR coatings has become an issue. For optical fibers, the ever-increasing demand for optical power delivery is a problem. When launching light into a fiber, the optical beam is focused down to a small spot with an NA (numerical aperture) below or matching the NA of the fiber. This spot must be

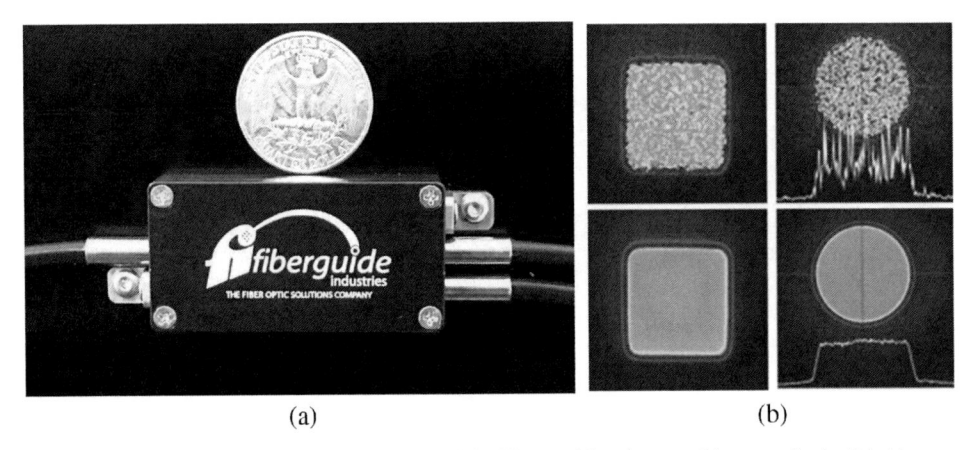

Figure 8.16 Fiber based de-speckler: (a) Fiberguide de-speckler product. (b) Upper: Without de-speckler; Lower: With de-speckler.

smaller than the diameter of the fiber core to allow for efficient coupling. The energy density quadruples with reduction in spot diameter leading to damage in the AR coating, rendering it unusable (Fig. 8.17).

This damage occurs at the glass interface and can be mitigated using end caps. Due to the much higher damage threshold for light traveling in glass, high-power light can be launched into fiber through an end cap. Where the size of the focused spot on the end cap can be larger at the glass-air interface and hence lower energy density while the spot at end cap/fiber (glass-glass) interface is smaller than the diameter of the fiber. The use of end caps on fibers has allowed the optical power delivered to increase by effectively

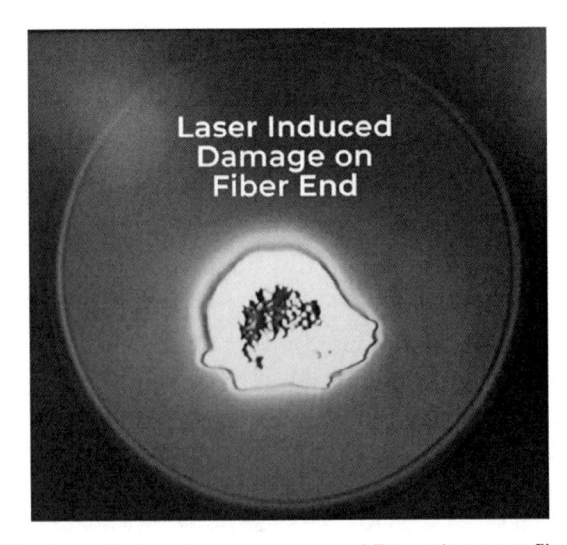

Figure 8.17 Laser-induced damage on an AR coating on a fiber end face.

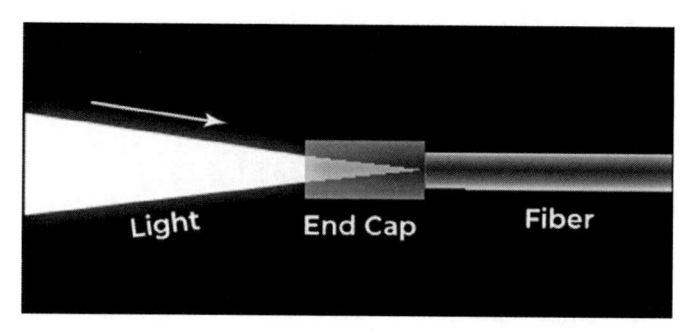

Figure 8.18 Showing light focused on the end of a fiber through an end cap.

allowing for a larger spot size at the launch face of the fiber (Fig. 8.18). However, AR films are still required to reduce reflections.

Conventional AR coatings have various limitations such as very narrow bandwidth, a relatively low damage threshold compared to that of glass, etc. Due to the ever-increasing demand for higher power delivery, the damage threshold for AR films has effectively been reached.

A better solution can be achieved by the use of meta-surfaces on the end of fibers that can replace standard AR coatings. These surfaces or motheye structures impart AR properties to the end faces of the fibers. They have superior damage threshold than coatings, and they work over a broad wavelength range with no angular dependence. Fibers with motheye end faces can be used in both high- and low-power applications.

8.2.3.1 Anti-reflection surfaces based on motheye structures

During the 1960s, it was discovered that corneal lenses on insects such as moths have periodic cone-like nano-structures that reflected very little light at any wavelength or incidence angle, as shown in Fig. 8.19 [27]. It was also discovered that the low reflectance was due to the continuous refractive index transition between air and the substrate (cones on the eye). The size of the

Figure 8.19 Close-up of an insect's eye.

structures affects the reflection properties. Typically, these structures are smaller than the wavelength of light. The lower wavelength limit is determined by the lateral size of the structures, and the upper wavelength limit is determined by the depth.

A new technique based on these structures has been developed to provide anti-reflection properties for glass and is now being applied to optical fiber end faces. These structures are smaller than the wavelength of light and have AR properties comparable to AR films as well as having a broad anti-reflection band greater than that of AR films but with much greater damage thresholds. When light encounters the nanoscale roughened surfaces with height and spacing that is smaller than the wavelength of light to be used, light propagating through the textured surface will encounter a gradual change in the refractive index as it enters the bulk material. This gradual change reduces the amount of reflected light. These structures have been recreated on glass using reactive ion etching which have shown anti-reflective properties.

Further work by Hobbs et al. has shown that random nanostructures rather than ordered structures produce even better AR properties [28]. The random anti-reflection surfaces provide anti-reflection properties that are better than films and provide a much broader anti-reflection bandwidth.

8.2.3.2 Random anti-reflection surfaces (RAR)

Random nano structures are created by the use of reactive ion plasma etching techniques on glass surfaces. These structures mimic motheyes and have anti-reflection properties. The damage threshold for AR-coated samples was 24 J/cm^2, and for plasma-etched RAR structures, it was 48 J/cm^2.

8.2.3.3 Random anti-reflection surfaces on optical fibers

Optical fibers with a 1500-μm core were etched to create RAR surfaces on their end faces as shown in Fig. 8.20.

Figure 8.20 Photograph of (a) an etched fiber in an SMA connector and (b) an SEM showing the structures in detail.

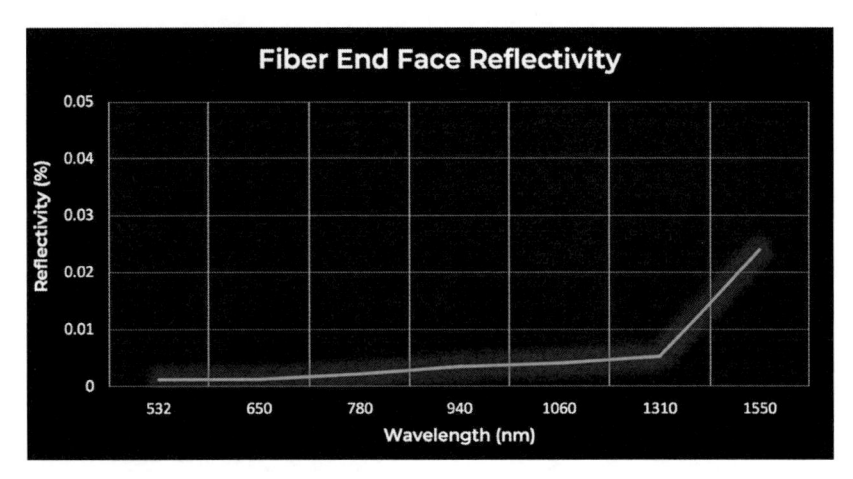

Figure 8.21 Graph showing reflectivity from an etched fiber.

The etching process produced fiber end faces with reflectivity of <0.1% over a broad wavelength band of 800 nm, as illustrated in Fig. 8.21.

These fibers (along with untreated and AR-coated fibers) were sent to Quantel for laser-induced damage threshold (LIDT) measurements using the same set up that was used for the fused silica windows. The results showed damage thresholds of 130 J/cm^2 for the untreated fiber, 59.1 J/cm^2 for the etched RAR fibers and 27.7 J/cm^2 for AR-coated fibers (Fig. 8.22). The AR coating was centered at 1064 nm with a reflectivity of 0.04%. AR technologies are needed to reduce the reflection from the glass-air interface, but AR films have a low damage threshold, hence the need for a new technology (RAR) to increase this threshold.

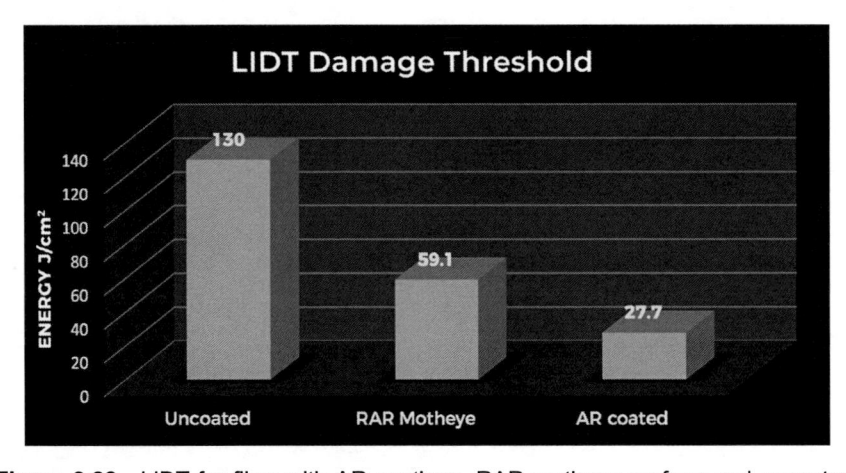

Figure 8.22 LIDT for fiber with AR coatings, RAR motheye surface and uncoated.

8.2.3.4 RAR (Meta) surfaces vs AR coatings

The results so far show that the RAR structures produce anti-reflection properties that are superior to those shown by conventional AR coating. RAR surfaces have a bandwidth of several hundred nm with a damage threshold that is twice that of AR films. This makes the RAR motheye structures eminently suitable for high-power applications where AR coatings cannot be used.

These surfaces can reduce reflectivities well below those achieved by conventional multilayer AR coatings with a very broad operating range of wavelengths and no angle dependence. They can be used in applications where detecting very low levels light over a broad band are essential (such as flow cytometry).

8.2.4 Assembly design considerations

Every fiber optic assembly has a mechanical component as well as an optical component. This could be a single fiber cable terminated with industry standard connectors on both ends and jacketed in a flexible plastic sheathing to a multi-fiber design consisting of multiple inputs and/or outputs, each with different cross-sectional areas and geometries, each requiring a custom-machined end fitting and a heavy-duty outer jacket to protect the assembly from being crushed. The mechanical design is highly dependent on the environment in which it will be used. Most of the industry-standard connectors are rated for a temperature range of −40°C to +80°C. Most fiber optic epoxies, fiber coatings and sheathing materials also support this range, so this could be considered a typical assembly environment.

Probably the most important factor when designing a fiber optic assembly is to choose the correct fiber construction to match the operating temperature and environmental conditions. Fibers have a variety of claddings that allow them to operate in a wide range of temperatures and conditions. Table 8.3 outlines many of the most common fiber types available and their temperature ranges.

Table 8.3 Common fiber materials and operating temperatures.

Core	Cladding	Buffer	Secondary Buffer	Operating Temperature
Fused Silica	Fused Silica	Acrylate		−40°C to +85°C
Fused Silica	Silicone	Nylon		−40°C to +100°C
Fused Silica	Fused Silica	Silicone	Nylon	−40°C to +100°C
Fused Silica	Hard Polymer	Tefzel		−60°C to +125°C
Fused Silica	Fused Silica	Silicone	Tefzel	−40°C to +200°C
Fused Silica	Fused Silica	Polyimide		−190°C to +325°C
Fused Silica	Fused Silica	Aluminum		−269°C to +400°C
Fused Silica	Fused Silica	Copper		−190°C to +600°C
Fused Silica	Fused Silica	Gold		−269°C to +700°C

Temperatures outside the typical range require considerations of a number of different parameters. One important factor is the choice of epoxies to bond the fiber and other mechanical parts of the assembly. When choosing an epoxy, thought must be given to the effect it might have on the fiber. Many high-temperature epoxies exhibit high amounts of shrinkage when curing. This may cause stresses in the core of the fiber, which can lead to poor modal quality or an increase in the numerical aperture (NA) of the fiber. High-temperature curing of the epoxy can also cause stress fractures in the fiber face. Gradual ramping of curing temperature is usually required to prevent this.

When choosing an epoxy, thought should be given to the glass transition (T_g) value of the epoxy being considered. A higher T_g allows use at higher temperatures. If the gap between the T_g of the epoxy and the operating temperature of the assembly is too great, then the fiber may piston at the face of the termination. The epoxy might even break down completely and cause the fiber to come loose from the connector. Cryogenic operating temperatures (<−150°C) tend to work best with epoxies that have lower T_g values. Glass transition temperatures below room temperature tend to cause less stress on the fiber as materials contract in very-low-temperature environments.

Very high temperatures (>300°C) require special designs, assembly techniques and materials. Gold-, aluminum-, and copper-coated fibers can be terminated with ceramic epoxies which will allow them to work up to their respective maximum temperatures. Gold-coated fibers can also be silver soldered (brazed) into a mechanical package. This allows the assembly to be hermetically sealed as well as work at temperatures as high as 700°C. This process can also be used for pressure feedthroughs that are able to work up to 10,000 psi at high temperature.

While temperature is of main concern other environmental conditions are also important. Factors like vacuum, high pressure, and caustic liquids or gases require careful examination of the fiber coating materials as well as the surrounding materials of the assembly. Ultra-high vacuum (UHV) assemblies have stringent outgassing requirements that require NASA approved adhesives and sheathing materials.

8.2.5 1D and 2D arrays

Fibers can be arranged in what are called one-dimensional (1D) and two-dimensional (2D) arrays. A 1D array is a single line of fibers, and a 2D array is a grid of fibers in both X and Y dimensions. In both cases, they are specified by the spacing between fibers. In the case of 1D array, there is a single defining dimension denoting the center-to-center position of the cores of the fibers. For 2D arrays, there are two dimensions, X and Y, that define the core positions.

One-dimensional arrays can be manufactured by several different techniques. The manufacturing technique will define the position accuracy

1.
SLIT ARRAY

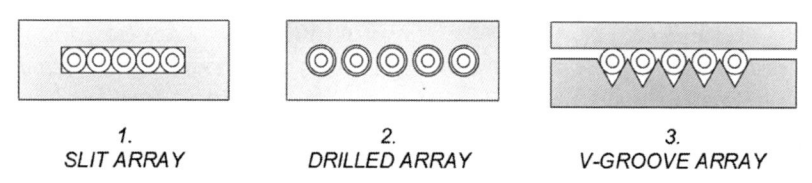

2.
DRILLED ARRAY

3.
V-GROOVE ARRAY

Figure 8.23 Various designs for a 1D array.

of the fiber cores. Fibers arranged in a machined slit (see Fig. 8.23.) will be defined by the fiber diameter tolerance and the machined opening size and tolerance. This type of array is common for spectrometer slits.

If more specific center positions are necessary, then machined holes can be drilled to position each fiber. Machining tolerances again come into play, so one can expect positional accuracy of ±25–40 µm. See Fig. 8.23.

For the highest position accuracy, a v-groove is used to hold the fibers (see Fig. 8.23). V-grooves can be made by grinding individual grooves into a glass substrate. A wide variety of materials, such as fused quartz, BK7, HK9L and other glasses, can be chosen to match other parameters of the assembly. This type of v-groove can be made to a very precise center to center tolerance of 1 µm or less. Glass v-grooves are usually specified to hold the fiber cores to ±1–2-µm positional accuracy after assembly. See diagram 3 in Fig. 8.23 for a view of an assembled v-groove.

Another technique for manufacturing v-grooves is by photolithography and KOH etching in silicon wafers. Photolithography is capable of sub-micron features in silicon, so this is the most accurate way to make a v-groove. Typical fiber positioning in an etched silicon wafer will attain ±0.5 µm or less in core-to-core measurements. Fiber position is dependent on factors other than just the v-groove substrate. Fiber tolerance, assembly technique, and epoxy choice play a role in how accurately the fibers can be placed in the termination.

Two-dimensional arrays range from a few fibers to thousands of fibers, depending on the applications. 2D arrays are used in optical switching and in sensing applications where spatial optical data is necessary, such as DNA sequencing, astronomy, and nuclear research.

Two-dimensional arrays can be made by similar methods as 1D arrays. A machined opening can hold multiple rows and columns of fibers. The fibers can be packed into this opening in two ways; a close (hexagon) packing or a square packing arrangement (see Fig. 8.24). These two types of arrays have many uses such as illumination and specialty spectroscopy applications, but the center-to-center tolerance is highly dependent on fiber tolerance and traditional machining tolerances.

The most accurate way to place fibers in a 2D array is by way of individual holes for each fiber. One such method creates holes in a glass substrate by a process that uses light to alter the structure of the glass (using a

4.
CLOSE PACKED ARRAY

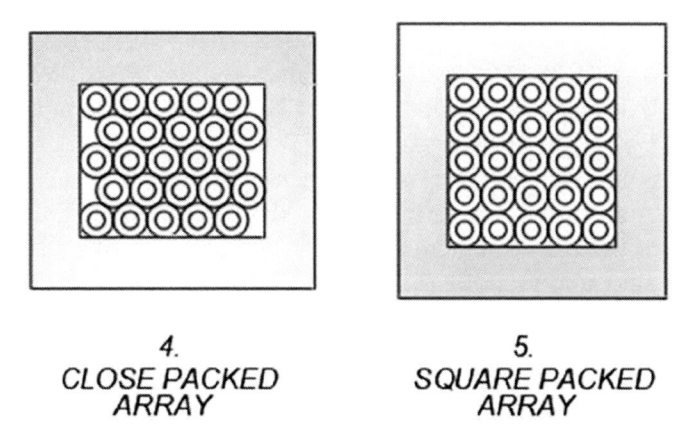

5.
SQUARE PACKED ARRAY

Figure 8.24 2D arrays.

femtosecond laser) to enable selective etching to remove the modified material. Sub-micron resolution has been attained with this method by Femtoprint SA.

The most common method of creating holes for 2D arrays is by photolithography and a KOH etching process. This is the same process that was developed to create integrated circuits. Fiber-to-fiber positioning of ±1–2 μm are regularly achieved with this method. This is a cost-effective way to create multiple die on a single wafer. The desired hole pattern is created with a photolithography mask, the wafer is etched, and then the individual die are diced out of the wafer.

The highest-precision arrays are made with single-mode (SM) fibers and polarization-maintaining (PM) fibers. Arrays made with multimode fibers are dependent on the fiber tolerance which can be as much as 1–2 percent of its diameter, while SM fibers have a tolerance of as little as ±1 μm or even submicron levels. See Fig. 8.25 for examples of a precision SM array and a PM array with the stress rods in alignment.

6.
SINGLE MODE ARRAY

7.
POLARIZATION MAINTAINING ARRAY

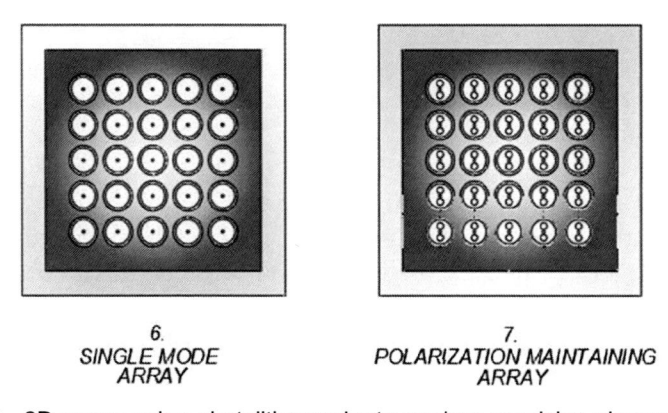

Figure 8.25 2D arrays using photolithography to produce precision placement of fibers.

8.2.6 High-power design considerations

When designing a high-power laser delivery cable, there are three main components that must be considered: the fiber material, the input connector, and the mode stripper (if necessary).

The first thing to consider is the fiber and the fiber-to-laser interface. Common to many high-power fibers is a step-index fused silica core and a fluorine-doped fused silica cladding. A high-purity fused silica core is capable of handling huge amounts of laser energy. The challenge to getting the energy into the fiber is the air-silica interface, which occurs at the input connector. The quality of the polish will maximize the amount of power that the interface will handle. Laser polishing of the fiber after the normal polishing step has been shown to increase the power handling by as much as 10–15%. This is done with a CO_2 laser that scans across the face of the fiber and re-melts the top surface, removing any subsurface micro-cracking or scratches left behind from the polishing media.

The failure mode for continuous wave (CW) lasers is thermal, caused by microscopic irregularities in the air-silica interface that absorb energy. For pulsed lasers, the failure mode can either be thermal or a dielectric breakdown at the atomic level, depending on the laser's characteristics. In either case, there is a maximum amount of power per unit of area that can be coupled into the fiber referred to as the laser damage threshold (LDT). For CW lasers, this is expressed in W/cm^2, also known as irradiance. For pulsed lasers it is J/cm^2 and is known as fluence. The reason for the difference is that CW lasers run, as the name suggests, as a continuous stream of unbroken energy, while pulsed lasers operate as a series of pulses, or bursts of repeating energy.

Determining if a laser will damage a fiber involves calculating the irradiance or the fluence for the laser by dividing the CW power or the energy per pulse, by the area of the beam where it makes contact with the fiber. The LDT for a CW laser and at the air-silica interface is 1.5 MW/cm^2 (at 1064 nm) and for a pulsed laser, 16.0 J/cm^2 (at 1064 nm and 1 ns). As a rule, lower wavelengths will require larger fibers or lower powers to couple into the fiber. For a pulsed laser, it may require lower power or a combination of lower power and longer pulses.

There are a few common connectors that are used for high-power assemblies. One is the SMA, another is the D80, but what make them work well are a few common characteristics. The connectors are precision machined and feature an epoxy free termination at the tip. This is done by what is called an air-gap or a cantilevered fiber tip. Epoxies and other organic materials, if left on the fiber face, will cause the fiber to burn when a laser strikes it. So, an air gap removes epoxy from the front where a laser might contact it and cause it to burn or heat up to the point that it fails and causes outgassing on the fiber face. See Fig. 8.26. for a view of an air-gap interface.

Figure 8.26 High-power connector with an air-gap interface.

The failure mode for these connectors is thermal overload due to physical size and construction constraints when materials (organics) break down. The connectors may utilize extra features to remove excess heat such as heat sinks or water-cooling to prevent the breakdown to the extent possible.

Input connectors can be designed with mode stripping capabilities. Mode strippers are designed to remove energy from the cladding of the optical fiber and dissipate it in the form of heat. For a variety of reasons such as poor alignment or poor beam quality, energy can be coupled into the cladding instead of all of it going into the core of the fiber. When this happens, it can cause heating in the connector and other beam quality issues. A mode stripper is a predictable way to remove the energy from the cladding at the input connector. Depending on the amount of heat to be removed, mode stripping may be done with an air-cooled heat sink or with a water-cooled housing.

When the amount of power needed is higher than the laser damage threshold of the base fiber, the designer will use an end cap as the air-fiber interface. This allows the user to couple higher power into the fiber core by reducing the power density at the air-silica interface. An end cap is designed to optimize the laser launch conditions with the fiber size and numerical aperture. End-cap size to fiber size ratios can range from 2:1 to 10:1 or even greater as the need requires. See Fig. 8.27 for an end-cap example.

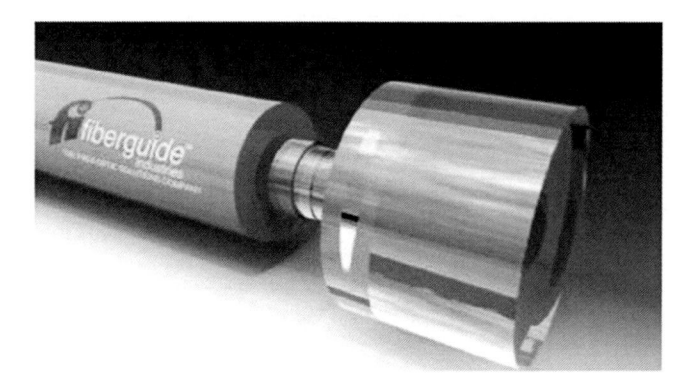

Figure 8.27 An end cap on a fiber.

References

[1] A. C. van Heel, "A new method of transporting optical images without aberrations," *Nature* **173**(4392), 39–39 (1954).
[2] H. H. Hopkins and N. S. Kapany, "A flexible fibrescope, using static scanning," *Nature* **173**(4392), 39–41 (1954).
[3] E. Snitzer, "Cylindrical dielectric waveguide modes," *JOSA* **51**(5), 491–498 (1961).
[4] K. C. Kao and G. A. Hockham, "Dielectric-fibre surface waveguides for optical frequencies," in *Proceedings of the Institution of Electrical Engineers*, **113**(7), 1151–1158, IET (1966).
[5] J. Hecht, *City of light: the story of fiber optics*, Oxford University Press (1999).
[6] F. Kapron, D. B. Keck, and R. D. Maurer, "Radiation losses in glass optical waveguides," *Applied Physics Letters* **17**(10), 423–425 (1970).
[7] A. Langner and G. Schötz, "Plasma outside deposition (pod) of fluorine doped silica for high-power laser applications," in *Fiber Lasers XI: Technology, Systems, and Applications*, **8961**, 896115, International Society for Optics and Photonics (2014).
[8] O. Humbach, H. Fabian, U. Grzesik, U. Haken, and W. Heitmann, "Analysis of oh absorption bands in synthetic silica," *Journal of non-crystalline solids* **203**, 19–26 (1996).
[9] S. Poole, D. N. Payne, and M. Fermann, "Fabrication of low-loss optical fibres containing rare-earth ions," *Electronics Letters* **21**(17), 737–738 (1985).
[10] B. J. Ainslie, S. Craig, and S. Davey, "The fabrication and optical properties of Nd3+ in silica-based optical fibres," *Materials Letters* **5**(4), 143–146 (1987).
[11] J. B. MacChesney and J. R. Simpson, "Multiconstituent optical fiber," (1987). US Patent 4,666,247.

[12] D. J. DiGiovanni, "Fabrication of rare-earth-doped optical fiber," in *Fiber Laser Sources and Amplifiers II*, **1373**, 2–8, International Society for Optics and Photonics (1991).

[13] M. Saha and R. Sen, "Vapor phase doping process for fabrication of rare earth doped optical fibers: Current status and future opportunities," *physica status solidi (a)* **213**(6), 1377–1391 (2016).

[14] M. Saha, A. Pal, and R. Sen, "Vapor phase doping of rare-earth in optical fibers for high power laser," *IEEE Photonics Technology Letters* **26**(1), 58–61 (2014).

[15] G. Oulundsen, K. Farley, J. Abramczyk, and K. Wei, "Matching active and passive fibers improves fiber laser performance," *Laser Focus World* **48**(1), 81–85 (2012).

[16] X. Cheng and Y. Jaluria, "Effect of draw furnace geometry on high-speed optical fiber manufacturing," *Numerical Heat Transfer: Part A: Applications* **41**(8), 757–781 (2002).

[17] K. Schuster, S. Unger, C. Aichele, F. Lindner, S. Grimm, D. Litzkendorf, J. Kobelke, J. Bierlich, K. Wondraczek, and H. Bartelt, "Material and technology trends in fiber optics," *Advanced Optical Technologies* **3**(4), 447–468 (2014).

[18] S. R. Schmid and A. F. Toussaint, "Optical fiber coatings," *Specialty Optical Fibers Handbook, edited by Mendez, A and Morse, TF*, 95–122 (2007).

[19] K. Schuster, J. Kobelke, K. Rose, M. Helbig, M. Zoheidi, and A. Heinze, "Innovative fiber coating systems based on organic modified ceramics," in *Optical Components and Materials VII*, **7598**, 75980H, International Society for Optics and Photonics (2010).

[20] V. Bogatyrev and S. Semjonov, "Metal-coated fibers," in *Specialty Optical Fibers Handbook, edited by Mendez, A and Morse, TF*, 491–512 (2007).

[21] P. J. Lemaire and E. A. Lindholm, "Hermetic optical fibers: Carbon-coated fibers," *Specialty Optical Fibers Handbook, edited by Mendez, A and Morse, TF*, 453–490 (2007).

[22] A. M. Ortiz and R. L. Sáez, "Multi-core optical fibers: Theory applications and opportunities," *Selected Topics on Optical Fiber Technologies and Applications* (2017).

[23] T. Hayashi and T. Nakanishi, "Multi-core optical fibers for the next-generation communications," *SEI Technical Review* **86**, 23–28 (2018).

[24] A. Van Newkirk, E. Antonio-Lopez, G. Salceda-Delgado, R. Amezcua-Correa, and A. Schülzgen, "Optimization of multicore fiber for high-temperature sensing," *Optics Letters* **39**(16), 4812–4815 (2014).

[25] P. Russell, "Photonic crystal fibers," *science* **299**(5605), 358–362 (2003).

[26] J. W. Goodman, "Speckle phenomena in optics: theory and applications," *SPIE* (2020).

[27] C. Bernhard and W. H. Miller, "A corneal nipple pattern in insect compound eyes," *Acta Physiologica Scandinavica* **56**, 385–386 (1962).

[28] D. S. Hobbs, B. D. MacLeod, and J. R. Riccobono, "Update on the development of high performance anti-reflecting surface relief microstructures," in *Window and Dome Technologies and Materials X*, **6545**, 65450Y, International Society for Optics and Photonics (2007).

Devinder Saini is currently the vice president of technology at Fiberguide industries in Caldwell Idaho responsible for product development and new technologies. Prior to joining Fiberguide in March 2015 he served as vice president and chief scientist of OxySense, Inc. since its founding in 2001. Prior to and in parallel with working for OxySense he was vice president and chief scientist for FCI environmental Inc. He holds 9 patents and has authored numerous articles on fiber optic sensing. He holds a Ph.D. in Physics from the City University in London England, a Master's degree in Materials Science and a Bachelor's degree in Physics and Astrophysics. He specializes in the development and manufacture of fiber optics-based sensors and instrumentation.

Kevin Farley received his PhD from Rutgers University in 2006 in the field of Materials Science Engineering with a specialty in Fiber Optics. He spent 11 years working for Nufern as a Research Scientist working on Rare Earth Doped fibers, beam delivery, and all forms of specialty fibers. He is listed on 3 patents as well as authored and co-authored numerous publications. He is currently the Director of Fiber Production and R&D at Fiberguide Industries.

Brian Westlund has worked in the fiber optic industry since 1979. The last 33 years he has been the applications engineering manager at Fiberguide Industries. As an application engineer he has designed and built many complicated and challenging fiber optic assemblies for varied environments and applications.

Chapter 9
Diffractive- and Micro-structured Optics

Tasso R. M. Sales
Viavi Solutions, Inc.,1050 John Street, Rochester, NY, 14586, USA

G. Michael Morris
Apollo Optical Systems, Inc., 925 John Street, Rochester, NY, 14586, USA

9.1 Introduction

Diffractive- and micro-structured optics technology provide new degrees of freedom for the design and optimization of optical systems. Using this technology one can create: large-aperture, lightweight optical elements; correct optical aberrations and achromatize optical systems; eliminate or reduce the need for exotic materials, such as flint glasses; combine multiple optical functions; and reduce significantly optical system weight, complexity and cost.

Important applications of diffractive- and micro-structured optics that have been developed, and in many cases have matured, include areas such as visible and infrared imaging systems, laser beam shaping for optical sensors, solid-state (LED) lighting, and display and image projection systems.

A diffractive optical element is an element whose fundamental operating principle is based on the wave nature of light. The basic physics of diffractive optical elements have been studied for hundreds of years and can be found in numerous modern textbooks [1,2,3]. A fundamental principle of wave optics is based on the superposition (or addition) of waves, wherein the notion of constructive and destructive of waves plays a key role, and results in bright and dark regions, bands or spots of light.

One can think of a diffractive optical element as an optical beam (or wavefront) converter, which converts an incident wavefront into the desired output wavefront. Two important attributes of the desired output wavefront are its shape and its diffraction efficiency, i.e., the fraction of the incident wavefront energy that is contained in the desired output wavefront.

Shortly after the invention of the laser and holography in the 1960s, researchers began to develop holographic optical elements. A hologram is a recording of an interference pattern produced by the superposition of spatially coherent light beams, and is one of the fabrication methods that has and can be used to create diffractive optical elements [4,5]. There has been a tremendous amount of work directed toward the development of methods to improve the diffraction efficiency of holograms, which can be divided into three basic approaches – (1) use of volume materials, which relied on the constructive interference from Bragg planes within the material; (2) bleached holograms, which involved the conversion of amplitude (absorptive)-type holograms into phase-type holograms; and (3) blazed surfacerelief- type holograms.

In the late 1960s to the early 1990s, methods of fabrication and applications of computer-generated holograms (CGHs) were developed. Rather than using the superposition of coherent optical beams to record the interference pattern, the interference pattern is calculated mathematically using a computer and then sent to some sort of plotter to create a CGH. Generally, CGHs are binary in nature – either an amplitude- (black and white) or phase- (0 and π) type hologram. One of the major uses of CGHs is for optical testing [6].

In the 1990s a new approach for the fabrication of diffractive optical elements was developed, which was called "Binary Optics" [7]. The binary-optics approach to fabrication is to utilize photolithographic equipment and processes developed for the semiconductor industry to produce surface-relief (or phase-type) optical elements. While multi-level phase elements can be fabricated using these techniques, the nomenclature of "binary optics" has been used to indicate that each step in the fabrication process is binary in nature. The binary-optics fabrication process will be discussed in greater detail in Section 9.2.

While the application of CGHs in optical testing generally does not require high diffraction efficiencies, most applications involving diffractive optical elements do require that the efficiency be as high as possible in the desired wavefront or diffracted order.

It is instructive to consider the maximum diffraction efficiencies that can be produced with various types of planar (thin) holographic gratings as illustrated in Fig 9.1 [8]. An amplitude transmission of "1" represents clear, and "0" represents opaque portions of the grating.

In Table 9.1, the corresponding transmission functions, together with the diffracted amplitude and irradiance in of the first diffracted order, are listed for each of the holographic gratings illustrated in Fig 9.1. In the case of the amplitude (or absorptive) sinusoidal grating, the sinusoidal portion of the transmission function can be written as

Diffractive- and Micro-structured Optics   305

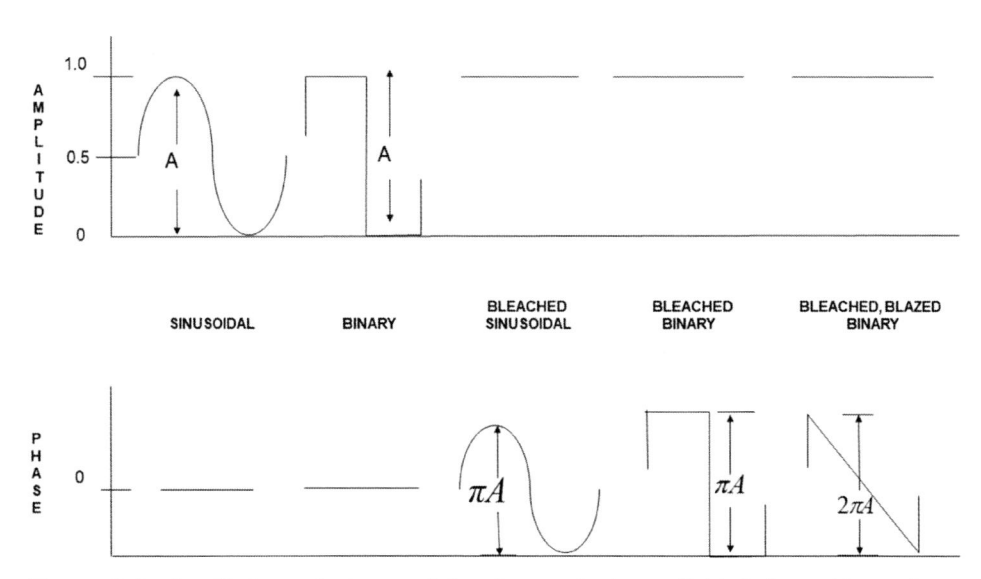

Figure 9.1 Amplitude and phase of five types of planar (thin) holographic gratings: sinusoidal and binary amplitude-type gratings; and sinusoidal, binary and blazed phase-type gratings.

$$\sin\left(\frac{2\pi x}{d}\right) = \frac{\exp^{+i2\pi x/d} + \exp^{-i2\pi x/d}}{2i} \quad (9.1)$$

in which the two exponential terms represent the $+1$ and -1 diffracted orders, and the ½ in the transmission represents the amplitude of the zero-order (or un-diffracted) wave amplitude. The complex field (or wave) amplitude of each of the diffracted orders is therefore $A = 4i$. Taking the squared magnitude of the diffracted amplitude and setting $A = 1$ gives an irradiance (or diffraction efficiency) of 1/16, or 6.25%. of the incident irradiance for each of the two diffracted orders.

The transmission function for the binary amplitude in Fig 9.1 can be represented using a Fourier series expansion of a square wave, where m

Table 9.1 Transmission functions, diffracted amplitude and diffracted irradiances for the first diffracted order for the holographic gratings shown in Fig 9.1.

Hologram Type	Transmission Function	Diffracted Amplitude	Diffracted Irradiance
Sinusoidal	$\frac{1}{2} + \frac{A}{2}\sin(\frac{2\pi x}{d})$	$\frac{A}{4i}$	6.2%
Binary	$\frac{1}{2} + \frac{2A}{\pi}\sum_{n=1}^{\infty}\frac{1}{n}\sin(\frac{2\pi nx}{d})$	$\frac{A}{i\pi}$	10%
Bleached sinusoidal	$\exp[i\frac{\pi A}{2}\sin(\frac{2\pi x}{d})]$	$J_1(\frac{\pi A}{2})$	34%
Bleached binary	$\frac{2i}{\pi}\sin(\frac{\pi A}{2})\sum_{n=1}^{\infty}\frac{(-1)^n-1}{n}\sin\frac{(2\pi nx)}{d}$	$\frac{2}{\pi}(\frac{\pi A}{2})$	41%
Blazed, bleached binary	$\frac{2}{\pi}\sin(\pi A)\sum_{n=1}^{\infty}\frac{(-1)^n}{n^2-A^2} \times [A\sin(\frac{2\pi nx}{d}) + i\cos(\frac{2\pi nx}{d})]$	$\frac{i\sin(\pi A)}{\pi(1-A)}$	100%

represents the m-th diffracted order. Therefore, the irradiance in the first diffracted order is $1/\pi^2$, or $\sim$10% of the incident irradiance.

The transmission function for the bleached sinusoidal grating is a moment-generating function for Bessel functions, wherein the first term in the expansion is a Bessel function of order 1, which gives a diffractive irradiance (or diffraction efficiency) of 34%. By converting a sinusoidal amplitude grating to a sinusoidal phase grating, the diffraction efficiency in the first order goes from 6.2% to 34%., with a similar increase for the case of amplitude and bleached binary gratings.

However, it is particularly interesting to note that for the case of blazed, bleached binary grating (using terminology in [8]), it is theoretically possible, again setting $A = 1$, to achieve a diffraction efficiency of 100% in the first diffracted order. In modern terminology, this case would simply be called a "blazed" grating with a phase jump of $2\pi A$.

The idea that, in principle, a blazed grating can obtain a diffraction efficiency of 100% is certainly not new. In fact, the idea that a blazed grating phase profile could give 100% diffraction efficiency was described Lord Rayleigh (aka J. W. Strutt) in the section on the "Wave Theory of Light" in the Encyclopedia Britannica (1888) [9], which states that

> "... If it were possible to introduce at every part of the aperture of the grating an arbitrary retardation, all the light might be concentrated in any desired spectrum. By supposing the retardation to vary uniformly and continuously we fall upon the case of an ordinary prism; but there is then no diffraction spectrum in the usual sense. To obtain such it would be necessary that the retardation should gradually alter by a wavelength in passing over any element of the grating, and then fall back to its previous value, thus springing suddenly over a wavelength. It is not likely that such a result will ever be fully attained in practice; but the case is worth stating, in order to show that there is no theoretical limit to the concentration of light of assigned wavelength in one spectrum, ..."

So, the physics "Ah Ha" moment, i.e., being able, at least in theory, to achieve a diffraction efficiency of 100%, has been known for over 100 years! However, the realization that has occurred during the past 20 to 30 years has actually been the development of sophisticated manufacturing techniques that produce surface-relief diffractive- and micro-structured optical elements with efficiencies approaching 100% for visible-light applications!

In Section 9.2, we review the fabrication methods that have been developed to produce high-efficiency surface-relief masters, which, as described in Section 9.3, can be used in high-volume manufacturing processes to produce hundreds of thousands to millions of high-efficiency, surface-relief diffractive and micro-structured optical elements. A particularly important

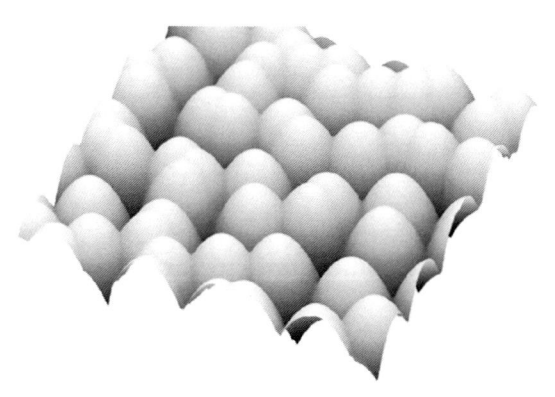

Figure 9.2 Mathematical model of a high-efficiency Engineered Diffuser® used for laserbeam shaping.

development that has occurred is the development of grayscale lithographic processes based on single-point laser exposure of photolithographic materials, which are capable of producing continuous, surface-relief structures with depths up to approximately 80 microns. In addition to the creation of blazed surface-relief diffractive optical elements, these grayscale "laser-writer" systems enable one to produce arrays of refractive microlens-based components.

In Section 9.4, we discuss the features and applications of high-efficiency diffractive lenses in broadband imaging applications.

An important new class of refractive microlens-based components, called Engineered Diffusers® [10], and shown in Fig. 9.2, consists of a highly complex arrays of refractive microlenses, which have different radii of curvature, diameters and depths.

Engineered Diffusers® provide a highly efficient means for shaping light distributions from lasers and solid-state (LED) sources, which, unlike diffractive optical elements, are insensitive to changes in illumination wavelength and illumination wavefront. High-volume applications of Engineered Diffusers® are discussed in Section 9.5.

9.2 Fabrication of Surface-Relief Masters

There are different methods of producing micro-optical masters, depending on the type of pattern that is being created or the specific symmetry required by the pattern. Rotationally symmetric patterns, for example, are well-suited for diamond-turning machines (see Section 9.2.1). Similarly, binary patterns are most efficiently produced by mask exposure processes. For continuously varying grayscale patterns, direct laserwriting is probably the most well-established approach [11]. Mask exposure processes have also been used for grayscale processes [12]. Multi-mask exposure has also been employed to create stepwise approximations of a continuous profile, particularly for diffractive elements [13]. In this section, we briefly cover some of the well-established methods of creating surface-relief masters for micro-optical applications.

9.2.1 Single-point diamond turning (SPDT)

Single-point diamond turning (SPDT), or diamond machining, uses a cutting tool with a diamond tip mounted on a precision lathe that is capable of producing highquality optical components that in many cases do not need to be post-polished; see Fig. 9.3.

A good review of the "Fabrication of optics by diamond turning" is given by R. L. Rhorer and C. J. Evans in Chap. 41 of [14]. SPDT can be used to produce small production lots, quick-turn prototypes and production and prototype inserts for injection molding and embossing-type manufacturing processes. There is a wide variety of materials that can be used to produce diamond-turned polymer optical components and mold inserts for visible light applications. These materials include: acrylic (PMMA), polystyrene, cyclic olefin co-polymer (COC), nickel, brass, copper, aluminum and many other materials. Typical surface finishes that can be achieved are less than ¼ to ½ wave peak-to-valley surface figure over a 25-mm diameter, and surface micro-roughness as low as 1–1.5 nm RMS in nickel, and 3–5 nm RMS with PMMA. SPDT is also capable of producing a number of different types of optical surfaces, including diffractive, aspheric, spherical, plano, Fresnel, conical, and toroidal surfaces.

Of particular interest here is SPDT of high-efficiency, blazed diffractive optical elements for visible-light applications. In Fig. 9.4, the surface profile of a section of a diffractive lens, obtained using a Zygo New View non-contact surface profilometer, is shown. This particular lens was designed for use with a

Figure 9.3 Diamond turning of a diffractive lens mold insert in nickel.

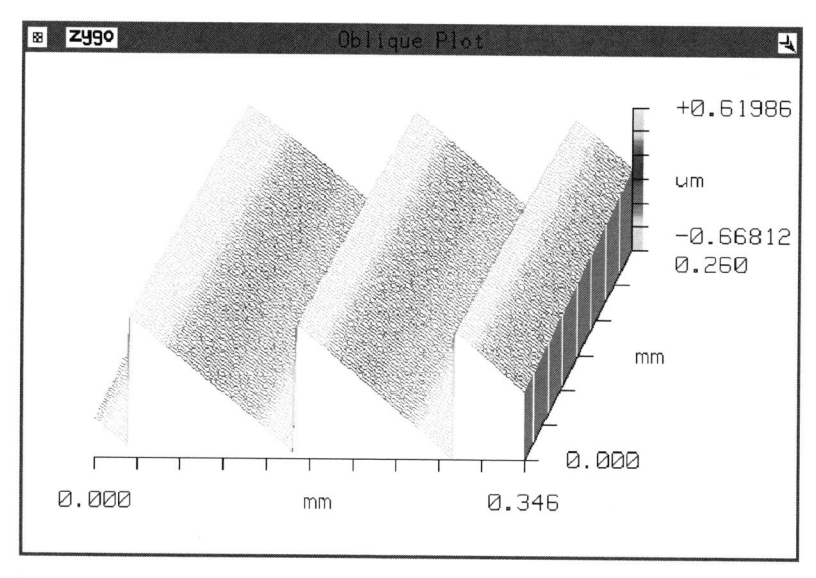

Figure 9.4 Non-contact surface profile trace of a section of a diffractive lens.

refractive lens to form a hybrid diffractive/refractive achromatic lens, as will be discussed in more detail in Section 9.4.

Fabrication of a diffractive optical element with high diffraction efficiency require smooth blaze (or surface) profiles within each zone and extremely sharp edges at the edge of each zone, which generally is not a compatible machining requirement. For this case, a proprietary diamond tool with a tip radius of approximately 1.5 microns was used to form the diffractive surface. In practice, diffraction efficiencies in the range of 97–99% at the design wavelength can be obtained with hybrid diffractive/refractive lenses.

Note that for the case of hybrid diffractive/refractive lenses, the diffractive zones across lens and step heights at each zone boundary are much greater than the wavelength of light; hence, scalar diffraction can be used to accurately predict the diffractive efficiency. However, for other types of diffractive optical elements, such as a low-f-number diffractive lens, in which the diffractive zone spacings are on the order of the wavelength of light, the diamond tool radius and zone shadowing can have a significant impact on the diffraction efficiency and must be modeled using rigorous electromagnetic theory [15].

9.2.2 Optical and E-beam lithography

In general terms, photolithography refers to the creation of a surface-relief pattern in a photosensitive polymer or emulsion (photoresist). Depending on the process and/or application, the photoresist pattern can be transferred into other materials such as UV-cure polymers through micro-replication processes, metallic materials though electroplating, or hard substrates such as fused silica and silicon through reactive ion etching. Some applications require binary patterns often due to the inability to produce grayscale solutions. Whenever possible, grayscale phase profiles enable design flexibility as well as improved performance, particularly in terms of efficiency. Binary patterns are limited in the types of patterns it can project but in the case of very small features it may be the most viable option. Binary patterns with small features, around 5 μm or less, are produced via electron-beam lithography. Larger features can be produced with optical lithography and sometimes with grayscale surface geometry. The processing conditions and challenges are significantly different in both cases.

While there are many variations and different implementations, the basic elements of the photolithographic process are listed below:

- Wafer preparation,
- Resist coating,
- Baking,
- Exposure,
- Development,
- Metrology.

The wafer preparation involves cleaning and possibly incorporation of adhesion layers. Ultrasonic cleaning is an effective method to prepare wafers for the next process steps, but manual cleaning with acetone and IPA or piranha baths can also be employed [16]. The use of an adhesion promoter, such as HMDS, is another option but that depends on the type of resist being used and the ultimate purpose of the wafer. Probably the most popular approach for resist application is spin coating as it provides very uniform resist films with uniform thickness over most of the wafer and is a quick process. Thicknesses of around 100 μm and less are routinely produced with a single application process. Thicker layers are also possible, depending on the photoresist, as well as the use of multiple coating steps. After coating, the resist undergoes a baking step to harden the material and expel some of its solvents prior to the exposure step. Resist baking has a direct impact on the response of the material to exposure and is a critical process step. Different resist manufacturers usually provide standard spin coating recommendations as well as baking recipes that can be used as starting point for establishing proper baking conditions. Resist storage after baking is also an important consideration, and it is most important to ensure consistently stable environmental conditions to make sure the exposure process starts with consistent material. In the exposure step, the photoresist is subjected to UV radiation that changes its chemical properties and enables the formation of the surface pattern. Basically, when exposed to radiation within a suitable wavelength or spectral range, portions of the resist may become more soluble or less soluble, depending on whether the resist is of positive or negative type. When developed with a generally alkaline solution, the portion of positive resist that was exposed with more radiation washes away more. Conversely, more negative resist remains at those locations of higher exposure. The development step then reveals the patterned photoresist. The final step consists of confirming that the intended pattern was created compared to the original target. This is done through direct or indirect measurements. Direct surface topography metrology can be suitable in the case of deterministic patterns if surface slopes are gentle enough to be captured by the instrument. For randomized patterns, direct measurement can also be instructive but often an indirect, functional, measurement is more revealing. In this case, one makes inferences about the pattern based on measurements of the optical distribution it creates such an intensity distribution or a diffraction pattern.

Electron-beam lithography refers to the patterning method based on the use of electron beams as the exposure tool. Negative resists are typically used in this case, such as PMMA (polymethyl methacrylate). E-beam systems are routinely used in the production of photomasks and generally when there is a need for high resolution, down to ~10 nm. Optical exposure is generally utilized for larger features, $\geq$ ~2000 nm, with the exposure beam constituting a focused laser beam on the ultraviolet/blue side of the spectrum. E-beam systems tend to be more expensive compared to optical exposure systems and involves longer exposure times over comparable pattern sizes.

9.2.3 Laser pattern generation

A versatile method of producing grayscale surface profiles is based on direct laserwriting, which is a maskless lithographic exposure process. As indicated previously, diffractive elements with grayscale profile provide the best possible diffraction efficiency. Whenever possible, as dictated for feature sizes and the capabilities of the exposure system, the grayscale surface profile is obviously preferable as it enables the most general design capabilities and highest efficiency.

Basic photolithographic processes are again utilized from substrate preparation through resist coating, although here there may be the need to consider thicker resist coatings than is typical of diffractive optical fabrication. Photoresist thickness up to about 150 mm are common to cover fabrication of deeper refractive-type optical elements. We should note, for completeness, there are other methods of producing grayscale profiles other than direct laserwriting. Grayscale masks previously mentioned and thermal reflow of photoresist are alternative methods albeit with limited fabrication capabilities [17].

A simplified diagram of the laserwriting process is shown in Fig. 9.5. The photoresist coated wafer is exposed by scanning a focused laser beam across the surface, on a point-by-point basis. In the diagram, the photoresist is

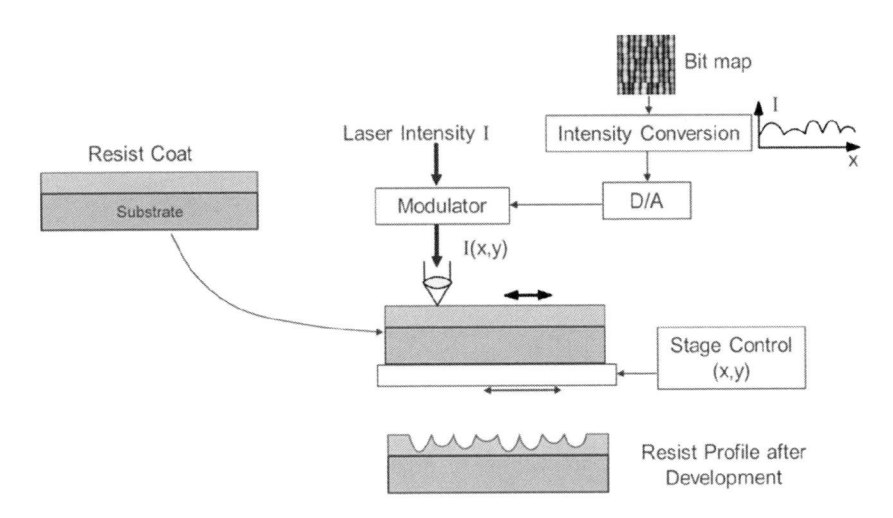

Figure 9.5 Basic diagram of laserwriting process.

assumed to be positive, so that high (low) exposure leads to more (less) removal of material upon development [18]. There are of course many details that need to be considered, as it relates to the size of the focused beam, pattern feature sizes, process transfer functions, write times, etc. Beam size and raster spacing, i.e., distance between consecutive scan lines, correlate directly with feature sizes and surface roughness. Larger beam size compared to raster spacing leads to smoother surfaces but also limits the minimum feature sizes that can be exposed. Write times are also dependent on raster spacing, so the specific write conditions need to be carefully evaluated to determine optical parameters dependent on requirements. It is also important to keep in mind subsequent processes such as etching, replication, embossing, molding, and electroforming, as these may impose further requirements that may need to be considered. Exposure is the part of the process that takes the longest, up to several days, depending on pattern size and process conditions. There is no strict rule on the ratio of feature size to beam size, but typically it should be in the range of 4-8. Naturally, as the feature size increases compared to the beam size the relief-profile can be created with higher definition. Non-imaging applications are generally less demanding as surface accuracy is only required in a statistical sense. For imaging applications, however, where wavefront quality is critical, a good understanding of the process transfer function is paramount to corrective measures can be taken at each step to ensure the final pattern meets the desired performance targets.

Other laserwriting geometries have been developed to address unique configurations or to simplify the exposure process, be it in terms of a particular pattern geometry, throughput, or manufacturing method. Examples include R-Θ laserwriting system, which is particularly suitable for rotationally symmetric components, such as diffractive and Fresnel-type lenses. This is similar to the diamond turning process (Section 9.2.1), except that it uses a lithographic process instead of mechanical pattern formation. Laserwriting systems have also been developed to create patterns on curved surfaces. An example is the drum laserwriting system described in Section 9.3.2 for seamless patterns. Creation of masters for roll-to-roll web processes is generally done through direct engraving or machining. The photolithographic process enables better controlled surfaces for optical applications as well as non-optical texturing. Drum masters for holographic diffusers through the recording of laser speckle is yet another method of producing drum masters [19].

Laser pattern generation is now routinely used as a flexible tool to create general surface profiles, refractive or diffractive, binary or grayscale. It enables capabilities that are often not possible with alternate technologies used to create similar products. An illustrative example is provided by optical diffusers, which are components traditionally used for homogenization and more recently for efficient beam shaping. Homogenizing diffusers are used to wash out illumination artifacts or hide sources while providing broad enough field of illumination but without much concern as to where or how light is distributed. Ground or etched glass is an example of an inexpensive diffuser that spreads light with a Gaussian profile. Consistent performance is often an issue with these types of diffuser. Holographic diffusers, created through the photolithographic recording of laser speckle [20] enables better control of directionality while still limited to Gaussian intensity profiles. Opal diffusers constitute a different class of diffusers where scatter centers are spread over a volumetric region of the material, not as a surface relief pattern. We mention it here for completeness and because it produces a Lambertian scatter profile, although in a rather inefficient manner: about 20% of the input illumination is transmitted. Lambertian scatterers are often used for calibration purposes. Diffractive diffusers [21] are another class of diffusers that rely on the coherent properties of the source to control the distribution of light. A unique feature of diffractive diffusers is that the phase depth is limited to one or a few multiples of π making it a very shallow relief profile while spreading light through the control of feature sizes. For these reasons, diffractive diffusers often need to be carefully matched to the operating wavelength. As spread angles get wider, feature sizes in diffractive diffusers get smaller and fabrication becomes more challenging. In those cases where there is a wavelength mismatch to the phase depth or features are not properly manufactured, one observes a strong hot spot where a significant fraction of the incident energy is concentrated on the zero-diffraction order. This is a well-known issue with diffractive diffusers which limits broad use of these types of components. Refractive diffusers [22], on the other hand, rely on surface slope to spread light and lead to much deeper surface patterns. Presently, when deep-sag photolithography is no longer a major challenge, refractive diffusers have become a more widespread diffuser solution, more flexible in producing general intensity profiles with

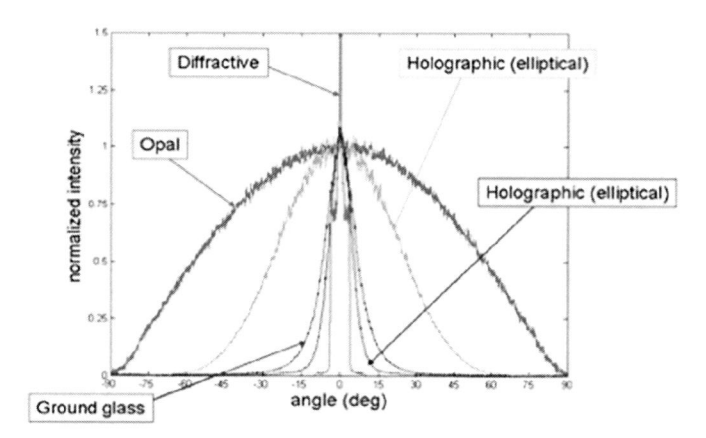

Figure 9.6 Intensity profiles for various types of common diffusers. In this example, the diffractive diffuser is not matched to the measurement wavelength, hence the hot spot.

controlled angles, and without the issue of hot spots. An illustration of measured intensity profiles from a variety of diffusers is shown in Fig. 9.6. A discussion of refractive diffuser applications is covered in Section 9.5.

9.2.4 Reactive ion etching

Reactive ion etching (RIE) constitutes a type of dry etching process, which is often utilized to create surface-relief structures in hard materials, such as fused silica and silicon. The pattern to be transferred (the mask) is first produced on materials such as photoresist, polymers, or metals, depending on the details of the etching system and the nature of the pattern to be transferred. A detailed description of the physics and chemistry behind dry etching is beyond the scope of this brief presentation but can be found in the large body of publications on the subject [16]. For our present purposes, we would like to indicate that RIE is an effective method of creating micro-optical elements in materials that are suitable for operation in the ultraviolet region of the spectrum, high power applications, adverse environmental conditions and general conditions where reliability is a fundamental requirement. At the same time, because of the more involved process, etched optical components tend to be significantly more expensive than most other alternatives.

In the etch process, a combination of chemical and physical processes within a vacuum chamber attacks the patterned mask, e.g., photoresist, and slowly transfers the mask pattern into the underlying substrate material. The mask and the substrate can have differing etch rates that can vary significantly depending on materials, gases, and conditions used, somewhere in the range of minutes to hours to transfer 1 μm of material. The ratio of etch rates defines the etch selectivity. When the selectivity equals one, the mask pattern is directly transferred into the substrate at the same scale. For selectivity greater (less) than one, the final pattern is stretched (compressed) compared to

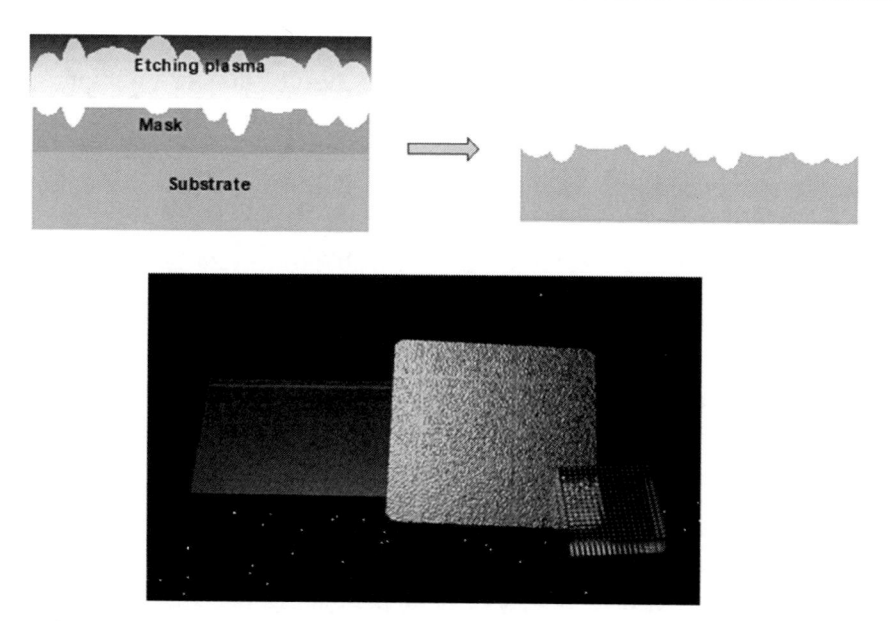

Figure 9.7 Etch transfer from photoresist mask and examples of etched silicon and fused silica.

the initial mask. As an aside, wet etching is another common approach to patterning, which is a much simpler process based on chemical baths that can be processed on a bench. Wet etching, however, is highly isotropic and while there are structures that can be produced with this method, reactive ion etching with its strongly anisotropic character enables a much wider variety of patterns to be created. In practice, depending on process conditions, plasma etching can also incorporate a stronger isotropic contribution to the pattern transfer that can be controlled to some extent.

Very general patterns can be transferred through RIE, with the primary considerations being limited to processing conditions, such as etch times, wafer sizes, and suitability of the etch chemistry given the available materials. Diffractive elements, which are generally shallow ($\sim$5–10 µm) in depth, undergo RIE without much difficulty and often with high throughput. In addition, since DOEs create light distributions as a function of feature size through interference and diffraction, there is no fundamental limit as to the field of illumination or energy distributions that can be produced through RIE. The only possible obstacle might be observed in the presence of a strong zero order, often an unavoidable aspect of diffractives with very small features. This "hot spot" (the zeroth diffraction order) generally arises due to small errors in manufacturing and can be very difficult to eliminate, or to even reduce, when the projected pattern covers field angles of $\sim$10–20 degrees or larger at visible wavelengths.

Refractive micro-optical elements typically do not show strong zero-order hot spots, but may encounter other limitations during the RIE process, depending on the material. Refractive structures tend to be much deeper, generally greater than 10–15 μm and as a result start to run into more significant etch times, especially for glass etching, e.g., fused silica. In addition, the surface transfer of slopes is more limited so that steeper grayscale profiles do not transfer as effectively, leading to a limitation in the types of illumination profiles that can be generated. Depending on the illumination spectrum, these limitations can be overcome using different materials. Silicon, for example, can be used to illuminate a very wide field of view, primarily enabled by its high index of refraction, if the wavelength of operation is at least 1 μm, preferably higher for best transmission. At the same time, these materials do require anti-reflection coatings to provide transmission comparable to glass over the same operational wavelength band. Here again, the etch process that is selected is an important consideration because the isotropic contribution to the etch transfer may significantly affect the final profile transfer is highly dependent on properties of the selected etch chemistry.

9.2.5 Nickel electroform tooling

Photolithographic processes based on photoresist constitute an effective method to create the original master, but photoresist is often a fragile material that needs to be kept in a controlled environment as it may change over time under exposure to changing storage conditions. Ideally, to preserve the surface profile one should transfer the photoresist pattern onto a more durable material, preferably usable in a manufacturing process. Electroforming into a metal is one such process and nickel (Ni) is often a material of choice due to the high-fidelity quality of the surface transfer, durability, and low cost. An excellent overview of the process is discussed in [23]. Here, we briefly describe the overall electroforming process.

The master to be electroformed is initially metalized with a seed layer to make it conductive – this will become the cathode in the electric circuit. The anode is generally composed of the plating metal. For the case of micro-optical elements, Ni is the most common material of choice. The electrolyte bath contains salts incorporating the metal to be deposited. The current across the circuit releases Ni ions into the solution as ionized Ni within the bath is deposited onto the plate. The Ni layer grows continuously over the master plate with high fidelity, making it suitable as a durable micro-optics master. The Ni thickness depends on the processing requirements for the final product, but typically it is somewhere in the range of under 100 microns to several millimeters thick. Once a first Ni master is prepared, multiple copies of the first origination can be created with minimal fidelity degradation.

The electroforming process is versatile enough to enable a wide variety of geometries, not only flat masters. See, for example, Fig. 9.8 for the case of a Ni electroform from a cylindrical drum master used for roll-to-roll film production.

Diffractive- and Micro-structured Optics 317

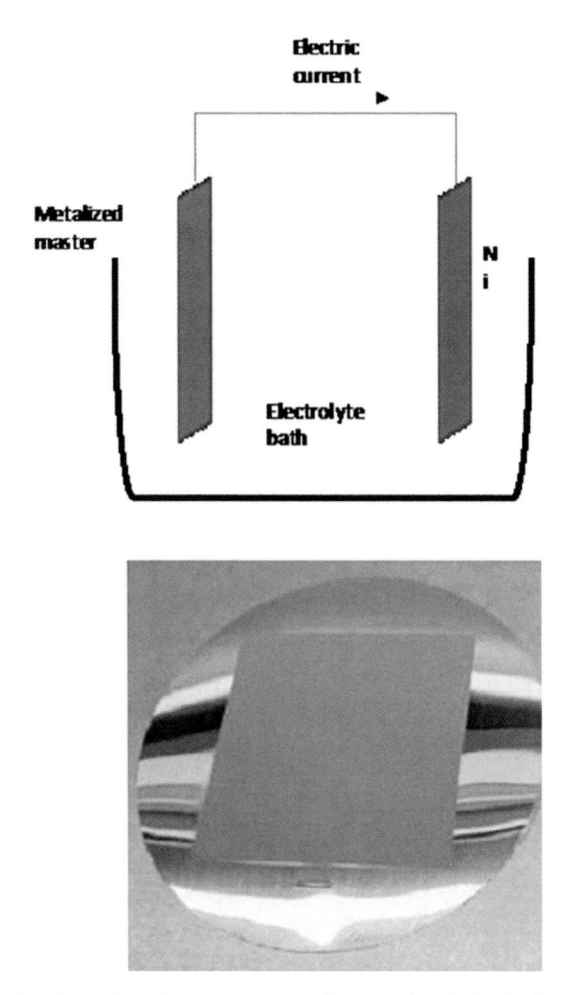

Figure 9.8 Basic electroforming process, and example of electroformed Ni master.

9.3 High-Volume Manufacturing Processes

In the late 1990s and early 2000s, a major focus was to develop and refine highvolume manufacturing processes for the production of polymer-based diffractiveand micro-structured optical surfaces. Four basic concepts for high-volume fabri cation of micro-structured optical surfaces are illustrated in Fig. 9.9.

An excellent overview of the replication methods that have been developed for surface-relief microstructures is given by M. T. Gale [24].

During these past 20 years, replication methods outlined by Gale have continued to evolve, be refined and are now capable of producing hundreds of thousands to millions of precision micro-structured optical products.

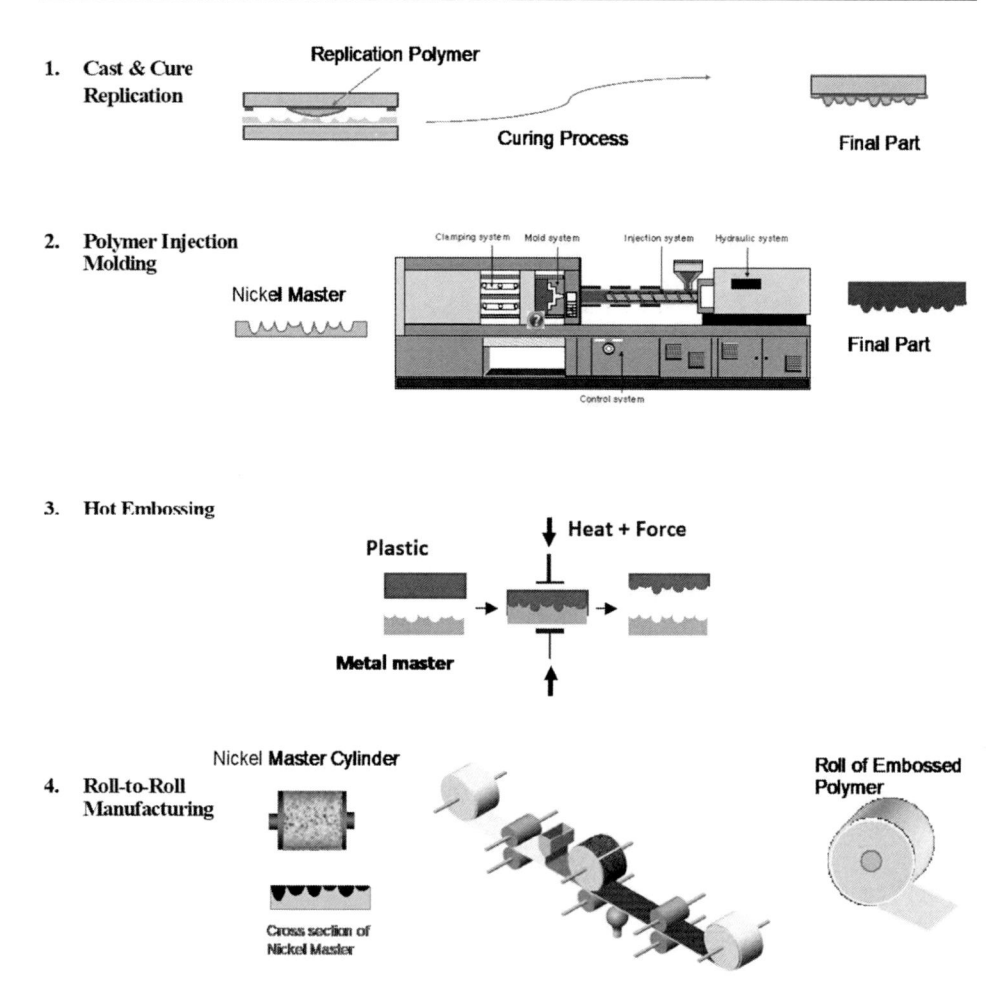

Figure 9.9 Four basic concepts for high-volume manufacturing of precision micro-structured optical surfaces: (1) Cast-and-cure polymer replication; (2) Polymer injection molding; (3) Hot embossing or stamping; and (4) Roll-to-roll manufacturing.

9.3.1 Polymer-on glass wafers

Of particular note are polymer-on-glass wafer-based components in which a thin layer of a UV-curable material is applied to a large-area, thin glass wafer and then diced into individual optical beam-shaping components, as shown in Fig. 9.10. The thickness of the resulting structured polymer layer is generally in the range of 10 to 100 microns, and the thickness of the glass substrate is typically in the range of 200 to 700 microns. The number of diced components that can be obtained from a given wafer obviously depends on the size of the individual component. For highvolume applications, such as a laser-beam pattern generator for 3D imaging and sensors for the consumer products industry, as discussed below in Section 9.5.1, a 300-mm polymer-on-glass

Diffractive- and Micro-structured Optics 319

Figure 9.10 Polymer-on-glass replicated wafers, 100-mm and 300-mm diameter (left), and diced polymer-on-glass wafer, 300-mm diameter (right).

wafer typically yields on the order of 5,000 to 10,000 diced components per wafer.

9.3.2 Roll-to-roll manufacturing

Also noteworthy during these past twenty years was the development and production of seamless Ni drum master tooling for the production of UV-curable surfacerelief polymer materials on flexible substrates, such as PET films.

The fabrication of seamless drum master tooling begins with the fabrication of a seamless patterned photoresist surface that is generated using a custom-designed cylindrical laser-writer system; see Figs. 9.11 and 9.12. In this process, a 200-mmdiameter metal cylinder, which is approximately one meter in length, is coated with a layer of photoresist with a thickness typically between 50 to 100 microns. The photoresist-coated metal cylinder is then placed in air-bearing rotation stages and rotated at several hundred rpm

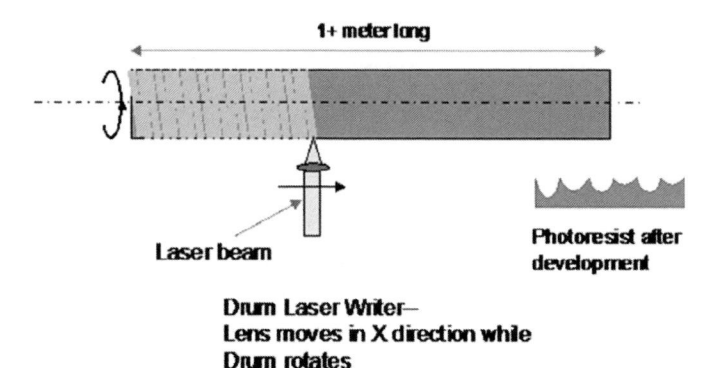

Figure 9.11 Concept for a seamless drum laser-writer system. A cylinder coated with a UV-curable photoresist is exposed by an intensity-modulated laser beam that traverses the length of the cylinder as it is being rotated. Development of the exposed photoresist produces the desired seamless structured surface in photoresist.

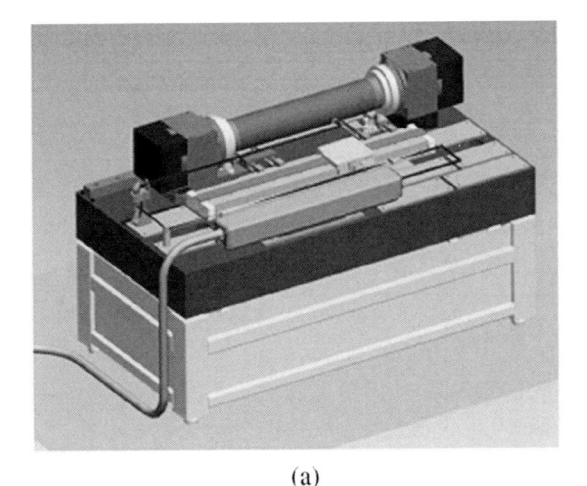

Figure 9.12 (a) Schematic and (b) photograph of a drum laser-writer system designed and developed at RPC Photonics, Inc. in Rochester, NY.

while being exposed using an intensity-modulated laser beam at mega-hertz rates, as it is slowly scanned along the length of the photoresist-coated cylinder.

Upon development of the photoresist, one obtains a seamless precision structured surface in photoresist, such as shown in Fig. 9.13.

An intermediate tool is formed by encasing the structured photoresist cylinder in a rubber-like material in which the interior surface of the rubber tool contains the negative of the surface-relief photoresist pattern. A Ni electroform is then formed on the interior surface of the rubber tool in a manner analogous to that described in [24].

The result is a cylindrical seamless surface-structured nickel tool, which has the same surface-relief pattern as the original laser-written photoresist master; see Fig. 9.14. Mandrels are then inserted into the ends of nickel

Figure 9.13 A fully patterned, seamless photoresist cylindrical drum for a projection screen application.

Figure 9.14 Seamless structured-surface nickel drum production tool used for roll-to-roll manufacturing of precision light-control films.

tooling and placed in a high-speed/volume roll-to-roll manufacturing line, as illustrated schematically in Fig. 9.9.

9.4 Diffractive Lenses in Broadband Imaging Systems

During these past 20 years, diffractive lenses have seen widespread implementation and use in broadband imaging systems. Diffractive lenses can significantly reduce the size, weight and cost of a variety of optical systems, which currently utilize only refractive and/or reflective optical components.

In this section the objective is to briefly review some of the fundamental properties of surface-relief diffractive lenses and then discuss how diffractive lenses can be and are used in important applications, such as 35-mm camera lenses, and large-aperture diffractive telescopes.

9.4.1 Diffractive lens fundamentals

A typical surface-relief diffractive lens consists of concentric annular zones, as shown in Fig. 9.15.

It is sometimes useful to think of a diffractive lens as a "modulo 2π" lens, i.e., at any given point along the lens radius, all the multiple 2π phase changes

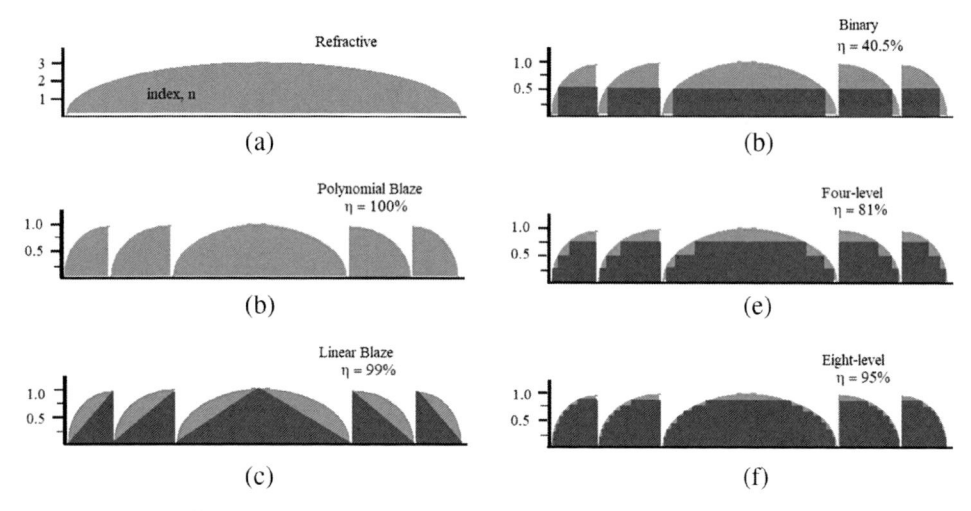

Figure 9.15 Diffractive lens cross-sections: (a) conventional refractive lens with index of refraction, n; (b) diffractive lens with continuous quadratic surface blaze profile; (c) diffractive lens with linear blaze profile; (d) phase reversal (or Wood) lens; (e) four-level approximation to a quadratic blaze profile; and (f) eight-level approximation to a quadratic blaze profile.

have been removed. The radius of the jth-zone boundary, r_j, is determined by the following relation:

$$\phi(r_j) = 2\pi j \tag{9.2}$$

where $\phi(r)$ is the desired phase profile for the lens; these zones are called fullperiod zones. In general, the phase function associated with a circularly symmetric lens is given by:

$$\phi(r) = Ar^2 + Gr^4 + Hr^6 + \ldots \tag{9.3}$$

where $A = \pi/(\lambda_0 F_0)$, G and H are associated with aspheric coefficients of the emerging wavefront, λ_0 is the design (or center) wavelength, and F_0 is the focal length of the lens when the incident wavelength $\lambda = \lambda_0$.

To construct the appropriate surface-relief profile from a spherical lens surface, one starts at the center of the lens and moves radially out from the center. For most cases, the surface of the lens is traced until the optical phase difference reaches a value of 2π or, equivalently, the optical path difference equals one wave. At this point, the surface-relief structure jumps abruptly back to the center height:

$$h_{\max} = \frac{\lambda_0}{n(\lambda_0) - 1} \tag{9.4}$$

where $n(\lambda_0)$ is the index of refraction of the lens material at the design wavelength λ_0 [25]. One continues tracing along the surface of the refractive lens in Fig. 9.15(a) until the optical phase difference again reaches a value of 2π and jumps back to a height, h_{max}, and continue this process until the edge of the refractive lens is reached. This construction process gives rise to a blazed zone-plate-type structure that consists of a series of zones that represent the required phase transformation. While the resulting diffractive lens produces essentially the same output wavefront as the associated refractive lens at the design wavelength λ_0, it has a number of additional properties when $\lambda \neq \lambda_0$.

The spacing of the various zones in a diffractive lens determines the possible foci, i.e., the set of possible diffraction orders, and shapes the emerging wavefront. The blaze (or surface) profile within each zone determines how the incident light is distributed among the various foci.

In most applications, one tries to maximize the amounts of light that goes into a given diffraction order, typically the first diffraction order. By using a quadratic blaze profile illustrated in Fig. 9.15(b) and the scalar Fresnel diffraction approximation, one can (theoretically) direct all the incident light at design wavelength λ_0 into the primary focus. For this case, the diffraction efficiency, λ_0 is said to be 100%. A so-called "phase reversal (or Wood)" lens, shown in Fig. 9.15(d), directs 40.5% of the incident light into the $+1$ diffraction order, and 40.5% into the -1 diffraction order. A four-step approximation to the quadratic phase profile, Fig. 9.15(e), yields a diffraction efficiency $\eta = 81\%$ in the first diffraction order.

An important effect that must be considered in the design of broadband optical systems employing diffractive lenses is the wavelength dependence of diffraction efficiency. As noted above, one can think of a diffractive lens as a modulo 2π lens. Clearly, the 2π phase steps at the zone boundaries are associated with only one particular wavelength, usually the design wavelength λ_0. Other illumination wavelengths experience a different phase step at the zone boundaries; hence, there is no longer perfect constructive interference. As illustrated in Fig. 9.16, the result is a reduction in diffraction efficiency [25].

A reduction in diffraction efficiency in the principal diffraction order results in an increase in the amount of light energy in other diffraction orders, which can contribute to veiling glare in the image plane. This effect reduces the modulation transfer function (MTF) of the optical system [26]. Therefore, in the design of diffractive lenses for broadband optical systems, careful consideration must be given to the diffraction efficiency of the diffractive lens over the spectral band of interest.

A useful figure of merit to describe the effects of non-unity diffraction efficiency of diffractive lens MTF is the integrated efficiency, η_{int}. In broadband (or polychromatic) applications, the integrated efficiency, η_{int}, for a wavelength band ranging from λ_{min} to λ_{max} is

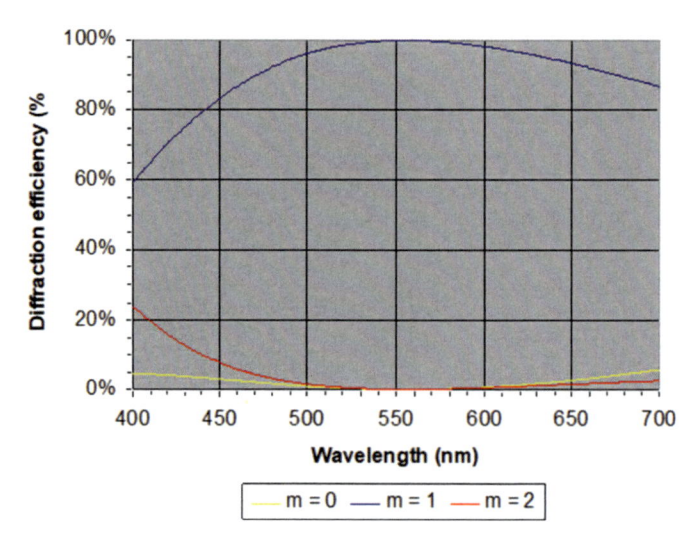

Figure 9.16 Diffraction efficiency as a function of the illumination wavelength for diffraction orders $m = 0$, 1 and 2. The design wavelength $\lambda_0 = 550$ nm. Note the theoretical diffraction efficiency $\eta = 100\%$ when $\lambda = \lambda_0$.

$$\eta_{int} = \frac{\int_{\lambda_{min}}^{\lambda_{max}} \eta(\lambda) d\lambda}{\lambda_{max} - \lambda_{min}} \quad (9.5)$$

where $\eta(\lambda)$ is the diffraction efficiency for the principal diffraction order at wavelength λ. The integrated efficiency serves as a limiting value and overall scale factor for the MTF [26].

With commercially available optical design software, there are two basic approaches that one can use to describe a diffractive lens. One approach is to specify and optimize a phase (polynomial) function that describes the wavefront to be produced by the lens. The other approach is to model the diffractive lens as a thin refractive lens with an ultra-high index of refraction [27,28].

The power of a thin lens, $\Phi_{TL}(\lambda)$, can be described by the lens equation

$$\Phi_{TL} = [n(\lambda) - 1]c \quad (9.6)$$

where $n(\lambda)$ is the index of refraction of the lens material at wavelength λ, and c denotes the lens curvature. A diffractive lens is equivalent to a thin lens with an infinite index of refraction. Of course, it is impossible to set the index of refraction equal to infinity in a computer. It is found, however, that by taking the model index of refraction, $n_{dif}(\lambda_0)$, of the diffractive lens at some design wavelength λ_0 to be approximately 10,000, one can generate results that are virtually indistinguishable from that predicted using scalar diffraction theory or the phase polynomial description of a diffractive lens [25].

The power of a diffractive lens is given by

$$\Phi_{\text{dif}} = \frac{\lambda}{\lambda_0 F_0} \tag{9.7}$$

where F_0 is the principal focal length of the lens when the illumination wavelength $\lambda = \lambda_0$. Equating Eqs. (9.6) and (9.7) for the case where $\lambda = \lambda_0$, one finds that

$$c = \frac{1/F_0}{[n_{\text{dif}}(\lambda_0) - 1]} \tag{9.8}$$

Using Eqs. (9.6)–(9.8), and assuming that the lens is operating in the first diffraction order and is immersed in air, it follows that the model index of refraction, $n_{\text{dif}}(\lambda)$, for a diffractive lens at wavelength λ is

$$n_{\text{dif}}(\lambda) = \frac{\lambda}{\lambda_0}[n_{\text{dif}}(\lambda_0) - 1] + 1 \tag{9.9}$$

If the diffractive lens is on a substrate of curvature c_s, one should take the two surfaces of the model lens to have the following curvatures:

$$c_1 = c_s + \frac{1/F_0}{2[n_{\text{dif}}(\lambda_0) - 1]} \tag{9.10}$$

and

$$c_2 = c_s - \frac{1/F_0}{2[n_{\text{dif}}(\lambda_0) - 1]} \tag{9.11}$$

With the thin-lens model, one can use aspheric coefficients to model higher-order terms in the phase polynomial. The thin-lens model also permits easy access to first-order lens properties and aberration data.

9.4.2 Diffractive/refractive (hybrid) achromatic lenses

A first-order layout for a diffractive/refractive (hybrid) achromatic lens is illustrated in Fig. 9.17.

Key parameters for the design of an diffractive/refractive achromat include [29,30] the lens powers for the constituent refractive, Φ_{ref}, and diffractive lens elements, Φ_{dif}, and the associated Abbe ν-numbers, ν_{ref} and ν_{dif}, are as follows:

$$\Phi_{\text{ref}} = \frac{\nu_{\text{ref}}}{\nu_{\text{ref}} - \nu_{\text{dif}}} \Phi_{\text{Total}}, \quad \Phi_{\text{dif}} = \frac{\nu_{\text{dif}}}{\nu_{\text{dif}} - \nu_{\text{ref}}} \Phi_{\text{Total}} \tag{9.12}$$

Figure 9.17 Configuration for a diffractive/refractive (hybrid) achromatic lens.

$$\nu_{\text{ref}} = \frac{n(\lambda_0 - 1)}{n(\lambda_{\text{short}}) - n(\lambda_{\text{long}})}, \quad \nu_{\text{dif}} = \frac{\lambda_0}{\lambda_{\text{short}} - \lambda_{\text{long}}} \qquad (9.13)$$

in which Φ_{Total} represents the total optical power of the achromat.

Let us consider the design and polychromatic MTF performance of a diffractive/ refractive achromat for use in the visible spectrum and compare its performance to that of a convention all-glass refractive (doublet) achromat, in which we take the design wavelength λ_0, λ_{short} and λ_{long} to be the Fraunhofer D_1-, F- and C-spectral lines (589.6 nm, 486.1 nm and 656.3 nm), respectively. Let us also assume that the crown and flint glasses used for the convention doublet have Abbe ν numbers of 64 and 36, respectively; and the crown glass used for the refractive lens element in the diffractive/refractive hybrid has an Abbe ν number of 64 (BK7 glass). The resulting lens powers for the various constitute lens elements for the two cases are shown in Fig. 9.18.

Key features of the diffractive/refractive hybrid doublet are, unlike a conventional doublet, the optical power both constituent lens elements have the same sign, which enables the constitute lens elements to have lower surface curvatures and thereby enables one to create lower f-number doublets. The resulting hybrid achromat is thinner and generally has 2 to 3 times lower weight than a conventional doublet. In addition, the hybrid achromat can be formed using a single optical material and, therefore, can be readily fabricated

Diffractive- and Micro-structured Optics 327

Conventional Doublet

Crown Glass
$v_{crown} = 64$
$\Phi_{ref} = 2.29\, \Phi_{Total}$

Flint Glass
$v_{flint} = 36$
$\phi_b = -1.29\, \Phi_{Total}$

Crown   Flint

Hybrid Doublet

Crown Glass
$v_{ref} = 64$
$\Phi_{ref} = 0.95\, \Phi_{Total}$

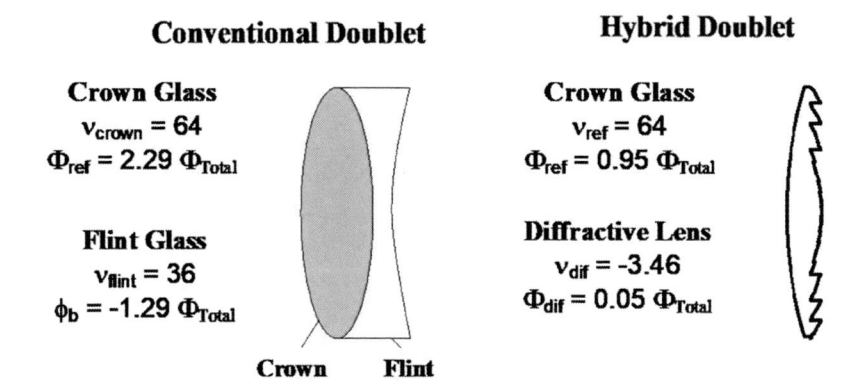

Diffractive Lens
$v_{dif} = -3.46$
$\Phi_{dif} = 0.05\, \Phi_{Total}$

Figure 9.18 Comparison of Abbe numbers and lens powers for a conventional achromat doublet and a diffractive/refractive hybrid doublet.

using high-volume manufacturing methods, such as polymer injection molding, thereby eliminating the need to use flint-glass materials, which tend to be more expensive and hard-to-manufacture glasses.

As an example, we compare the theoretical polychromatic MTF performance that can be obtained using a glass singlet, a glass doublet, and hybrid diffractive/refractive lenses with two different peak diffraction efficiencies at the design wavelength, λ_0, both without and with spherical aberration correction.

Lens parameters for the hybrid achromats used to calculate the MTF curves shown in Fig. 9.19 are as follows:

- Refractive substrate: BK7 singlet, 25.4-mm diameter;
- Diffractive lens: UV-cured polymer-on-glass replica on planar surface of BK7 singlet;
- f/5.6, 75-mm focal length;
- Achromatized range: 486.1 nm to 656.3 nm;
- Design wavelength, λ_0: 589.6 nm;
- Diffraction efficiencies λ_0: 87% and 98%, with associated integrated efficiencies of $\eta_{int} = 84\%$ and $\eta_{int} = 95\%$, respectively;
- Number of diffractive zones: 30;
- Minimum zone width: 117 μm.

Features to note in Fig. 9.19 include the effects of the diffraction efficiency on the polychromatic MTF in which the primary performance limitation of the hybrid designs is associated with secondary color [30]. It is also interesting to note the increase in performance that can be obtained by including aspheric correction of the spherical aberration in the hybrid achromats, which is easily incorporated during the diffractive lens manufacturing process with small shifts in zone radii.

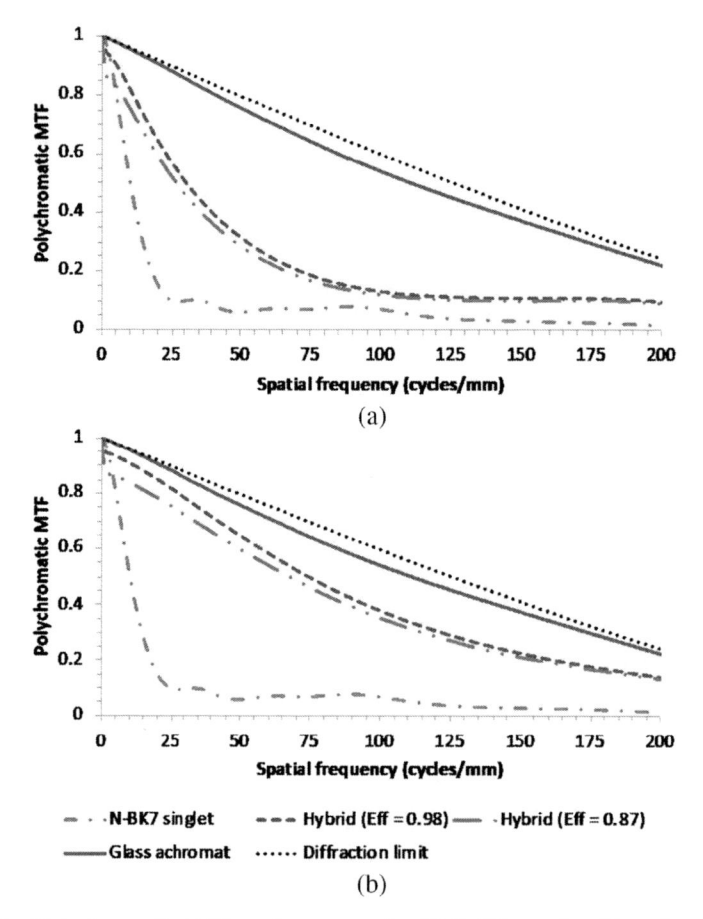

Figure 9.19 Polychromatic MTF performance of a f/5.6, 75-mm focal length–glass singlet, hybrid achromats with peak efficiencies of 87% and 98%, and a conventional all-glass doublet: (a) without and (b) with spherical aberration correction in the hybrid achromats.

9.4.3 Multi-layer diffractive optical elements

As noted above, the integrated efficiency, η_{int}, serves as a limiting value and overall scale factor for the MTF of an imaging system employing diffractive lenses. The primary effect is to lower the contrast of the image, but the portion of light that goes into other diffraction orders (see Fig. 9.16) can produce other deleterious effects as well, such as spurious, or "ghost", artifacts in the image plane that need to be evaluated during the optical design process.

In 2000, Canon, Inc. introduced a new multi-layer diffractive optical element that is capable of providing nearly 100% diffraction efficiency over the entire visible spectrum, thereby significantly improving image contrast and eliminating any artifacts associated with light going into other diffracted orders, and applied it to a commercially available photographic lens – a super telephoto lens (Model EF-400mm, F4 DO IS USM).

The Canon EF400mm FY DO IS USM lens, which contains a multi-layer diffractive lens [31,32], is 26% shorter and 31% lighter than a conventional all refractive lens with the same optical specifications.

The multi-layer DOE consists of two single-layer multi-order diffractive (MOD) lenses. As an example, the first MOD lens with minus optical power diffracts light form the -13th order to the -7th order in the visible region. The second MOD lens with positive optical power diffracts light from the 8th order to the 12th order in the visible region. In this design, the -9th order of the first MOD lens, which has the highest diffraction efficiencyat 560 nm, and the second MOD lens +10th order have approximately the same design wavelength, and the combination diffracts light into the first order $(-9 + 10 = 1)$, which results in high diffraction efficiency in the composite diffraction order [31]. Alternative design methods for multi-layer DOEs have been reported by Xue and Cui [33], and the effects of manufacturing errors and material selection on the diffraction efficiency have also been investigated [34,35,36].

A more detailed description of MOD lenses is given in Section 9.4.4.

9.4.4 Multi-order diffractive lenses

As noted above, diffractive lenses offer system designers new degrees of freedom that can be used to enhance the performance of optical systems: zone spacings are used to shape the emerging wavefront; substrate curvature and aperture location can be used to compensate and balance field aberrations; and as will be described in this section, the zone-step height offers an additional degree of freedom to control the spectral properties of the emerging wavefront [37,38,39].

Unlike a conventional diffractive lens in which the optical power of the lens is directly proportional to the illumination wavelength, [see Eq. (9.7)], a MOD lens, as illustrated in Fig. 9.20, is capable of bringing multiple wavelengths across a given spectral band to the same focal length, F_0, with high diffraction efficiency.

Here, we consider the MOD lens case in which the construction of the diffractive zone structure is defined such that the optical path difference at the j-th zone boundary is equal to $(F_0+jp\lambda_0)$, where λ_0 is the design wavelength, F_0 is the focal length when the illumination wavelength $\lambda = \lambda_0$, and p is an integer that represents the maximum phase modulation as a multiple of 2π, as illustrated in Fig. 9.21.

The optical phase introduced by the MOD lens construction illustrated in Fig. 9.21 is given by

$$\Phi_r = 2\pi\alpha p\left(j - \frac{r^2}{2p\lambda_0 F_0}\right), \; r_j \leq r \leq r_{j+1} \qquad (9.14)$$

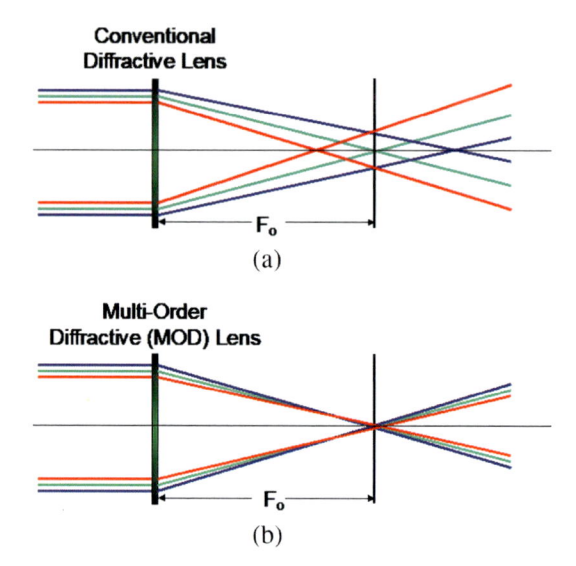

Figure 9.20 (a) The optical power of a conventional diffractive lens varies linearly with the wavelength of light, and thus is highly dispersive with longer (red) wavelengths focusing closer to the lens than shorter (blue) wavelength, whereas in (b) a MOD lens is designed to bring multiple wavelength within a given spectral band to the same focal point with high diffraction efficiency.

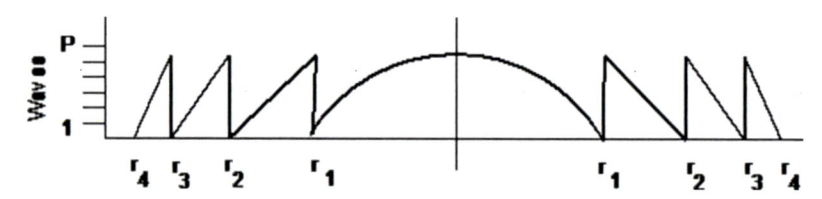

Figure 9.21 Diffractive lens construction for a multi-order diffractive (MOD) lens.

where α is defined as the fraction of 2π phase delay that is introduced for illumination wavelength λ and is given by

$$\alpha = \frac{\lambda_0}{\lambda} \left[\frac{n(\lambda) - 1}{n(\lambda_0) - 1} \right] \quad (9.15)$$

where n is the index of refraction of the material in the grating region. The maximum height of the surface relief structure is given by

$$h_{max} = \frac{p\lambda_0}{n(\lambda_0) - 1} \quad (9.16)$$

Within the context of scalar diffraction theory [37], the focal length of the m-th diffraction order at wavelength λ of a MOD lens is

Diffractive- and Micro-structured Optics 331

$$F(\lambda) = \frac{p\lambda_0 F_0}{m\lambda} \tag{9.17}$$

and the corresponding diffraction efficiency, η_m, of the m-th diffracted order is

$$\eta_m = \text{sinc}^2(\alpha p - m) \tag{9.18}$$

where $\text{sinc}(\pi x) = \sin(\pi x) = (\pi x)$. The diffraction efficiency given by Eq. (9.18) is unity when the argument of the sinc function is equal to 0. Note that, depending of the choices for p and m, this condition can allow for high diffraction for several wavelengths within a given spectral band.

For example, consider the case of a MOD lens operating in the visible wavelength range with $p = 10$ and $\lambda_0 = 555$ nm. Figure 9.22 illustrates the wavelength dependence of the diffraction efficiency for a range of diffracted orders with material dispersion neglected. The peaks in the diffraction efficiency occur at precisely those wavelengths that come to a common focus, see Eq. (9.17), i.e.,

$$\lambda_{\text{peak}} = \frac{p\lambda_0}{m} \tag{9.19}$$

It is important to note a well-known property of operation when higher diffracted orders are used. The wavelength bandwidth of the diffraction efficiency around a given diffracted order narrows with increasing values of p, see Eq. (9.18) and Fig. 9.23 [40,41]. Also note that the material dispersion, see Eq. (9.15), is an important parameter that must be retained in a detailed optical design.

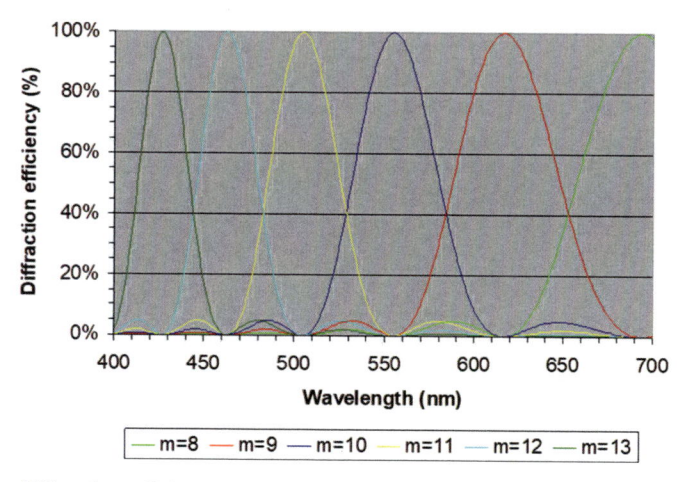

Figure 9.22 Diffraction efficiency of the *m*-th diffracted order versus wavelength for a MOD lens with $p = 10$ and $\lambda = 555$ nm.

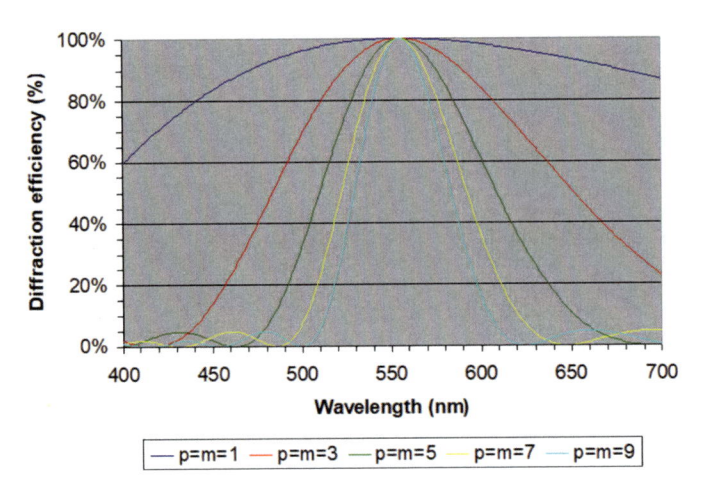

Figure 9.23 Diffraction efficiency versus wavelength for a MOD lens when $p = m$.

Using Eq. (9.19), one can choose the parameters p and m that provide high diffraction efficiency for certain wavelength bands over a given spectrum, with the center wavelength of each of the bands coming to focus at a distance F_0 behind the lens.

9.4.5 Multifocal diffractive ophthalmic lenses

Simultaneous vision bifocal (or multifocal) contact or intraocular lenses provide a method of vision correction for presbyopes. Unlike spectacles, where the user can change the power of the refractive correction by directing his or her gaze through a differently powered portion of the lens, simultaneous vision (used in both contact lenses and intraocular lenses) works by providing two or more images to the retina at all times. With proper design, only one of these images is in focus at any object vergence. There are two primary methods to achieve the production of multiple (or simultaneous) images:

(a) The aperture of the lens is divided into segments, with each segment possessing a single corrective power (for distance, near, and/or intermediate vision).

(b) A diffractive structure is used to split the incoming light into two wavefronts (one wavefront corresponding to the distance image, and one wavefront corresponding to the near image). A feature of the diffractive (or wavefront-splitting) approach, in contrast to the aperture-splitting approach, is that the resulting images are independent of pupil diameter.

Herein, we will focus on surface-relief diffractive structures that are designed to produce two primary focal points. All diffractive bifocals are based on the concept of utilizing a periodic structure that diffracts the incident light into more than one diffracted order. The different diffracted orders correspond to lenses with different powers. It would be desirable, in the case of a bifocal lens, to have all of the incident light diffracted into only two orders. However, it has been shown that this is theoretically impossible with a pure phase structure (i.e., a transparent medium with a surface height modulation) [42]. For the bifocal cases of interest, the total efficiency in the two orders of interest is limited to a value of around 80–85%. The remaining light is diffracted into other orders. This additional out-of-focus light reduces the contrast of the in-focus image.

Of course, it is also possible to combine the aperture- and wavefront-splitting approaches using various combinations of refractive and diffractive zones within the pupil aperture to produce different multifocal designs, including bi-focal, trifocal and quadfocal designs [43].

9.4.5.1 Blazed diffractive bifocal lenses

The Eqs. (9.14)–(9.18) also apply to non-integer values of p. As an example of a diffractive lens with two principal focal points, let us choose $p = \frac{1}{2}$, which when using Eq. (9.16) gives a maximum step height of

$$h_{max} = \frac{\lambda_0/2}{n(\lambda_0) - 1} \qquad (9.20)$$

In other words, the phase step at each zone boundary is π rather than 2π, as discussed above. In this case, again with $p = \frac{1}{2}$, we find using Eq. (9.18) the diffraction efficiencies in the various diffractive orders in Table 9.2:

We find 40.5% of the incident energy in the $m = 0$ diffraction order and 40.5% of the incident energy in the $m = 1$ diffraction order. As illustrated in Fig. 9.24, when the $p = \frac{1}{2}$ diffractive lens is combined with a refractive lens produces a bifocal hybrid lens in which the $m = 0$ order is used for distance vision, and the $m = +1$ order provides the optical power needed for reading.

The remaining light is distributed among the other diffraction orders; these highly defocused images serve to reduce the contrast of the in-focus image.

This form of a hybrid diffractive/refractive lens was first proposed for ophthalmic use by M. H. Freeman [44] and serves as the basis for virtually all of the commercially available bifocal intraocular lenses (IOLs) on the market today.

Table 9.2 Diffraction efficiencies at various diffracted orders for a bifocal lens.

Diffracted Order	Diffraction Efficiency
$m = 0$	40.5%
$m = +1$	40.5%
$m = -1$	4.5%
$m = +2$	4.5%
$m = -2$	1.62%

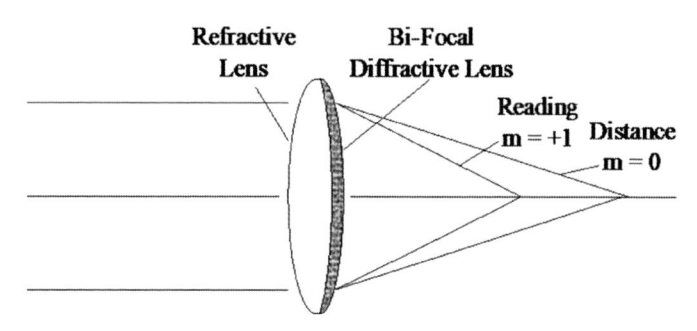

Figure 9.24 Configuration for a bifocal diffractive/refractive lens.

Note that the value of p can be adjusted to change the relative amount of energy between the $m = 0$ and $m = +1$ orders. As an example, the diffraction efficiencies for the 0 and +1 orders for an acrylic diffractive lens in air as a function of the blaze depth are shown in Fig. 9.25. The vertical lines indicate the efficiencies (at the design wavelength of $\lambda_0 = 555$ nm).

In Fig. 9.26, the diffractive zone pattern can be seen in the microscope image of the surface of bifocal diffractive acrylic lens with a depth (or zone step height) equal to 0.64 μm.

Finally, it is also noted that instead of choosing $p = ½$ as discussed above, other fractional values of p can be chosen to create two principal focal points in diffraction orders other than $m = 0$ and $m = 1$.

9.4.5.2 Apodized bifocal diffractives

Apodized bifocal diffractive designs are based on C.-S. Lee and M. J. Simpson [45] and J. A. Davison and M. J. Simpson [46]. The fundamental principle is

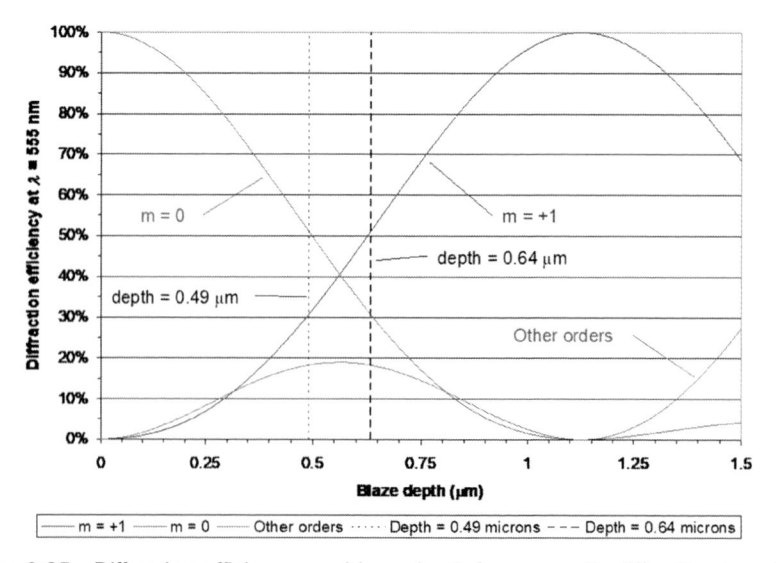

Figure 9.25 Diffraction efficiency vs. blaze depth for an acrylic diffractive lens in air.

Figure 9.26 Microscope image of diffractive bifocal lens surface showing the diffractive zones extending to the edge of the lens. (Courtesy of Apollo Optical Systems, Inc., Rochester, New York)

the same as the blazed diffractive bifocal, with a nominally even energy split between distance and near (40.5% in each order), but the height of the diffractive steps is gradually reduced as the radial distance from the center of the lens increases. This has the effect of putting more energy in the zero order (i.e., the distance image) for larger pupil diameters.

In the Lee-Simpson designs, the blaze depth at a radial point r on the surface is multiplied by an apodization factor. Modifying the depth in this way changes the diffraction efficiency across the aperture of the lens. Therefore, these lenses combine aspects of both aperture segmentation and wavefront splitting.

The diffractive zone pattern of a Lee-Simpson apodized bifocal lens design can be seen in the microscope images of the surfaces shown in Fig. 9.27.

9.4.5.3 Commercially-available ophthalmic multifocal diffractive lenses

During the past 20 years, there has been a great deal of time, effort and money devoted to the development of diffractive optics designs and clinical trials for both contact lens and intraocular ophthalmic applications. To our knowledge, currently there are no commercially available multifocal diffractive contact products available in the market. However, there are some bifocal and trifocal intraocular (IOL) lenses in the market today that utilize diffractive optics designs.

AcrySof® ReSTOR® (Alcon Laboratories, Inc., Ft Worth, TX) was the first multifocal diffractive IOL to receive approval for use in the United States

Figure 9.27 Microscope image of Lee-Simpson apodized bifocal lens surface showing the diffractive zones transitioning to a refractive outer ring. (Courtesy of Apollo Optical Systems, Inc.)

since 2005. It features an apodized diffractive surface design in the center of the lens, similar to that shown in Fig. 9.27. It is been noted in the literature that patients have acceptable distance and near vision with this lens, although intermediate vision may be be somewhat compromised.

The Tecnis® Multifocal IOL (Johnson & Johnson Vision Care, Inc., Moorestown, NJ) was approved for use in the United States in 2009. The diffractive surface design in the Tecknis® IOL extends across the entire surface of the lens, similar to that shown in Fig. 9.26. Like the ReSTOR® IOL, patients have reported acceptable distance and near vision, but intermediate vision is somewhat compromised compared with other designs [47].

To address the issues associated with the clarity of vision at intermediate distances, two companies, Alcon Laboratories, Inc. and Zeiss Inc., recently introduced new diffractive trifocal IOL designs that produce acceptable performance at distant, intermediate and near ranges.

The Alcon Laboratories, Inc. AcrySof® IQ PanOptix® trifocal IOL uses diffractive optics in its central 4.5-mm pupil to create an intermediate add power of $+1.65D$ and a near add power of $+2.35D$. Alcon's technology, which focuses 88% of the light on the retina, is designed to provide images independent of pupil size [48,49,50].

The Zeiss Medical Technology Division of, Zeiss, Inc. introduced its Zeiss AT Lisa® line of trifocal IOLs, which uses diffractive optics to produce an intermediate add power of $+1.66D$ and a near add power of $+3.33D$ [51].

9.4.5.4 Large-aperture diffractive telescopes

An area that has garnered a great deal of international attention in the past several years is the development of large-aperture, diffractive space telescopes. A key motivation for use of a large-area diffractive lens as the primary space telescope lens is because it is/can be ultralight-weight, flat, foldable, and hence packable and deployable. The challenge, of course, is that the optical power of a conventional (modulo 2π) diffractive lens is highly dispersive, see Eq. (9.7) and Fig. 9.20(a). Therefore, the basic question is: given a large-aperture diffractive primary lens, what additional secondary optics are required to form a high-quality, broadband real image at the detector plane?

In the 1970s and 1980s, several authors investigated the effects of dispersion in imaging systems consisting of combinations of diffractive (holographic) lenses. Katyl [52] suggested compensating systems consisting of dispersive lens groups to correct for primary chromatic aberrations in diffractive systems. Numerical examples using optical design software were reported by Latta [53] to illustrate the correction of lateral and longitudinal chromatic aberrations. Conditions for the elimination of longitudinal dispersion in two- and three-lens systems were derived by Bennett [54] in the context of geometrical optics. He found that systems consisting of two diffractive lenses could not produce an achromatic real image. Sweatt [27] concluded that even for systems consisting of three diffractive lenses, simultaneous dispersion correction and real image formation cannot be realized. Hufnagel [55] patented an achromatic holographic optical system based on a Schupmann-type lens system [56] that consisted of a combination of holographic and conventional lenses, and Faklis and Morris [57] provided a paraxial Fresnel diffraction analysis for a broadband imaging system consisting of three lenses of arbitrary dispersion, and found a general solution for the wavelength dependence of the lenses that simultaneously correct the imaging system for both longitudinal and lateral paraxial chromatic aberration, a special case of which is the Schupmann-type lens system proposed by Hufnagel [55].

The layout of a Schupmann-type lens system analyzed by Faklis and Morris [57] is illustrated in Fig. 9.28, in which a large-aperture diffractive lens with focal length $F_1(\lambda)$ is focused onto an achromatic field lens with focal length $F_2(\lambda)$, which is used to image the large-aperture diffractive lens onto the third lens group, which contains a diffractive lens with negative focal length, $-F_1(\lambda)$, wherein the focal lengths of the three lens groups, $F_1(\lambda)$, $F_2(\lambda)$, and $F_3(\lambda)$ are

$$\frac{1}{F_1(\lambda)} = \frac{\lambda}{\lambda_0 F_0} \qquad (9.21)$$

$$\frac{1}{F_2(\lambda)} = \frac{1}{B} + \frac{1}{C} \qquad (9.22)$$

$$\frac{1}{F_3(\lambda)} = \frac{1}{C} + \frac{1}{D} + \frac{B^2}{C^2}\left[\frac{1}{A} + \frac{1}{B} - \frac{1}{F_1(\lambda)}\right] \qquad (9.23)$$

There are now numerous reports in the literature that describe the development large-aperture diffractive space telescopes, many of which are based on the Schupmann-type lens system illustrated in Fig. 9.28. "Eyeglass" is a large-aperture space telescope, which consists of two spacecrafts separated by several kilometers. The large-aperture diffractive primary lens (25-100 m) is mounted on one spacecraft, and the corrective optics and image plane are mounted on the other spacecraft [58,59,60].

The Membrane Optical Imager Real-time Exploration (MOIRE) program sponsored by the Defense Advanced Research Projects Agency (DARPA) also used a large-aperture diffractive lens as primary telescope primary lens, which was fabricated using binary-optics, reactive ion etching and polymer replication processes, described above in Section 9.2 [61,62].

Recently, researchers at the University of Arizona (Tucson, AZ) and Northrop Grumman Aerospace Systems (Redondo Beach, CA) developed a new approach for the design of lightweight diffractive telescopes. In this case, the large-aperture primary lens consists of a combination of a MOD lens and a weak conventional diffractive lens, which is used to compensate for the small amount of chromatic aberration arising from the MOD lens [63]. The authors refer to the combined diffractive lens structure as a "Multiple-Order Diffractive Engineered" (MODE) lens. A primary goal of the proposed Nautilus telescope program, which is based on MODE lenses, is to create a diffractive space telescope to survey exo-planets for biosignatures [64].

Lastly, we note that an extensive review of the international efforts on the development of the space-based diffractive telescopes can be found in the recent article [65].

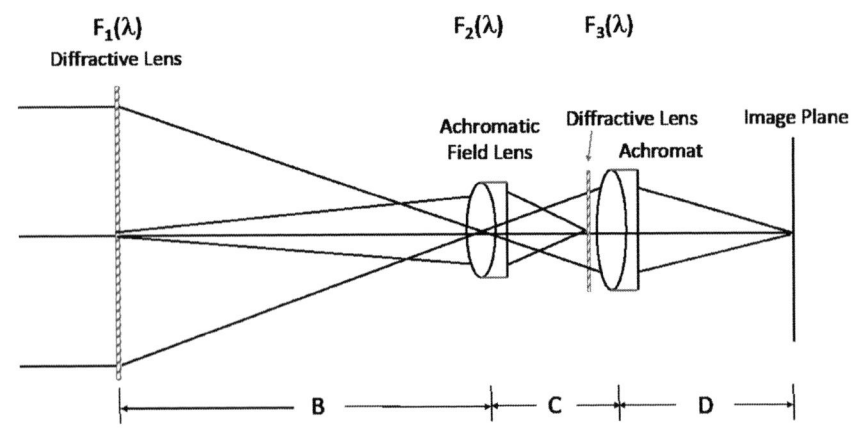

Figure 9.28 Broadband imaging system consisting of a combination of diffractive and refractive lens elements.

9.5 Markets for Micro-Structured Optics

In this section, we discuss a few select applications with significant market impact. One should not assume, however, that other important applications are not emerging, such as in automotive, medical, defense, space, etc. Micro-optics, as well as the embryonic nano-photonics, continue to grow and no doubt over the next few years and decades will impact various market segments.

9.5.1 Gesture recognition and 3D imaging/sensing systems

Three-dimensional (3D) sensing is a growing application that greatly benefits from micro-optical components. 3D sensing refers to systems capable of scanning the environment to detect the presence and distance of nearby objects. The technology has widespread application in consumer electronics, automotive, medical, gaming, virtual and augmented reality, etc. 3D sensing systems are incorporated in many cellphones for face recognition and gaming systems where, through gesture recognition algorithms, the player's own motion becomes part of the game, the Microsoft Kinect sensor being an early example. Also, 3D sensing is used in robotic navigation and is starting to make inroads into automobiles for both in-cabin and surround detection. As the technology continues to evolve and new uses of 3D sensing emerge, this market is expected to grow significantly and become intrinsic to a large set of applications.

In broad terms, any 3D sensing system can be separated into two basic channels: the illumination channel and the detection channel. Micro-optical elements play a major role in the illumination channel so here we will concentrate on that portion of the system. The exact type of micro-optical component to be used depends on the detection system, and there are a few different approaches that are currently employed, namely stereoscopic illumination, structured light illumination, and time-of-flight systems. In the stereoscopic approach [66], the sensing channel incorporates two cameras offset by some distance. The two images detected from the same scene are then combined to create a depth map, in much the same way as human eyes enable depth perception. Structured light systems project specific light patterns that in combination with some detection algorithm enable the creation of the 3D map based on the distortions of the projected illumination pattern. There is a variety of structured light schemes [67]. Time-of-flight (ToF) systems are based on the time it takes for photons to propagate from the illumination channel and make their way back to the sensing channel, tracing its origins to earlier LIDAR technology [68]. There are of course pros and cons for any of these approaches. The ToF approach is probably the most popular approach currently in use due to its simpler implementation, but stereoscopic vision and structured light have been adopted and used in some commercial systems.

The most common structured light implementation is based on dot projection where a micro-optical element, typically in the form of one or multiple DOEs, generates an array of dots. An important challenge for diffractive elements in this type of application is that the angular range that needs to be illuminated (field of view, FOV, or field of illumination) is very wide, often covering more than 50–60 degrees full width. This requires very small features. For example, a 50-degree FOV at 940 nm requires a grating period of about 2.2 μm and smaller as angles get wider. For such wide angles one is generally limited to binary phase profiles, which provide maximum theoretical efficiency of 80% and impose certain symmetries on the projected pattern. More significantly, for wide FOVs it becomes very difficult to avoid the presence of a strong hot spot – the zero-order diffraction spot. To successfully implement DOEs in structured light applications, the common approach is to use two DOEs so as to reduce the angular requirements on either element or to combine a VCSEL array with a simpler DOE and in this way multiply the spot distribution as necessary [69]. The combined contribution of the two elements, (DOE + DOE) or (VCSEL + DOE), makes the presence of zero diffraction order more manageable and, while manufacturing tolerances are still tight, allows the use of wide-angle dot patterns in many important applications. Similar patterns can also be used in the stereographic detection scheme with the processing algorithms constituting the main change in this approach.

For the ToF, flood illumination is more common wherein the scene is illuminated in a continuous fashion, as opposed to the discrete dot patterns of structured light. There are no requirements on the light source other than those dictated by the detection scheme. However, in some applications, such as consumer electronics where battery lifetime is critical, highly efficient illumination where only a specific region of the scene is probed at any given time becomes very important. Engineered Diffusers® [22] constitute the preferred components in ToF illumination systems because they can provide efficient illumination with controlled energy distributions, either in intensity (angle) space or irradiance (screen) space. A simplified diagram of the ToF system is shown in Fig. 9.29.

VCSEL arrays are typically used in 3D sensing systems due to their low cost and efficiency [70]. These lasers are generally not well collimated, with typical divergence angles in the range of 15–25 degrees, depending on manufacturer. When utilizing a diffuser to provide the necessary FOV the lack of collimation will cause the ideal illumination to convolve with the laser divergence distribution thereby reducing the concentration efficiency at the scene being illuminated. However, the need for compact packaging often

Diffractive- and Micro-structured Optics 341

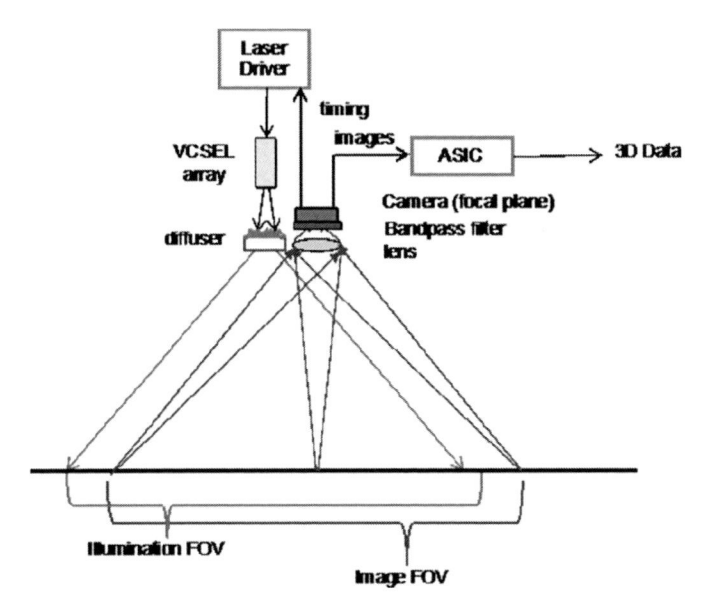

Figure 9.29 Basic elements of ToF 3D sensing system showing the illumination and detection channels.

takes precedence over the potential gain in efficiency, which would require the addition of collimating optics into the system.

The proper diffuser selection thus enables not only the best use of the laser energy but also control of the illumination pattern. Engineered (microlens-based) diffusers [22] constitute one such solution, or in some cases, periodic microlens arrays [17] are used. Depending on the application, uniform illumination may be required in angle space (intensity) or at a flat surface (irradiance), or something more complex. There are many possible reasons to select specific illumination patterns, some due to intricacies of the detection algorithm or regulatory reasons, such as eye-safe illumination levels. Uniform illumination in irradiance requires intensity profiles with a so-called "batwing" shape, wherein more energy is directed towards wider angles. Depending on the FOV, there may be significant energy within certain regions that pose a potential threat in terms of eye safety. Figure. 9.30 shows some of the typical illumination schemes used in ToF applications. The first row shows the case where the illumination is uniform (flattop) in intensity or angle space. At a plane away from the source (image plane), the energy distribution will undergo a $\cos^3$ transformation, from radiometric considerations [71]. At the sensor, again due to radiometric considerations, it becomes $\cos^7$. The second row represents the case where unform illumination is required in irradiance, at the image. In this case the diffuser needs to project a $\cos^{-3}$ from the source (batwing profile). The final row presents the case of uniform illumination at the sensor, leading to a $\cos^{-7}$ and a $\cos^{-4}$ pattern from the

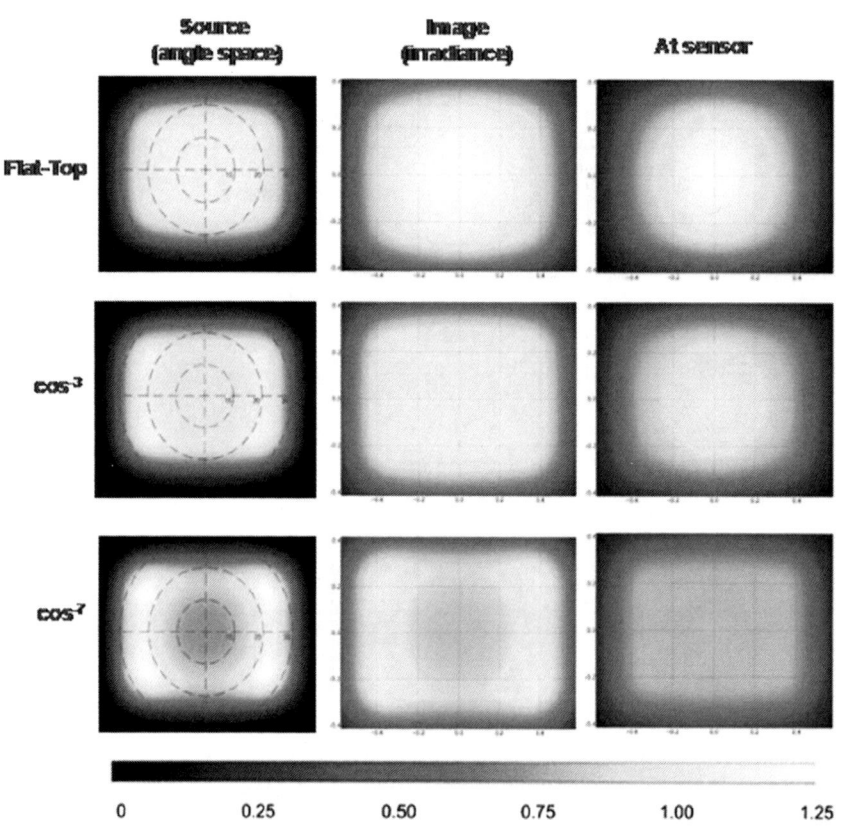

Figure 9.30 Illustrative illumination schemes for 3D sensing from uniform illumination in angle space (first row) to uniform illumination at the sensor plane (third row).

source and at the image place, respectively. (This assumes the image plane behaves as a Lambertian screen.)

9.5.2 Display and image projection systems

Micro-optical elements also play an important role in "light management" for display and projection systems. Light management used to be a term heard with frequency in the days of projections television, those large but very bulky TV sets that used to be the stars of showrooms before the advent of LCD TVs took over and completely transformed the market. The concept, while not as much a buzzword these days, remains true in the sense of controlling the distribution of light to make the best use of photons and provide the most efficient illumination distribution in a way that meets the important requirements of the application.

In typical displays, one finds the imaging system, usually comprised of traditional optical components up to the last or nearly last element, which comprises the projection screen. The function of the screen is to spread the image within viewer space and uniformly and efficiency as possible. Screens generally take the form of a diffusive component that can cover a reasonably wide angular space. In earlier projection screen systems, the screen was generally comprised of a lenticular onedimensional cylindrical array together with some volume diffusing element. The lenticular is primarily responsible for the wide-angle FoV while the volume diffuser would provide enough narrow-angle field. In volume diffusers, particulates dispersed in the bulk material provide vertical gain as well as minimize diffraction and color artifacts caused by the periodic nature of the lenticular array. The periodic nature of the lenticular constitutes a potential source of image artifacts, such as moiré fringing and/or color banding. To improve resolution as well as image contrast, screens based on microlens scatter elements offer improved performance [72]. To minimize image artifacts, however, the microlens array needs to be randomized. A comparison of the effect of periodic versus randomized diffuser screens is shown in Fig. 9.31.

Scanning electron microscope pictures of the microlens screen with both periodic and random geometries are shown in Fig. 9.32.

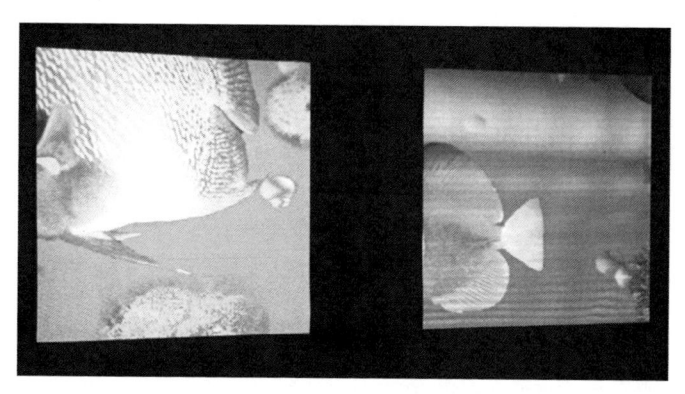

Figure 9.31 Image projected through a randomized microlens array (left) and periodic microlens array (right).

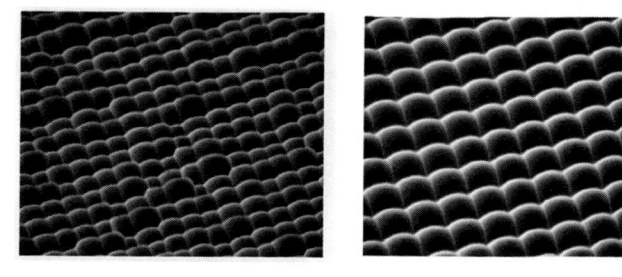

Figure 9.32 Examples of microlens diffuser screens with randomized geometry (left) and periodic arrangement (right).

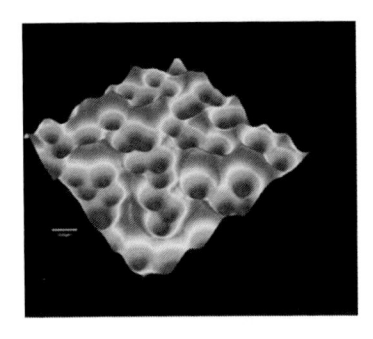

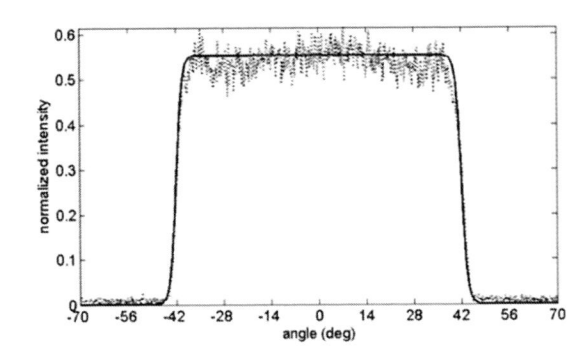

Figure 9.33 Engineered microlens diffuser (left) that produces flattop uniform illumination (right).

Randomized diffusers with deterministically designed surface structures [73] are well suited for display applications as well as image projection applications. For example, for high-power laser projectors diffusers can be used as "speckle busters", wherein a rotating diffuser placed in the path of the beam averages multiple independent speckle patterns to provide uniform-looking images, devoid of the graininess that is typical of the pattern observed under coherent illumination. Because of the high powers involved, the micro-optic diffuser needs to be produced in a hard material that can withstand high energy densities, fused silica often being the material of choice. To ensure the speckle pattern itself is devoid of diffraction artifacts that might lead to objectionable image defects, the diffuser surface needs to be strongly randomized to eliminate any residual hints of periodic structures. Examples are shown in Fig. 9.33.

9.5.3 Solid-state lighting

The development of light-emitting diodes (LEDs) started back in the 1960s, and were primarily used as indicator sources on instruments and portable devices. Since then LEDs have evolved considerably and are now covered under the broader "solid-state lighting" (SSL) term that includes not only the LED source itself but the whole system, including the optical components, electronics, and thermal management. Efficiency in converting electrons to photons had been the primary damper to the widespread use of SSL, but many of those obstacles have been overcome and capabilities continue to evolve to such an extent that these days, LED sources have become preferred to incandescent or fluorescent lighting for general lighting.

Micro-optics has been utilized to enhance extraction efficiency in LED sources where typical emissive materials have a high enough index that the transmission cone is limited due to total internal reflection, which leads to excessive heating of the package. Extracting as many photons as possible helps improve the optical efficiency while reducing the need to resort to more

Diffractive- and Micro-structured Optics 345

Figure 9.34 Common LED light fixture for general lighting.

involved heat dispersal solutions. There are limited options for suitable encapsulants so structuring some of the surfaces in the LED package helps improve efficiency [74].

Outside of the source construction, micro-optics can play a role in the distribution of emitted light from the LED. Here, efficiency is also a crucial parameter where the available energy must be optimally utilized by making sure light is directed as much as possible to the desired region of illumination. Particularly in outdoor situations, optimal directivity of illumination helps reduce light pollution and enhance the aesthetic contrast, which can be of interest in architectural lighting designs. LED light fixtures are now commonly found in nearly every home improvement store incorporating simple diffusers, either volume-type or simple Gaussian-type surface diffusers. For these applications, requirements on light distribution are very loose, the only requirement being good enough efficiency and sufficient diffuse power to spread light over a specific region. In most cases, the LEDs themselves emit with a near-Lambertian distribution so that the primary function of the diffuser may be to just hide the source and provide a uniform appearance.

In specialized applications, however, light propagation may require better control, and in this case, suitably designed micro-optics becomes important. As an example, we briefly discuss the case of LED-based billboard lighting, a prototypical example of off-axis incidence or other "wall-washing" applications in which it is desirable to obtain a uniform irradiance pattern over a surface when illumination comes at oblique angles; see Fig. 9.35. This is a particularly challenging illumination problem because of its unique geometry. At least one or several source assemblies are lined up along the edge of the billboard. For illustrative purposes, we consider the case of one source assembly, as shown in Fig. 9.35. This is a typical example where a long-lasting, efficient source is highly desirable to minimize energy costs and maintenance needs. It is also a case where reducing light pollution is highly desirable.

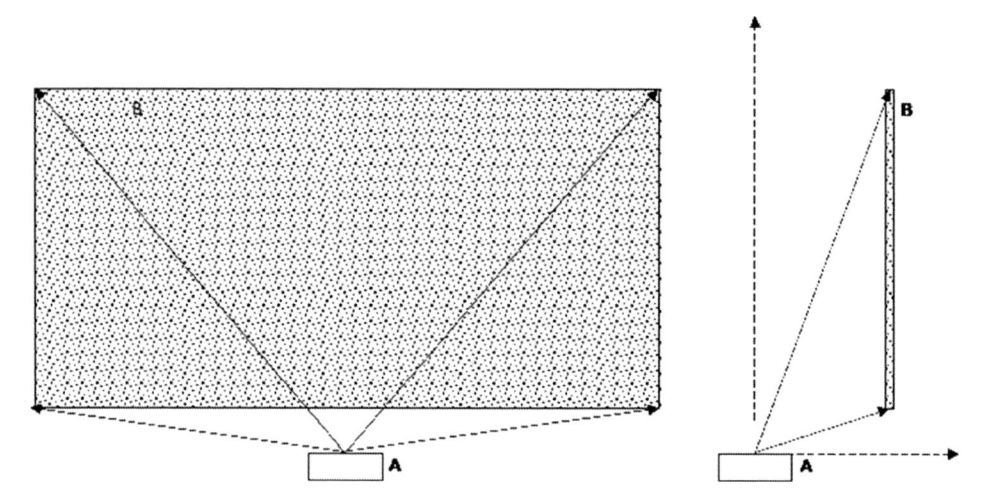

Figure 9.35 Basic billboard setup should the source (A) and the billboard screen (B) to be illuminated.

The challenge here is to illuminate the billboard screen as uniformly as possible with maximum efficiency. The angles involved in the billboard set shown above can be quite large, which makes it difficult to compensate for the natural change in irradiance across the target. As an example, for a 16-foot-high billboard where the source is 3 feet below the bottom edge, illumination angles are in the range of 25–75 degrees top to bottom. From the front view, with a 12-foot extent, coverage angles are about 33 degrees along the top and 84 degrees along the bottom. A possible solution to this problem is described by [75], where the basic beam-shaping element works to collimate a Lambertian LED source and concentrate nearly all the available energy towards the billboard target surface while taking advantage of the geometry of the problem to enable uniform irradiance over most of the surface. The final element, intrinsically asymmetric with a "teardrop" shape, as shown in Fig. 9.36, enables significant illumination improvement over common metal halide samples with conventional optical systems.

9.6 Summary

Over the last few decades, diffractive- and micro-structured optics design and fabrication have experienced extraordinary progress, particularly in manufacturing capabilities, successfully transitioning from university research into established commercial successes. This process continues to this day, and much more should be expected for the future as capabilities continue to evolve and new applications emerge. Initially, one could say somewhat elegant optical solutions have morphed into what now form the basis for critical components in a variety of markets. 3D sensing is probably one of the best

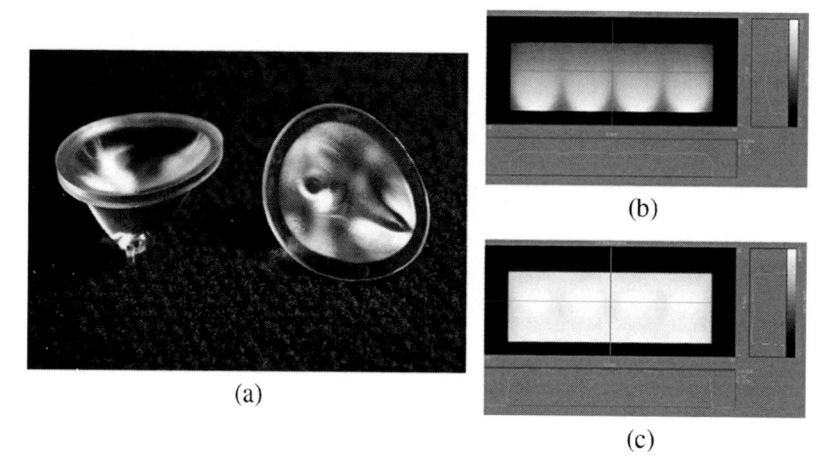

Figure 9.36 (a) "Teardrop" illumination optic for uniform billboard illumination. (b) Irradiance distribution simulation for conventional lamps. (c) Irradiance distribution using the teardrop element showing improved illumination uniformity.

examples of that success. Interestingly, early concepts based on binary photolithography continue to be as prominent today as they were promising decades ago, with important applications still relying on these types of solutions. Capabilities for high-precision patterning have been a critical aspect of this evolution, especially in regard to grayscale lithography, with direct laserwriting establishing as the method of choice to create general surface patterns that enable one to come very close to the maximum theoretical performance. While the area continues to develop with new approaches becoming more practical, such as two-photon exposure for 3D sculpting [76], we now observe order of magnitude improvements in processing times compared to earlier write times, not to mention improved capabilities and tools. Accurate modeling continues to be an active area of interest, not only from the fundamental point of view but the need to incorporate considerations, so as to allow strong computer-based modeling that can be directly translated into manufacturing. While some of that already exists in a variety of micro-optical components, these are strongest in relation to imaging-type structures. Work on random, non-imaging surfaces, is still on-going with significant improvements expected in the near future, which will further accelerate the pace of discovery, development, and transition to current and new high-volume manufacturing applications of diffractive- and micro-structured optical elements.

References

[1] E. Hecht, *Optics, 5th Edition*, Pearson Education (2002).
[2] B. D. Guenther, *Modern optics*, OUP Oxford (2015).

[3] J. W. Goodman, *Introduction to Fourier optics*, Roberts and Company Publishers (2005).

[4] R. Collier, C. Burckhardt, and L. Lin, "Optical holography," *Academic Press, New York* **19712**, 504–514 (1971).

[5] H. Caulfield, "Handbook of optical holography," (1979).

[6] M. Bass, "Handbook of optics, vol. ii, chap. 31," *McGraw-Hill, New York* (1995).

[7] M. Bass, "Handbook of optics, vol. ii, chap. 8," *McGraw-Hill, New York* (1995).

[8] B. Brown and A. Lohmann, "Computer-generated binary holograms," *IBM Journal of research and Development* **13**(2), 160–168 (1969).

[9] E. Britannica, *Wave Theory of Light*, New York, Charles Schribner's Sons (1888).

[10] Engineered Diffusers® is a registered trademark of Viavi Solutions.

[11] T. Hessler, M. Rossi, R. E. Kunz, and M. T. Gale, "Analysis and optimization of fabrication of continuous-relief diffractive optical elements," *Appl. Opt.* **37**, 4069–4079 (1998).

[12] T. J. Suleski and D. C. O'Shea, "Gray-scale masks for diffractive-optics fabrication: I. commercial slide imagers," *Applied Optics* **34**(32), 7507–7517 (1995).

[13] E. Hasman, N. Davidson, and A. Friesem, "Heterostructure multilevel binary optics," *Optics letters* **16**(19), 1460–1462 (1991).

[14] M. Bass, "Handbook of optics, vol. ii, chap. 41," *McGraw-Hill, New York* (1995).

[15] Y. Tamagawa and Y. Ichioka, "Efficiency of blazed diffractive optics produced by diamond turning," *Optical review* **5**(5), 291–294 (1998).

[16] L. Thompson, W. C. Grant, and M. Bowden, "Introduction to Microlithography, 2nd Edition," (1994).

[17] M. Hutley, "Refractive lenslet arrays," in *Micro-optics: Elements, Systems and Applications*, ed. by H. P. Herzig, 127–152, Taylor & Francis (1997).

[18] M. T. Gale, M. Rossi, J. Pedersen, and H. Schuetz, "Fabrication of continuous-relief micro-optical elements by direct laser writing in photoresists," *Optical Engineering* **33**(11), 3556–3567 (1994).

[19] S. Z. Wang, G. D. Savant, E. Kaiser, and K. S. Lee, "Seamless master and method of making same," (2004). US Patent 6,675,863.

[20] S. L. Yeh, "A study of light scattered by surface-relief holographic diffusers," *Optics communications* **264**(1), 1–8 (2006).

[21] D. R. Brown, "Beam shaping with diffractive diffusers," in *Laser Beam Shaping VI*, F. M. Dickey and S. C. Holswade, Eds., 249–271, Marcel Dekker, New York (2005).

[22] T. R. Sales, "Engineered microlens diffusers," *Laser Beam Shaping: Theory and Techniques*, 367–399 (2014).

[23] H. P. Herzig, *Micro-optics: elements, systems and applications*, CRC Press (1997).
[24] M. T. Gale, "Replication," in *Micro-optics: Elements, Systems and Applications*, ed. by H. P. Herzig, 153–178, Taylor & Francis (1997).
[25] D. A. Buralli, G. M. Morris, and J. R. Rogers, "Optical performance of holographic kinoforms," *Applied optics* **28**(5), 976–983 (1989).
[26] D. A. Buralli and G. M. Morris, "Effects of diffraction efficiency on the modulation transfer function of diffractive lenses," *Applied optics* **31**(22), 4389–4396 (1992).
[27] W. Sweatt, "Achromatic triplet using holographic optical elements," *Applied Optics* **16**(5), 1390–1391 (1977).
[28] W. Kleinhans, "Aberrations of curved zone plates and fresnel lenses," *Applied optics* **16**(6), 1701–1704 (1977).
[29] R. Kingslake and R. Johnson, "Lens design fundamentals, 2nd edition," (2010).
[30] T. Stone and N. George, "Hybrid diffractive-refractive lenses and achromats," *Applied optics* **27**(14), 2960–2971 (1988).
[31] T. Nakai and H. Ogawa, "Research on multi-layer diffractive optical elements and their application to camera lenses," in *Diffractive Optics and Micro-Optics*, DMA2, Optical Society of America (2002).
[32] H. Genda and M. Niwa, "Multilayer diffractive optical element," (2017). US Patent 9,696,469.
[33] C. Xue and Q. Cui, "Design of multilayer diffractive optical elements with polychromatic integral diffraction efficiency," *Optics letters* **35**(7), 986–988 (2010).
[34] L. Yang, Q. Cui, T. Liu, and C. Xue, "Effects of manufacturing errors on diffraction efficiency for multilayer diffractive optical elements," *Applied optics* **50**(32), 6128–6133 (2011).
[35] L. Gao, S. To, H. Yang, X. Nie, T. Liu, and C. Xue, "Effect of assembling errors on the diffraction efficiency for multilayer diffractive optical elements," *Applied optics* **53**(31), 7341–7347 (2014).
[36] C. Xue and S. To, "Analysis of materials selected for multilayer diffractive optical elements," *Optik* **125**(13), 3245–3248 (2014).
[37] D. Faklis and G. M. Morris, "Spectral properties of multiorder diffractive lenses," *Applied Optics* **34**(14), 2462–2468 (1995).
[38] D. W. Sweeney and G. E. Sommargren, "Harmonic diffractive lenses," *Applied Optics* **34**(14), 2469–2475 (1995).
[39] D. Faklis and G. M. Morris, "Polychromatic diffractive lens," (1996). US Patent 5,589,982.
[40] K. Miyamoto, "The phase fresnel lens," *JOSA* **51**(1), 17–20 (1961).
[41] J. C. Marron, D. K. Angell, and A.M. Tai, "Higher-order kinoforms," in *Computer and Optically Formed Holographic Optics*, **1211**, 62–66, International Society for Optics and Photonics (1990).

[42] R. Borghi, G. Cincotti, and M. Santarsiero, "Diffractive variable beam splitter: optimal design," *JOSA A* **17**(1), 63–67 (2000).

[43] A. Zhang, "Multifocal diffractive lens design in ophthalmology," *Applied Optics* **59**(31), 9807–9823 (2020).

[44] M. H. Freeman, "Multifocal contact lenses utilizing diffraction and refraction," (1987). US Patent 4,637,697.

[45] C.-S. Lee and M. J. Simpson, "Diffractive multifocal ophthalmic lens," (1997). US Patent 5,699,142.

[46] J. A. Davison and M. J. Simpson, "History and development of the apodized diffractive intraocular lens," *Journal of Cataract & Refractive Surgery* **32**(5), 849–858 (2006).

[47] J. Zvorničanin and E. Zvorničanin, "Premium intraocular lenses: the past, present and future," *Journal of Current Ophthalmology* **30**(4), 287–296 (2018).

[48] T. Kohnen, "First implantation of a diffractive quadrafocal (trifocal) intraocular lens," *Journal of Cataract & Refractive Surgery* **41**(10), 2330–2332 (2015).

[49] *AcrySof IQ PanOptix trifocal intraocular lens and AcrySof IQ PanOptix toric trifocal intraocular lens*, U.S. Food and Drug Administration (2019). https://www.fda.gov/medical-devices/recently-approved-devices/alconlaboratories-inc-acrysofr-iq-panoptixr-trifocal-intraocular-lens-model-tfnt00-and-acrysofr-iq.

[50] *Alcon Introduces AcrySof IQ PanOptix Trifocal IOL in the U.S., the First and Only FDA-Approved Trifocal Lens*, Alcon Inc. (2019). https://www.alcon.com/media-release/alcon-introduces-acrysof-iq-panoptixtrifocal-iol-us-first-and-only-fda-approved.

[51] *ZEISS AT LISA tri family*, ZEISS Medical Technology (2019). https://www.zeiss.com/meditec/int/product-portfolio/iols/multifocal-iols/atlisa-tri-family.html.

[52] R. H. Katyl, "Compensating optical systems. part1: broadband holographic reconstruction," *Applied optics* **11**(5), 1241–1247 (1972).

[53] J. N. Latta, "Analysis of multiple hologram optical elements with low dispersion and low aberrations," *Applied Optics* **11**(8), 1686–1696 (1972).

[54] S. J. Bennett, "Achromatic combinations of hologram optical elements," *Applied optics* **15**(2), 542–545 (1976).

[55] R. E. Hufnagel, "Achromatic holographic optical system," (1985). US Patent 4,550,973.

[56] A. Offner, "Field lenses and secondary axial aberration," *Applied optics* **8**(8), 1735–1736 (1969).

[57] D. Faklis and G. M. Morris, "Broadband imaging with holographic lenses," *Optical engineering* **28**(6), 286592 (1989).

[58] R. A. Hyde, "Eyeglass. 1. very large aperture diffractive telescopes," *Applied Optics* **38**(19), 4198–4212 (1999).

[59] I. M. Barton, J. A. Britten, S. N. Dixit, L. J. Summers, I. M. Thomas, M. C. Rushford, K. Lu, R. A. Hyde, and M. D. Perry, "Fabrication of large-aperture lightweight diffractive lenses for use in space," *Applied Optics* **40**(4), 447–451 (2001).

[60] R. A. Hyde, S. N. Dixit, A. H. Weisberg, and M. C. Rushford, "Eyeglass: a very large aperture diffractive space telescope," in *Highly Innovative Space Telescope Concepts*, **4849**, 28–39, International Society for Optics and Photonics (2002).

[61] J. A. Britten, S. N. Dixit, M. DeBruyckere, D. Steadfast, J. Hackett, B. Farmer, G. Poe, B. Patrick, P. D. Atcheson, J. L. Domber,et al., "Large-aperture fast multilevel fresnel zone lenses in glass and ultrathin polymer films for visible and near-infrared imaging applications," *Applied Optics* **53**(11), 2312–2316 (2014).

[62] P. Atcheson, J. Domber, K. Whiteaker, J. A. Britten, S. N. Dixit, and B. Farmer, "Moire: ground demonstration of a large aperture diffractive transmissive telescope," in *Space Telescopes and Instrumentation 2014: Optical, Infrared, and Millimeter Wave*, **9143**, 91431W, International Society for Optics and Photonics (2014).

[63] T. D. Milster, Y. S. Kim, Z. Wang, and K. Purvin, "Multiple-order diffractive engineered surface lenses," *Applied optics* **59**(26), 7900–7906 (2020).

[64] *The Nautilus Array*, University of Arizona. http://nautilus-array.space/.

[65] W. Zhao, X. Wang, H. Liu, Z. Lu, and Z. Lu, "Development of space-based diffractive telescopes," *Frontiers of Information Technology & Electronic Engineering* **21**, 884–902 (2020).

[66] L. Li, "Time-of-flight camera–an introduction," *Technical white paper* (SLOA190B) (2014).

[67] J. Geng, "Structured-light 3d surface imaging: a tutorial," *Advances in Optics and Photonics* **3**(2), 128–160 (2011).

[68] R. Horaud, M. Hansard, G. Evangelidis, and C. Ménier, "An overview of depth cameras and range scanners based on time-of-flight technologies," *Machine vision and applications* **27**(7), 1005–1020 (2016).

[69] A. Shpunt and B. Pesach, "Optical pattern projection," (2013). US Patent 8,384,997.

[70] R. Michalzik, *VCSELs: fundamentals, technology and applications of vertical-cavity surface-emitting lasers*, vol. 166, Springer (2013).

[71] R. W. Boyd, *Radiometry and the detection of optical radiation* (1983).

[72] T. R. Sales, "High-contrast screen with random microlens array," (2004). US Patent 6,700,702.

[73] G. M. Morris and T. R. Sales, "Structured screens for controlled spreading of light," (2006). US Patent 7,033,736.

[74] Y.-K. Ee, P. Kumnorkaew, R. A. Arif, H. Tong, H. Zhao, J. F. Gilchrist, and N. Tansu, "Optimization of light extraction efficiency of

iii-nitride leds with self-assembled colloidal-based microlenses," *IEEE Journal of Selected Topics in Quantum Electronics* **15**(4), 1218–1225 (2009).

[75] D. J. Schertler, "Optical element providing oblique illumination and apparatuses using same," (2015). US Patent 9,217,554.

[76] J. Fischer and M. Wegener, "Three-dimensional direct laser writing inspired by stimulated-emission-depletion microscopy," *Optical Materials Express* **1**(4), 614–624 (2011).

Tasso R. M. Sales was the CTO of RPC Photonics, now part of VIAVI Solutions. He has worked extensively on the design and modeling of diffractive optical elements. He has B.S. and M.S. degrees in Physics from UFPE, Brazil, and a PhD in Optics from the Institute of Optics at the University of Rochester. He has worked on a variety of topics ranging from statistical mechanics of complex systems to optical super-resolution through grating diffraction and optical diffusers, summarized in dozens of publications, patents, and conference presentations. He current focuses on overseeing micro-optical design and fabrication efforts at VIAVI Solutions as well as working directly with customers to bring their projects and ideas from concept to reality.

G. Michael Morris was the CEO and co-founder of RPC Photonics and is presently serving as the CEO of Apollo Optical Systems. Prior to RPC and AOS, he was a Professor of Optics at The Institute of Optics, University of Rochester. He is an OSA Past-President (2003) and Past President of the OSA Foundation (2009–2015). He received his BS degree from the University of Oklahoma, and his MS and Ph.D. degrees from the California Institute of Technology. He is the recipient of many awards, including the David Richardson Medal, Joseph Fraunhofer Award/Robert M. Burley Prize, and the Stephen D. Fantone Distinguished Service Award from OSA. His research has spanned a wide variety of topics in statistical optics, optical information processing, quantum-limited imaging, automatic pattern recognition, and diffractive- and micro-optics technology. His current research/development interests include optical beam-shaping components with a particular emphasis on 3D imaging and sensing systems for consumer electronics, robot vision, autonomous vehicles, and surveillance markets. He holds over 30 U.S. patents, and has published more than 70 referred journal articles, three book chapters, and numerous conference proceedings.

Chapter 10
Illumination Optics

Henning Rehn
FISBA AG, Rorschacher Str. 268, 9016 St. Gallen, Switzerland

Julius Muschaweck
JMO GmbH, Zugspitzstr. 66, 82131 Gauting, Germany

10.1 Introduction

An illumination system consists of the light source(s), some optical components and a target, the latter literally carrying the specs which are defined by the field of application. In our review we will discuss design, technology, and fabrication of light sources and components. As there are considerable differences between the many fields of application, we start with a general view and derive the technical specialties afterwards.

In some cases we will name representative suppliers without the intention to exclude others. With the help of such a reference, the interested reader will be able to gather information, understand the product and find similar suppliers easily.

We restrict the review to visible, incoherent light and related optical elements.

10.2 Fields of Application

We start with an overview on the various fields and markets of lighting and illumination optics:

- General lighting (office lighting, high bay, decorative lighting, hospitality, and much more). *Interestingly, mankind did and does always spend the same share of the gross domestic product for lighting purposes* [1]
- Automotive exterior lighting (low and high beam, daytime running lights)

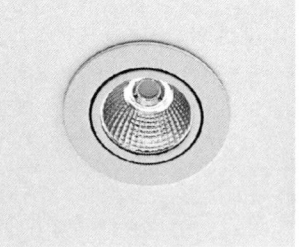

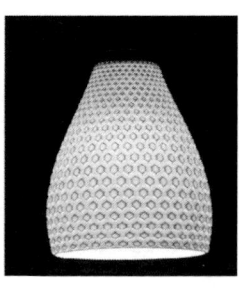

Figure 10.1 Indoor luminaires: Fluorescent trougher, LED downlight, and decorative ceiling light.

- Projection (various imagers and light sources)
- Lithography (though mainly UV)
- Entertainment and studio (as well as stage, theater, and architainment)
- Medical lighting (wherever light is needed on or in the body)
- Horticulture (plant growth with light recipes)
- Solar collection (concentrated and flat cells)
- Solar simulation (mimic aspects of sun irradiation)
- Airfield lighting (runway, landing, and taxi lights)
- Street lighting (streets, pedestrian ways, tunnels)

We characterize the needs of the field, suitable light source technologies and related optical components (their function and making are explained in Section 10.4).

10.2.1 Indoor lighting

In general lighting [2], the target étendue is usually much larger than that of the source. If one source does not produce enough light, one can easily use a plurality of them. The necessary luminance is rather low, efficacy and a pleasant spectral distribution are key features. Tolerances can be neglected to some degree.

Incandescent bulbs and their halogen cousins are currently replaced by LED luminaires and LED retrofits (an LED assembly that mimics a legacy source). In traditional luminaires, the sources work inside without much involvement of optics (there is at least one exception: for omnidirectional lamps, a light spreading element is placed on the LEDs [3], made of a transparent plastic by injection molding; – see Fig. 10.15).

For higher performance, such as in high bay lighting, lamps do greatly benefit from optics such as reflectors. In the office world, fluorescent troffers did not change much when transformed by LED tubes. But new designs developed—for example, lightguide plates with carefully engineered emission became popular over the last 10 years. There are components such as reflectors of various kind (halogen reflectors, downlights, troffers), LED domes to improve light extraction, total internal reflection (TIR) lenses for efficient collimation, and lightguides for light distribution and shaping.

10.2.2 Outdoor lighting

The outdoor market, mainly street lighting [4], is characterized by legal standards (a wide variety of situations or cases) to be followed (defined target illuminance distribution and avoidance of light pollution) and economic demands on the customer's side (mainly local authorities). Luminance requirements are basically moderate but may increase for efficient light shaping.

In the new millenium, a transition began from sodium and mercury lamps to LED because of lifetime, efficacy, spectral features and the chance to benefit from new illumination designs for, e.g., better homogeneity, and options for smart operation. From the beginning, the LED side has claimed to be the only technique to solve the light pollution (urban night sky brightness [5]) problem. However, for efficiency reasons, mainly *cold white LEDs* were chosen. Their light consists of a considerable fraction of blue light, and besides subjective factors [6] the stronger Rayleigh scattering of blue light created an effective *increase* in nightly skyglow [7].

Street lights comprise optical components such as metal reflectors (deep drawn or bent), LED reflectors (mostly made from injection molded and coated plastic, see Fig. 10.2), and TIR lenses.

10.2.3 Automotive lighting

In the world of automotive exterior lighting [9], the Economic Commission for Europe (ECE) and other standards define far-field distributions that have to be reached within some tolerances. We concentrate our discussion on the

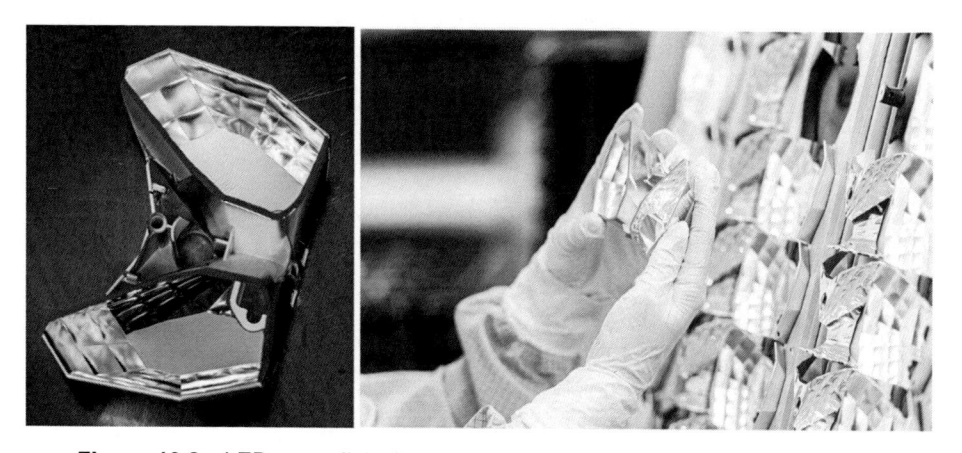

Figure 10.2 LED street-light freeform reflector [8] (photo courtesy of Siteco).

headlights. The necessary source luminance is moderate to high and is driven by the wish for a small luminaire size for aesthetic reasons.

The traditional way to realize the target distributions consisted of placing a light bulb in a reflector and provide some beam-shaping means, realized by the reflector shape [10,11] or some designed shape in the transparent plastic cover. Another technique is to provide a baffle in the beam and project it to the far field, a rather inefficient but conceptually easy way to produce the desired structures for the low beam.

Automotive lighting started with incandescent and halogen sources. In the early 1990's, we saw discharge sources (D lamps [12]) greatly improving lighting performance. Typical elements were reflectors and projection lenses (aspheres made by glass molding). Ten years later, LEDs began to enter the game. Their lifetime enabled permanently lit *daytime running lights* (using TIR lenses and lightguides), and their robustness and new design options lead to many headlamp designs. Recent developments include the pixelated or adaptive high beam [13] (which requires a sophisticated projection lens) and a laser-boosted supplemental high beam [14].

10.2.4 Medical lighting

There are two main applications of illumination tecnology for medical purposes. First, we have surgical (and dental) lights with low to moderate luminance requirements, strictly governed by standards. On the other hand, there are fiber light sources for endoscopy and related methods, always requesting the highest luminance source technologies can provide and more.

In the past, the medical lighting market was dominated by halogen and xenon lamps because of their good color rendering. Surgical lights usually consisted of such a lamp in a faceted elliptical reflector made of glass with a dichroic coating. For light guides, there were mostly xenon discharge sources in elliptical reflectors (made of glass as well), sometimes with very low efficiency, but providing sufficient, high-quality light.

Later LEDs conquered the market, and there is just one dominant design style (Figure 10.3), using one optical element: the TIR lens.

10.2.5 Airfield lighting

There are various luminaire types, such as runway, taxiway and obstruction lights. The field is still partially dominated by halogen lamps that resist the LED technology for a number of reasons, e.g., in cold countries, the heat generation in such a luminaire is considered a feature and not a bug, as it prevents the luminaire from being covered with snow.

Optical elements include reflectors (glass and metal), molded front glasses to realize an optical function, prisms and glass or plastic domes.

Figure 10.3 LED surgical light: LEDs of various colors are arranged side by side and collimated by TIR lenses.

10.2.6 Entertainment lighting

Here we include fields such as stage, show, architainment, studio and hospitality. Traditional luminaires comprise Fresnel lights and ellipsoidals operated with reflectorized tungsten halogen and metal halide lamps. Typical components are reflectors (smooth and faceted elliptical glass reflectors, retro reflectors [15]), dichroic filters, metal and glass gobos, glass and plastic Fresnel lenses, and classical projection lenses.

When high-luminance LEDs became available, they gained a considerable market share, and some novel luminaire types were created. To get the necessary luminous flux, the light of a large number of LEDs had to be combined, using arrayed collimation optics [16], dichroic mirrors, and homogenizers [17].

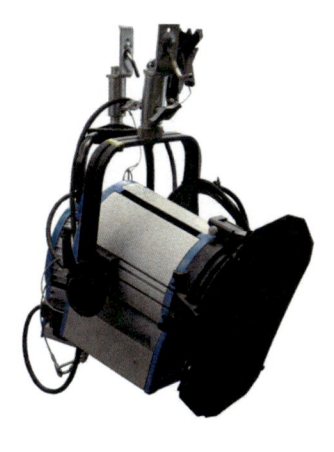

Figure 10.4 Tungsten Fresnel light, LED wash light, and Moving Head (left to right).

10.2.7 Data and video projection

When an object has to be projected to a screen, the system étendue (defined by the size of the microdisplay and the aperture of the lens) is kept low, mainly for economic reasons. Consequently, a high-luminance light source is needed. Spectral needs come in a different flavor; instead of a good color rendering, a lot of light in some primary color regions is required. In the late 1990s, microdisplay technology imager technology including liquid crystal display (LCD), digital light processing (DLP) [18], liquid crystal on silicon (LCOS), and ultra-high-pressure mercury (UHP)-type lamps [19] came almost simultaneously to maturity, causing a steep rise of the projection market. After 2010, laser-phosphor-based projectors [20,21] entered the market, and a few years later, LED projectors passed the 1000 ANSI lm threshold thanks to thin film LED technology and higher étendue imagers [22,23,24].

With the rise of projectors, large numbers of glass reflectors (elliptical or parabolic), high performance dichroic filters, integrators (rods or fly's eye condensers), and big quantities of lenses, prisms, mirrors, and dichroic coated elements were required.

But soon after this industry was fully developed, projectors were partially ousted by the rise of large flat panel displays. These were initially much behind projectors in terms of economic large area display but quickly gained market share.

10.2.8 Flat panel displays

Flat panel displays are used for television, computer screens, signage, commercials and more. After some technology and cost breakthroughs in the new millenium, they formed a multi-billion dollar industry.

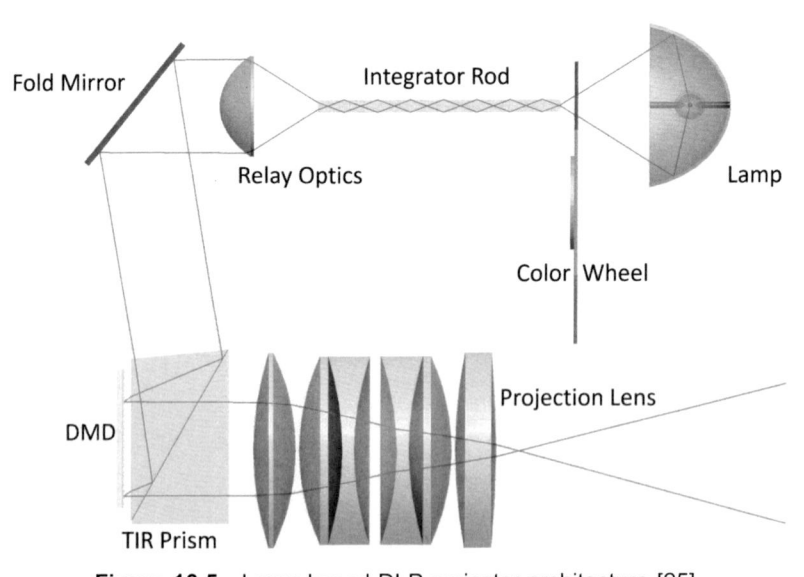

Figure 10.5 Lamp-based DLP projector architecture [25].

Figure 10.6 Fresnel lenses for solar collection.

There are active (plasma, OLED, micro-LED) and passive (mostly back-illuminated LCD) displays. Actie matrix technologies have advantages in terms of contrast, switching times, viewing angle, and luminance. Whereas plasma displays are nearly extinct due to their high energy consumption, OLEDs are on the rise and micro-LED displays are laying in wait to possibly take over the market.

All flat panel displays use various optical sheets such as diffusers, polarizers, brightness enhancement films, and color filters. An LCD display needs to be illuminated from the rear side by either a side lit lightguide or a "white box". The white box is less popular because it would add an inch or so to the overall thickness, but makes image improvement techniques such as local dimming possible. The ubiquitous lightguides, made from transparent plastic, are illuminated by LEDs from the side, often side emitting LEDs, and the light guide is equipped with some means for outcoupling.

10.2.9 Smartphones and smart watches

Illumination optics in a smartphone does not only comprise the display but also the flash LEDs for the rear camera [26]. These include some micro-Fresnel lenses made by injection molding. Smart watches may include sensors on the inner side for pulse and oxygen monitoring that are operated with an LED.

10.2.10 Solar

The long expected rise of *concentrated photovoltaics* (CPV) based on sophisticated optical knowledge [27] was to a large extent replaced by cheap flat solar cells. Optical elements of the field include reflectors (glass or metal), Fresnel lenses (usually made of plastic), and compound concentrators (mainly by glass molding), all with strong requirements on the material to resist concentrated solar radiation.

CPV systems always competed with the ordinary flat solar cells. But those have to be tested and certified, introducing a market for *solar simulation*, cell testing and weathering. Solar simulators are regulated by standards as well, and the degree of fulfillment defines the optical system. There are traditional light sources such as reflectorized xenon and metal halide lamps, and LEDs (white ones for simplicity or full spectrum solutions). Their optics include collimators and homogenizers [28].

10.2.11 Freeform optics

In illumination optics, some sophisticated light shaping tasks can be realized by one or more surfaces with a *lack of symmetry*. There is a variety of old and new methods, and sometimes old results are rediscovered by new methods. The most prominent methods are simultaneous multiple surface (SMS) [29,30] and tailoring, where the light of a point source (thus an academic case but often applicable for small sources) is converted to a prescribed distribution. This is fascinating science, in some years producing 90% of the illumination design publications [31,32,33,34,35] but there are not many business cases. Considerable progress in the field after 2010 led to at least two groups who established the whole chain from design to manufacturing and measuring freeform elements (Center for Freeform Optics [36], Rochester / Charlotte and FO-Plus [37], Jena). Applications are mainly in the imaging domain, e.g., AR goggles [38], miniaturized telescopes [39] (the ubiquitous TMA [40,41]) and spectrometers.

In illumination, there are logo projectors [42,43] (Fig. 10.7), lenses and reflectors in street lighting [8,44], and variable beam systems with various kind of freeform elements [45,46,47,48,49]. Some principles, like the Alvarez lens pair [45], can be applied in both illumination and imaging.

Point source tailoring for illumination is available in commercial software [50], but optimization approaches of others may work too in some cases. The

Figure 10.7 Logo projection demonstrator of the former OEC company. Actually it is not a projector but a beam shaping device based on a tailored lens surface.

methods allow for the design of freeform lenses and reflectors. Such elements would be most likely made of optical plastics, either by injection molding or diamond turning.

10.2.12 Trends after 2000

In the past, the core of all illumination was the light bulb, placed inside a luminaire and sometimes assisted by some optics. In the new millenium, we saw the maturity (or culmination?) of discharge lamp technologies, enabling high performance in the fields of projection for cinema, conference rooms, TV, and stage lighting, and in lithography. In the same period LED technology broke into nearly all illumination fields, causing the fall of incandescent and fluorescent light sources, up to the point of disintegration of the legacy light source industry—inevitable or not.

Developments in electronics and displays technologies called for specialized LEDs and did benefit considerably from their performance, such as in flat panel displays, smartphones, smart watches and biomonitoring, and more recently in AR goggles.

On the knowledge and design side, illumination design was established as a field inside optics, evident by the dissemination of the étendue concept and the rise of new design principles such as SMS and tailoring.

Recently, laser diodes reached a level of maturity that enabled their use for lighting purposes, such as for laser-phosphor sources. A list of current hot topics would include smart lighting with occupancy sensing and LiFi (data transmission via light modulation).

So we have technologies enabling new applications, but also applications pushing the technology on the light source and optical system side.

10.3 Light Sources and Their Fabrication

In the following, we take a deeper look at some trends and identify the relevant illumination systems and elements. Before discussing the making of optical elements for illumination, we may spend a moment at the very basic and essential item: light sources.

10.3.1 Legacy light sources

In the 20th century, electric light sources consisted of a glass vessel (the bulb), a provision to bring in some electric energy, and a means of light generation inside the vessel (filament or discharge arc). For an overview, see Refs. [51,52].

There is a wide range of legacy light sources with different features such as

- Electrical power: from a few milliwatts (incandescent signal lamps) up to 50kW (halogen lamps for stage, theater and light houses)
- Efficacy: from 5 lm/W up to 100 lm/W (metal halide lamps)

- Luminance: from 10 Mcd/m^2 to 10 Gcd/m^2 (projection lamps)
- Spectra: Black body radiation (incandescent), spectral lines (low pressure mercury), continuum (Xenon discharge)
- Color rendering (CRI): from very bad (low pressure mercury) up to perfect (incandescent)

Lamp fabrication was characterized by mastering

- glass production and bulb forming
- current feedthrough, realized by metal-glass joints (seals)
- metallurgy for electrodes, filaments, and feedthroughs (especially tungsten and molybdenium)
- bulb chemistry (gas fill, metal salts, halogens, getters, phosphor)

Lighting enterprises owned glass and metal factories. Lamp production grew from manual assembly to an automated process. For a long time, there were only three big players: General Electric, Philips, and OSRAM. As a result of a limited lifetime of the sources, and the high market entry barriers, lighting was a stable replacement business. Then a new technology reached maturity and broke into the world of lighting: the LED.

10.3.2 Rise of LED technology

In their early times, LEDs were tiny sources that could be used for signaling, but not to actually illuminate anything. Then, mainly in the 1990s, the technology saw significant progress in terms of performance (chip efficacy, phosphor conversion [53,54,55]), lifetime (orders of magnitude longer than legacy sources), current carrying capacity, resulting in higher luminance, and cost reduction [56]. As a result, legacy light sources were partially or completely forced out of the market and consequently even legally banned. The specific properties of LEDs introduced tremendous technical and commercial changes, as listed in Table 10.1:

Table 10.1 The specific properties of LEDs and the consequences.

LED Property	Consequence
Hemispheric emission	New optical architectures
Spectrum	Advanced color rendering metrics
Mostly no mains operation	Specific LED drivers needed
Less Heat	Other materials for optics and mechanics
Performance loss at higher temperatures	Heat management necessary
Tolerances and Binning	Complex supply chain
Lifetime	No replacement business any more

Illumination Optics

Figure 10.8 Halogen lamps are getting replaced by LED ones.

As a result, the plastic optics industry grew at the expense of glass component makers, and the thermal solutions providers expanded into lighting. From a logistics point of view, a light source is now merely an electronic component. Except for the retrofit market, the concept of a "bulb" has dissolved. Current discussions such as the *blue light hazard* and *human centric lighting* do not essentially affect optical components.

As the required luminance is mostly defined by the application, LEDs were adapted to what has been used for a long time. Many components were only slightly adapted, whereas some legacy concepts were modified such as reflector design, or adopted such as the use of phosphors to generate white light. There are three basic directions of LED development that are mutually exclusive to a large degree (Fig. 10.9). Each aspect is of different importance for the applications described in section 10.2.

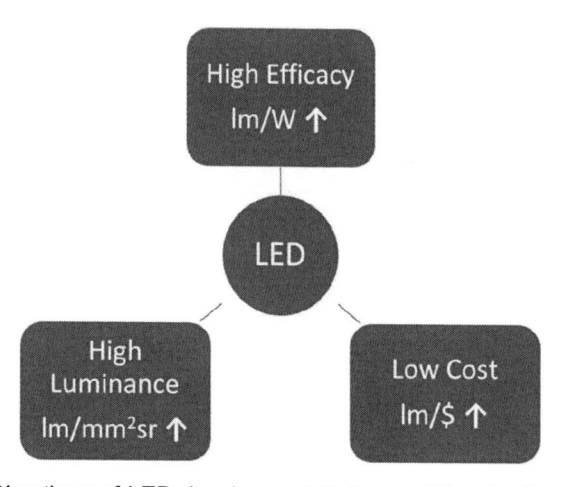

Figure 10.9 Directions of LED development that cannot be simultaneously fulfilled.

10.4 Optical Components for Illumination and Their Fabrication

There is a wide range of optical elements (much more than merely lenses) that realize the required function. Accordingly, materials and manufacturing methods do vary widely.

Illumination systems share some optical elements with their imaging relatives, but there are differences. In illumination, everybody expects the cost of elements to be lower. This can be especially true in fields with a large number of luminaires. Illumination often allows for wider tolerances than we know from imaging optics. Illumination engineers usually have a gut feeling what effort is necessary. For example, if source and target étendues are comparable, one has to take care.

Often one has to pay attention to the amount of light that passes the elements. There can be implications to overheating, internal transmission and mounting. Even if it's no high flux application, transmission may be key (e.g., in daylighting [57] and light detection [58]).

10.4.1 Standard lenses

The making of optical lenses is a well established technology, and in many projects a group of a few lenses may be cheaper and easier to get than a compound concentrator tooled. However, this may change in the future due to additive manufacturing of optics.

In contrast to imaging optics, illumination lenses have usually wider tolerances and are often made from materials other than optical glass—a big chance for plastic and silicone lenses.

Dome lenses on top of LEDs are a new class of illumination lenses, made mostly from silicone. For prototypes, plastic lenses can be made by diamond turning without post processing.

Glass asphere molding for illumination has its place established at the lower end of the accuracy scale. Large numbers of lenses are made of low

Figure 10.10 Molded lens array as the primary optics on some LEDs.

index and rather cheap glasses such as H-K51, Suprax, LIBA2000, and Pyrex instead of special molding glasses by the standard optical glass suppliers. The molding processes depend on volume: there are manual ones in low cost environments, but also automated production lines optimized for large numbers with medium precision. One example is the molding of projection aspheres for automotive high intensity discharge (HID) lamps.

There is usually no post-processing for illumination lenses beside some very basic centering (we may consider it as deburring). The same technology is used to make lens *arrays* if the source comes as an array as well.

10.4.2 TIR lenses and related collimators

TIR lenses are the best simple choice for collimated spot lamps if target étendue is much larger than that of the (LED) source and efficiency is key. The TIR lens design ensures efficiency but in most cases not conservation of effective luminance. TIR lenses are mostly made from molded plastic, but also from silicone and glass. There is a wide variety of stock lenses available from many suppliers all over the world, designed for many LEDs and diverse applications. The exit surface often provides an additional beam shaping or light mixing function (frosted, microlenses, cylindric, see Fig. 10.11).

10.4.3 Fresnel lenses

Fresnel lenses are rather popular in illumination because of their lower cost (you don't want to pay and handle a glass lens of a 500 mm diameter), low weight, and small sag which for example enables mobile phone flash solutions (Fig. 10.12).

Fresnel lenses are mostly made of plastics, for example by diamond turning for prototypes or compression molding for mass production. Even roll-to-roll production is sometimes used [59]. There are glass Fresnel lenses mainly for entertainment lighting, and hybrid ones (by casting of silicone onto glass).

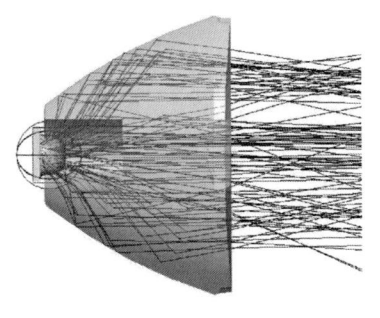

Figure 10.11 TIR lenses with different front face textures for beam shaping. TIR lens in a simulation. TIR lenses are the only means of beam shaping of the surgical light in Fig. 10.3.

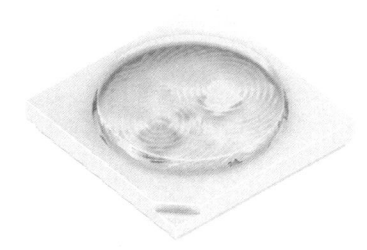

Figure 10.12 Large fresnel lens and mobile phone flash lens (photo courtesy of OSRAM OS).

Stock Fresnel optics is offered, for example, by NTK (Japan), Orafol (Germany), and Fresnel Technologies (US).

10.4.4 Compound concentrators

A compound parabolic concentrator (CPC) [60] or compound elliptical concentrator (CEC) [61] google image search reveals thousands of sketches and drawings, but very few photos. So, from the fabrication point of view, it's a rather academic topic. The use as a primary optics for LEDs is not popular as good performance requires the small face of the device to be perfectly filled by the source, which then needs to almost touch the CPC. There are a few examples of a CPC made from glass (Fig. 10.13) by either injection molding or using a synthetic quartz process [62].

Figure 10.13 Square CPC made of glass.

Illumination Optics 369

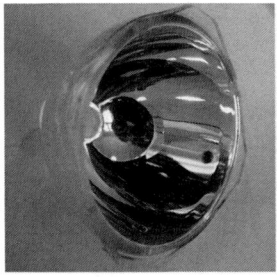

Figure 10.14 Glass reflector for video projection lamps, electroplated plastic reflector, deep drawn metal reflector (bright anodized).

10.4.5 Reflectors

Reflectors have been made for more than a hundred years, right from the beginning of the artifical lighting era. They can be produced by metal deep drawing, electroforming (a *mandrel* is galvanically coated with e.g., nickel), glass molding (with metallic or dielectric coating), or plastic injection molding with e.g., electroplating. Glass reflectors deliver the highest performance in terms of geometric accuracy and thermal stability.

10.4.6 Light guides

A lightguide is a means that transports light from a entrance surface to one or more exit surfaces. Lightguides can be made of transparent glass or plastic and benefit from the efficiency of total internal reflection (TIR) or hollow with reflecting side faces. One type of lightguides have a well specific face and serve to transport and possible to homogenize light. The purpose of another group of lightguides is to "lose" light on the way in a well defined manner. These can be flat or of a curved oblong shape.

Most TV, computer monitor and mobile device displays use a planar, thin PMMA light guide with white LEDs coupling light into the edges. Outcoupling is achieved by printed white dot patterns or embossed microstructures, whose density and/or size varies to compensate the inhomogenous local flux density, achieving good homogeneity.

In automotive exterior and interior lighting, light guides are everywhere, except for high and low beams. These light guides, mostly made from polycarbonate for thermal reasons, are injection molded parts, often long and thin, with groove structures on the back side to couple out the light into the desired directions. The primary reason for using light guides here is design: shapes of brake lights and daytime running lights create brand recognition.

Some LED retrofit bulbs comprise a lightguide (such as shown in Fig. 10.15) that collects the light from one or more LEDs and realizes omnidirectional emission by a suitably shaped exit region.

Figure 10.15 Left: glass light guides for the combination of the light of four high power LEDs to a target on the right. Right: LED bulb featuring a light guide for an omnidirectional light distribution.

10.4.7 Homogenizers

The principles of light mixing and homogenization were discussed in a previous paper [17]. There are two basic designs:

Rod integrators are special lightguides that produce a homogeneous *spatial* distribution of an arbitrary incident light distribution. They can be hollow or made of a transparent dielectric. The solid or dielectric type is made by glass molding (Fig. 10.16) or entirely by flat optics grinding and polishing technology. The guiding by TIR is perfectly efficient, but care has to be taken for the entrance and exit faces. It is not that easy to provide a highly resistant anti-reflection coating there. Molded glass rods of a hexagonal or even octagonal shape are becoming popular for LED Fresnel luminaires.

For the hollow variant, rectangular glass segments are put together (Fig. 10.17). There are products with special coatings for grazing incidence available [63]. A hint for mock-ups: there are highly reflecting sheets (such as Alanod Miro) that can be cut and folded and serve well as a light guide or rod integrator.

Fly's eye condensers (FECs) (Fig. 10.18, left) produce a homogeneous distribution in the *angular* domain. Combined with a condenser lens or

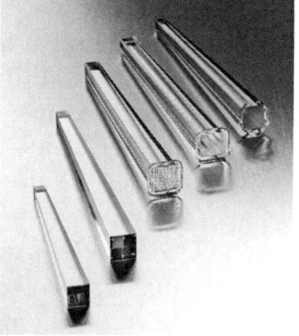

Figure 10.16 Faceted reflector. Integrator rods (both made of glass, photo courtesy of Auer Lighting).

Illumination Optics 371

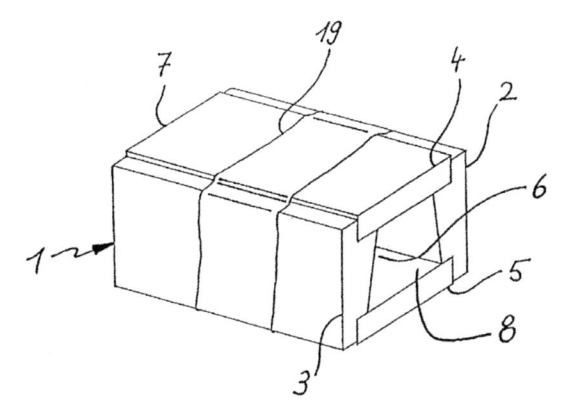

Figure 10.17 Rod integrator assembly from coated segments [64] and held together by some shrink tubing.

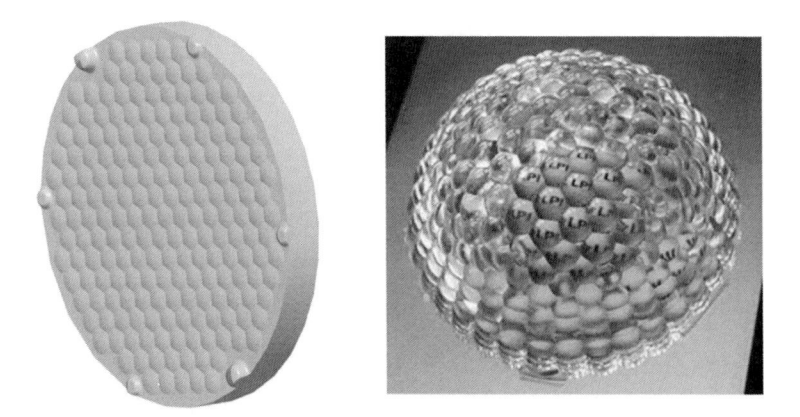

Figure 10.18 Model of a fly's eye condensor. Shell mixer [66] made by plastic molding (photo courtsey by LPI).

system, a target can be illuminated homogeneously. FECs are ubiquitous in LCD projectors, coming as pairs and usually made by glass molding. Nowadays, mostly for use with LEDs, there are tandem FECs from molded plastics [65] which carry lens arrays on both faces. There are versions with rectangular and with hexagonal lenslets, depending on the target shape. The shell mixer (Fig. 10.18, right) is a variant that surrounds a multi-color source making it effectively white.

Finally, we mention that optical surfaces such as of lenses and reflectors can by *faceted*, i.e., consiting of small flat areas to smoothen the image of a source in a target plane (Fig. 10.16).

10.4.8 Sheets

There are many different plastic sheets with optical functionality. *Deglaring films* ensure that ceiling luminaires obey the glare standards. The principle consists of rejecting (and recycling) low angle rays by microstructures such as tiny prisms (e.g., Jungbecker or BWF, both Germany). *Brightness enhancement films* are applied between backlight and liquid crystal element of an LCD screen. They reduce the emitted intensity distribution and thus increase the effective luminance in a restricted angular cone towards the viewer, at the expense of overall efficiency and luminance at high angles (e.g., 3M Vikuiti, US). The effect can be achieved by arranging *arrays of concentrators* on the sheet [67]. Light shaping diffusers and direction turning films are sheets that scatter or deflect light of a source to create a prescribed distribution (e.g., Luminit Corp., US).

10.5 Prototype Technologies

For the near future, we expect more 3D printing technologies applied to illumination optics as the needed accuracy is lower than in imaging, where it is much harder to get the optical quality right. Luxexcel (The Netherlands) is successfully printing lenses using an adapted ink jet printing process.

There are trials reported to make lenses by fused deposition modeling with transparent filaments. Actually, blanks are produced and post-processed by grinding/polishing or transparent coating with a resin.

Microlens arrays can be produced by precise ink jet printing individual lenses on a glass substrate and UV curing them (e.g., IOF, Germany).

There is even some direct printing of silicone lenses directly on a LED board. 3D printing of optics is just at its beginning, in our opinion. But some 3D printing vendors might disagree.

Gradient index optics, which offer degrees of freedom which are unavailable to optics made of homogeneous materials, are manufactured by layering, laminating, thermoforming and post-processing thin plastic films with varying refractive index (PolymerPlus, USA). Such elements can be potentially used instead of glass lenses.

For reflector prototypes, it can be easier. Many reflectors have been milled from bulk metal and used with some reflective anodization. Today, 3D printing can be used as well of course; with the right manual or computerized polishing and reflective coating by electroplating or sputtering, reflectors can be realized.

10.6 Outlook

Illumination optics has always reflected the needs of markets and technological possibilities. In the past, this meant mainly glass elements; in the times of LED, plastic optics is gaining considerable market share.

References

[1] J. Tsao, H. D. Saunders, J. R. Creighton, M. E. Coltrin, and J. A. Simmons, "Solid-state lighting: an energy-economics perspective," *Journal of Physics D: Applied Physics* **43**(35), 354001 (2010).

[2] M. Karlen and J. Benya, *Lighting Design Basics*, Wiley (2004).

[3] T. Schmidt, "LED lamp lens," *USD744157S1* (2014).

[4] W. v. Bommel, *Road lighting: fundamentals, technology and applications*, Springer, Cham Heidelberg New York (2015). OCLC: 922283804.

[5] S. Bara, A. Rodriguez-Aros, M. Perez, B. Tosar, R. Lima, A. Sanchez de Miguel, and J. Zamorano, "Estimating the relative contribution of streetlights, vehicles, and residential lighting to the urban night sky brightness," *Lighting Research & Technology* **51**, 1092–1107 (2019).

[6] M. Davidovic, L. Djokic, A. Cabarkapa, and M. Kostic, "Warm white versus neutral white LED street lighting: Pedestrians' impressions," *Lighting Research & Technology* **51**, 1237–1248 (2019).

[7] F. Falchi, R. Furgoni, T. Gallaway, N. Rybnikova, B. Portnov, K. Baugh, P. Cinzano, and C. Elvidge, "Light pollution in USA and Europe: The good, the bad and the ugly," *Journal of Environmental Management* **248**, 109227 (2019).

[8] K. Schroll, "Development of a Freeform 3-Zone Streetlight Reflector - How Much Degree of Freedom is Necessary?," in *Imaging and Applied Optics 2015*, OSA, (Arlington, Virginia) (2015).

[9] B. Wördenweber, J. Wallaschek, P. Boyce, and D. Hoffman, Eds., *Automotive lighting and human vision: with 22 tables*, Springer, Berlin (2007). OCLC: 180941546.

[10] D. E. Spencer, L. L. Montgomery, and J. F. Fitzgerald, "Macrofocal Conics as Reflector Contours," *Journal of the Optical Society of America* **55**, 5 (1965).

[11] G. Kloos, *Entwurf und Auslegung optischer Reflektoren: Theorie und Anwendungen*, Reihe Technik, expert-Verlag, Renningen (2007).

[12] P. Flesch and M. Neiger, "AC modelling of D2 automotive HID lamps including plasma and electrodes," *Journal of Physics D: Applied Physics* **37**, 2848–2862 (2004).

[13] M. Wood, "Pixelated Headights," (2019).

[14] R. Fiederling, J. Trommer, T. Feil, and J. Hager, "The Next Step — Pure Laser High-beam for Front Lighting," *ATZ worldwide* **117**, 32–37 (2015).

[15] T. T. True, "*LIGHT COLLECTION SYSTEM*," *US 4, 305, 099*, 10 (1981).

[16] Q. Zhang and Y. Li, "Led lluminating device for stage lighting and method for improving color uniformity of the device," *US 2012/0153852 A1* (2012).

[17] J. Muschaweck and H. Rehn, "Illumination design patterns for homogenization and color mixing," *Advanced Optical Technologies* **8** (2019).

[18] L. J. Hornbeck, "Projection displays and MEMS: timely convergence for a bright future," 2–2 (1995).

[19] G. Derra, H. Moench, E. Fischer, H. Giese, U. Hechtfischer, G. Heusler, A. Koerber, U. Niemann, F.-C. Noertemann, P. Pekarski, J. Pollmann-Retsch, A. Ritz, and U. Weichmann, "UHP lamp systems for projection applications," *Journal of Physics D: Applied Physics* **38**, 2995–3010 (2005).

[20] M. Matsubara, "Projector light source apparatus having collimator disposed between excitation light source and phosphor element," *US9046750B2* (2010).

[21] T. Miyazaki, "*LIGHT SOURCE UNIT AND PROJECTOR,*" *US 8,807,758 B2*, 27 (2014).

[22] J.-W. Pan, C.-M. Wang, W.-S. Sun, and J.-Y. Chang, "Portable digital micromirror device projector using a prism," *Applied Optics* **46**, 5097 (2007).

[23] T. Brukilacchio, "Light emitting diode illumination system," (2007).

[24] D. Li, B. Zhang, and J. Zhu, "Illumination optics design for DMD Pico-projectors based on generalized functional method and microlens array," *Journal of the European Optical Society-Rapid Publications* **15**, 11 (2019).

[25] "DLP System Optics," *Texas Instruments White Paper* (2010).

[26] U. Streppel, "Lens and optoelectronic lighting device," *WO 2016/180814 A1* (2016).

[27] R. Winston, J. C. Miñano, and P. Benitez, *Nonimaging Optics*, Academic Press (2005).

[28] N. Morgenbrod, "Solar simulator and method for operating a solar simulator," *US20130294045A1* (2011).

[29] J. C. Miñano and J. C. González, "New method of design of nonimaging concentrators," *Applied Optics* **31**, 3051 (1992).

[30] P. Benitez, J. C. Minano, J. Blen, R. Mohedano, J. Chaves, O. Dross, M. Hernandez, and W. Falicoff, "Simultaneous multiple surface optical design method in three dimensions," *Optical Engineering* **43**, 1489 (2004).

[31] H. Ries and J. Muschaweck, "Tailored freeform optical surfaces," *Journal of the Optical Society of America A* **19**, 590 (2002).

[32] Y. Ding, X. Liu, Z. Zheng, and P. Gu, "Freeform LED lens for uniform illumination," *Optics Express* **16**, 12958 (2008).

[33] K. Desnijder, P. Hanselaer, and Y. Meuret, "Ray mapping method for off-axis and non-paraxial freeform illumination lens design," *Optics Letters* **44**, 771 (2019).

Julius Muschaweck, a physicist, is the owner and CEO of his company, JMO. After receiving his M.D. from the Ludwig-Maximilians-University in Munich, Germany in 1989 and a stay as Visiting Scholar at the University of Chicago, he co-founded and ran OEC, a unique combination of optical engineering service and pioneering freeform optics research institute. In 2006, he moved on to OSRAM, where he became Senior Principal Key Expert for Optical Design, and in 2013 joined ARRI, the maker of professional movie cameras and lamp heads, as Principal Optical Scientist. He authored over 25 scientific papers and is the inventor of over 50 patents. He is a member of SPIE.

Henning Rehn, a physicist, graduated from Friedrich-Schiller-University Jena in 1991 and received a Dr. rer. nat. in Applied Optics there in 1995. After some years as a post doc scientist he started an industrial career at Carl Zeiss, Jena developing some of the early data projectors. In 2001, he moved to OSRAM for projector lamp development. Later, he also worked as a group leader in pre-development and LED based specialty products. In 2013, he became Principal Key Expert for optical design. In 2018, he moved to Switzerland to join FISBA and became a team leader of the optical design group. He is a member of SPIE.

[53] K. Höhn, A. Debray, P. Schlotter, R. Schmidt, and J. Schneider, "Wavelength-converting casting composition and its use," *US6,066,861* (2000).

[54] Y. Narukawa, M. Ichikawa, D. Sanga, M. Sano, and T. Mukai, "White light emitting diodes with super-high luminous efficacy," *Journal of Physics D: Applied Physics* **43**, 354002 (2010).

[55] R. Mueller-Mach, G. Mueller, M. Krames, and T. Trottier, "High-power phosphor-converted light-emitting diodes based on III-Nitrides," *IEEE Journal of Selected Topics in Quantum Electronics* **8**, 339–345 (2002).

[56] C. Branas, F. J. Azcondo, and J. M. Alonso, "Solid-State Lighting: A System Review," *IEEE Industrial Electronics Magazine* **7**, 6–14 (2013).

[57] M. Knoop, O. Stefani, B. Bueno, B. Matusiak, R. Hobday, A. Wirz-Justice, K. Martiny, T. Kantermann, M. Aarts, N. Zemmouri, S. Appelt, and B. Norton, "Daylight: What makes the difference?," *Lighting Research & Technology*, 147715351986975 (2019).

[58] H. Hinterberger and R. Winston, "Efficient Light Coupler for Threshold Čerenkov Counters," *Review of Scientific Instruments* **37**, 1094–1095 (1966).

[59] X. Zhang, K. Liu, X. Shan, and Y. Liu, "Roll-to-roll embossing of optical linear Fresnel lens polymer film for solar concentration," *Optics Express* **22**, A1835 (2014).

[60] R. Winston, "Dielectric compound parabolic concentrators," *Applied Optics* **15**, 291 (1976).

[61] M. Ploke, "Axialsymmetrische Lichtführungseinrichtung," *DE1472267A1* (1965).

[62] R. Schwarz and P. Kleinschmit, "Pyrogen hergestellte Kieselsäure und Verfaghren zu ihrer Herstellung," *DE 3016010* (1980).

[63] K. Hohenegger and P. Wierer, "*HIGH-REFLECTION SILVER MIRROR*," *US 5,751,474*, 7 (1998).

[64] B. Wagner, "Hohlintegrator," *US6625380B2* (2003).

[65] P. Schreiber, S. Kudaev, P. Dannberg, and U. D. Zeitner, "Homogeneous LED-illumination using microlens arrays," 59420K (2005).

[66] J. Chaves, A. Cvetkovic, R. Mohedano, O. Dross, M. Hernandez, P. Benitez, J. Minano, J. Vilaplana, and M. Hernandez, "Inhomogeneous source uniformization using a shell mixer Köhler integrator," in *Proc. SPIE*, **8550** (2012).

[67] J. Lee and D. Kessler, "Brightness enhancement film using an array of light concentrators," *US7581867* (2009).

[34] R. Wu, Z. Feng, Z. Zheng, R. Liang, P. Benitez, J. C. Minano, and F. Duerr, "Design of Freeform Illumination Optics," *Laser & Photonics Reviews* **12**, 1700310 (2018).

[35] J. ten Thije Boonkkamp and W. IJzerman, "Illumination freeform design using Monge-Ampère equations," in *Optical Design and Fabrication 2019 (Freeform, OFT)*, FT2B.1, OSA, (Washington, DC) (2019).

[36] "CeFO." *The Center for Freeform Optics*, https://centerfreeformoptics.org/.

[37] "FO Plus." (freeform optics plus), https://fo-plus.de/en/.

[38] S. R. Kiontke, "Monolithic freeform element," 95750G, (San Diego, California, United States) (2015).

[39] M. Beier, J. Hartung, T. Peschel, C. Damm, A. Gebhardt, S. Scheiding, D. Stumpf, U. D. Zeitner, S. Risse, R. Eberhardt, and A. Tünnermann, "Development, fabrication, and testing of an anamorphic imaging snap-together freeform telescope," *Appl. Opt.* **54**, 3530 (2015).

[40] J. C. Papa, J. M. Howard, and J. P. Rolland, "Three-mirror freeform imagers," in *Optical Design and Engineering VII*, 43, SPIE, (Frankfurt, Germany) (2018).

[41] Y. Zhong, H. Gross, A. Broemel, S. Kirschstein, P. Petruck, and A. Tuennermann, "Investigation of TMA systems with different freeform surfaces," 96260X, (Jena, Germany) (2015).

[42] H. Ries and J. A. Muschaweck, "Tailoring freeform lenses for illumination," 43–50, (San Diego, CA) (2001).

[43] A. Bruneton, A. Bäuerle, R. Wester, J. Stollenwerk, and P. Loosen, "High resolution irradiance tailoring using multiple freeform surfaces," *Optics Express* **21**, 10563 (2013).

[44] E. Chen and R. Wu, "Design of freeform reflector array for oblique illumination in general lighting," *Optical Engineering* **54**, 065103 (2015).

[45] L. Alvarez, "Two-element variable-power spherical lens," *US3305294* (1967).

[46] R. E. Albrecht, "Zoom flash with wave-lens," *US5666564*, 7 (1997).

[47] U. Hartwig and H. Rehn, "Luminaire, camera or camcorder having same and optical element for a luminare," (2012).

[48] H. Deng and H. Rehn, "Lighting apparatus with zooming function," *EP2835577A1* (2013).

[49] D. Crosby, G. Storey, R. Taylor, and O. Reading, "Adjustable refractive optical device," *US 9,335,446 B2* (2016).

[50] "LightTools."

[51] D. DiLaura, "A Brief History of Lighting," *Optics and Photonics News* **19**, 22 (2008).

[52] S. Kitsinelis, *Light sources: basics of lighting technologies and applications* (2017). OCLC: 1062326378.